PHYSICS BEFORE AND AFTER EINSTEIN

Physics Before and After Einstein

Edited by

Marco Mamone Capria

University of Perugia, Department of Mathematics and Informatics, Perugia, Italy

Amsterdam • Berlin • Oxford • Tokyo • Washington, DC

ISBN 1-58603-462-6
Library of Congress Control Number: 2005923350

Publisher
IOS Press
Nieuwe Hemweg 6B
1013 BG Amsterdam
Netherlands
fax: +31 20 687 0019
e-mail: order@iospress.nl

Distributor in the UK and Ireland
IOS Press/Lavis Marketing
73 Lime Walk
Headington
Oxford OX3 7AD
England
fax: +44 1865 750079

Distributor in the USA and Canada
IOS Press, Inc.
4502 Rachael Manor Drive
Fairfax, VA 22032
USA
fax: +1 703 323 3668
e-mail: iosbooks@iospress.com

Physics Before and After Einstein
M. Mamone Capria (Ed.)
IOS Press, 2005

Preface

It is a century since one of the icons of modern physics submitted some of the most influential scientific papers of all times in a few months; and it is fifty years since he died. There is no question that Albert Einstein with his work on relativity and quantum theory has marked the development of physics indelibly. To reappraise his lifework forces one to rethink the whole of physics, before and after him.

The aim of the present book is to contribute to this daunting task. Though not an encyclopedic work, it tries to provide a perspective on the history of physics from the late 19th century to today, by taking the series of groundbreaking and sometimes provocative contributions by Einstein as the demarcation line between the "old" and the "new" physics. The treatment is not meant as celebratory, but to provide accurate information (both historical and conceptual) and critical appraisal.

Since it is clearly impossible to deal within a relatively small compass with all the topics which would have been suitable for special treatment, a choice – sometimes a painful one – had to be made. It would be wanton to assume that all readers should find the selection presented here as ideal; however the editor is confident that it is neither so arbitrary nor so conventional as to seriously detract from the interest of the whole.

Although this book has been inspired by an historical occasion (a double one, in fact), it is not an occasional one. The authors have a long record of working on themes related to Einstein or Einstein's work and its consequences. What they have to say should be of interest to a wide range of scholars and students in physics, history of science, and epistemology.

However, this is not a book for specialists only. An effort has been made to make the bulk of the book understandable to lay persons with some knowledge of undergraduate mathematics and physics. Most parts of physics can surrender a considerable part of their cultural import to any non-specialist seriously intentioned to grasp the main concepts. The authors have tried to avoid unnecessary technical jargon and to paraphrase in words the main formulas, though some sections or chapters may turn out to be more hard going than others. However, it is a fact that Einstein's work has been instrumental in introducing a previously unheard-of degree of mathematical sophistication into theoretical physics. Ninety years after the event it would be a deceptive simplicity that achieved at the cost of concealing this crucial aspect.

In closing this preface, the editor is very pleased to thank Dr. Einar H. Fredriksson, of IOS Press, for his constant interest and encouragement, and to acknowledge the care and forbearance of Anne Marie de Rover and the other staff of IOS Press.

Marco Mamone Capria
mamone@dipmat.unipg.it

March 2005

Physics Before and After Einstein
M. Mamone Capria (Ed.)
IOS Press, 2005
© 2005 The authors

Contents

Preface   v
  Marco Mamone Capria

Chapter 1: Albert Einstein: A Portrait   1
  Marco Mamone Capria

Chapter 2: Mechanics and Electromagnetism in the Late Nineteenth Century:
  The Dynamics of Maxwell's Ether   21
  Roberto de Andrade Martins

Chapter 3: Mechanistic Science, Thermodynamics, and Industry at the End of the
  Nineteenth Century   49
  Angelo Baracca

Chapter 4: The Origins and Concepts of Special Relativity   71
  Seiya Abiko

Chapter 5: General Relativity: Gravitation as Geometry and the Machian   93
  Programme
  Marco Mamone Capria

Chapter 6: The Rebirth of Cosmology: From the Static to the Expanding Universe   129
  Marco Mamone Capria

Chapter 7: Testing Relativity   163
  Klaus Hentschel

Chapter 8: Einstein and Quantum Theory   183
  Seiya Abiko

Chapter 9: The Quantum Debate: From Einstein to Bell and Beyond   205
  Jenner Barretto Bastos Filho

Chapter 10: Special Relativity and the Development of High-Energy Particle
  Physics   233
  Yogendra Srivastava

Chapter 11: Quantum Theory and Gravitation   253
  Allan Widom, David Drosdoff and Yogendra N. Srivastava

Chapter 12: Superluminal Waves and Objects: Theory and Experiments.   267
  A Panoramic Introduction
  Erasmo Recami

Chapter 13: Standard Cosmology and Other Possible Universes   285
  Aubert Daigneault

Physics Before and After Einstein
M. Mamone Capria (Ed.)
IOS Press, 2005

Chapter 1

Albert Einstein: A Portrait

Marco Mamone Capria

> Only a free individual can make a discovery.
> *A. Einstein*

Whatever one thinks of the common belief that the personality of a scientist has little to do with the creation and success of his theories, it is a prudent guess that there may be at least some exceptions to it. Albert Einstein is arguably one of the best candidates to be such an exception. But there are also other reasons for being interested in his life and opinions. His career departs in many ways from common ideas on how a young person grows to become a great scientist, indeed one of the symbols of the depth and power of his science. He was also one of the few great scientists to devote a considerable number of his thoughts, writings and actions to social and political issues, often going against the current (mis)conceptions of his and our time. Although pictures of his face have been reproduced millions of times, his physics is not the only thing about him that is not as widely known as it deserves to be.

This chapter is divided into three parts. The first gives a chronological outline of Einstein's life; the second sketches his personality; and the third one presents, mainly using his own words, some of his opinions on a wide range of topics.

I

The life of the most famous scientist of the twentieth century began in Ulm on March 14, 1879, as a son of cultivated, "entirely irreligious" Jewish parents.

As a child he was a late speaker (he started at about 2½). Up to the age of seven, he used to repeat to himself (softly) all the phrases he uttered. At about the age of five he had his first violin lesson.

He showed an early liking for arithmetical problems, though he was not particularly precise in his computations. His first geometry book made an exhilarating impression on him, which he vividly described in his "Autobiographical Notes" of 1949 [28, pp. 8–11]. He taught himself calculus between the ages of 12 and 16.

Einstein's relationship with school was far from smooth. At 9 he entered the Luitpold Gymnasium in Munich, but there he found an authoritarian atmosphere which made him feel uneasy. Although his marks were generally good, his relationship with his professor of Greek was strained. At one time, this teacher unwittingly earned himself posthumous fame by angrily telling Albert that nothing would ever become of him. Eventually Albert presented a certificate from his family doctor to the school's principal, left school without finishing it, and in 1895 went to live with his parents, who had moved to Pavia (Italy) the previous year.

In 1895 he failed his entrance examination to the Federal Institute of Technology (*Eidegenössische Technische Hochschule*, ETH) in Zürich, because of low marks in French, chemistry and biology (but not in mathematics and physics). He was advised to attend the cantonal school at Aarau for one year, and then try again to get into the Federal Institute of Technology, which he did, getting very good marks except in geography and drawing (both artistic and technical). In finding this interim solution for the brilliant sixteen-year-old, student the ETH physics professor, Heinrich Weber, was helpful.

In 1897 he finally entered the ETH, where Marcel Grossmann, who was to become a lifelong friend and a valuable collaborator at a crucial stage of Einstein's development, was a fellow student; as was Mileva Marič, a Serbian, his elder by four years, who was to become his wife in 1903. Around that year he also met Michele Angelo Besso, an engineering student, who played a decisive role in his life, first by introducing him to Mach's criticism of Newtonian mechanics. Another acquaintance was a physics student, Friedrich Adler, son of Victor, the leader of the Austrian Social Democrats.

Albert was rather remiss in attending classes (he by far preferred to work in the laboratory, and, especially, studying by himself the works of Kirchhoff, Helmholtz, Hertz and others), but he could rely on the careful notes taken by Grossmann – and on Grossmann's generosity in sharing them. One of the teachers at the ETH was the eminent mathematician Hermann Minkowski, who gained an unfavourable impression of Einstein; indeed, he described him as "a lazy dog".[1] Forty years later, remembering his days as a student at ETH, Einstein contrasted himself and Grossmann as follows: "He, the irreproachable student, I myself, unorderly and a dreamer. He, on good terms with the teachers and understanding everything, I a pariah, discontent and little loved" [5, p. 291].

Mileva had met with strong opposition from Einstein's parents, and his mother never really accepted her. She bore Albert two legitimate children, Hans Albert (1904–73), who was to become a professor of hydraulic engineering, and Eduard (1910–65), who at first looked very promising as a literary and musical boy, but then at university showed symptoms of mental illness. Only in 1987, a sheaf of intimate letters between Albert and Mileva was published, which documented many novel facts concerning their relationship, the most unexpected of which was that when they were not yet married, they had a daughter, Lieserl, and that for reasons which are not completely clear Albert did not recognize her.

In 1900 both Einstein and Grossmann got their diplomas. Einstein's marks were consistently good: "5 for theoretical physics, experimental physics and astronomy; 5.5 for theory of function; 4.5 for a diploma paper (out of a maximum of 6)" [24, pp. 44–45]. In October he sent his first manuscript to *Annalen der Physik*, on the "Consequences of the capillarity phenomena".

Notwithstanding his good record, he was the only one of four graduates in 1900 (Jacob Herat, Louis Kollros, Grossman in mathematics, and Einstein in physics) who was *not* chosen as an assistant at the ETH. Actually, a graduate in mechanical engineering was preferred to him by the same Professor Weber who had encouraged the sixteen-year-old Albert at the time of his first (failed) entrance examination. The basic reason

---

[1] In 1907–1908 it was the same Minkowski, who in 1902 had moved to Göttingen, who provided the geometric interpretation of special relativity, an essential step in the difficult path that eventually led to the birth of the general theory (cf. Chapter 5).

for this rejection was that his teachers, and most crucially Weber, did not appreciate his independence of thought and slightly irreverent demeanour, and were by that time determined to punish him for that.[2]

Thus it happened that the man who was to be arguably the greatest physicist of his century was denied access to an academic career. The graduates who *were* chosen are known today essentially because of their connections to Einstein.

Albert sent letters to several institutes all over Europe, and to famous physicists like Ostwald and Kamerlingh Onnes, asking them for a job, but he received no replies (his return address at the time was in Milan).

For a few months he earned a little money as a secondary school teacher. The turning point of his life was, however, his being hired in June 1902 as a technical expert third class at Bern's patent office, thanks to Grossmann's father, who recommended him to the director. In 1901 he had obtained Swiss citizenship. At this point, had Einstein been a less strong-willed youth, or lacking the emotional support of his fiancée (then wife) and of a handful of affectionate and admiring friends, his bent to speculative thinking in physics could well have been stifled forever. On the other hand, had he not been given any job at all, he "would not have died, but might have spiritually wasted away", as he wrote in 1936 [24, p. 224].

The same year he married, he started to hold meetings with his friends Conrad Habicht and Maurice Solovine in their own private "Akademie Olympia". They read and discussed together classics of epistemological thought such as Mill's *System of Logic*, Hume's *Treatise*, Kant's *Critic of the pure reason*, and, most important for his future physical speculations, Mach's *The Science of Mechanics* and Poincaré's *Science and Hypothesis*. In 1916 he wrote in his obituary of Mach: "I can say with certainty that the study of Mach and Hume has been directly and indirectly a great help in my work". So it is a documented fact that the father of one of the revolutions of twentieth-century physics was a thinker with a deep and precocious philosophical awareness. In his time, however, this was not so rare among scientists as it is today (cf. [22]).

In 1905 Einstein had four papers published in *Annalen der Physik*: they include the two relativity papers, a paper usually referred to as 'the one on the photoelectric effect' (in fact a much more wide-ranging contribution, the one of the four he qualified as "very revolutionary"), and one on Brownian motion.

He became a *Privatdozent* at the University of Bern in 1908, and the following year he left the Patent Office, as he had been appointed professor of theoretical physics at the University of Zürich. His friend Friedrich Adler, who had been proposed to occupy that chair instead of him, refused the offer in his favour, since, as he wrote to the university committee, "I must quite frankly say that my ability as a research physicist does not bear even the slightest comparison to Einstein's" [14, p. 75].[3]

Thus Einstein's academic career began relatively late (at 29), but from then on it was consistently brilliant: in 1911 he became full professor at the German university in

---

[2] As Einstein recalled when he was already famous: "[My teachers] disliked me for my independence and passed me over when they wanted assistants" [9, p. 61].

[3] People who think unselfish behaviour in the context of academic competition as impossible (rather than merely unlikely) have argued for what can be termed as 'extenuating circumstances' to Adler's 'crazy' decision (cf. [16, pp. 123–124]). It is interesting in this connection to mention that when, a couple of decades later, physicist Otto Hahn advanced the proposal that "thirty distinguished professors should jointly protest the treatment of their Jewish colleagues", Planck replied that "if you bring together thirty such gentlemen today, then tomorrow one hundred and fifty will come out against them because they want their positions" [5, p. 248].

Prague and attended the first Solvay conference in Brussels; then, for three semesters, he was professor at the ETH (it is during this period that a crucial collaboration with his friend Grossmann took place); in 1914 he went to Berlin, where he held a professorship with no teaching duties until 1932.

In 1913 he went to Vienna for the Congress of German Scientists and Physicians, and took the opportunity of paying a visit to Mach.[4]

In the meantime, his marriage to Mileva had deteriorated, and shortly after their move to Berlin they separated. Their divorce was granted in 1919, and in the same year he married Elsa Einstein Löwenthal, twice his cousin (both first and second cousin), whom he had begun to meet in 1912. Elsa was three years older than him, a divorcée, and had two daughters from her first marriage.

During the First World War, in 1916, Adler, who was a pacifist, shot and killed the Austrian prime minister, Count Stürgkh. Einstein offered to testify in his favour, though this did not happen. Adler was sentenced to death, but, thanks to his influential father, this was soon commuted to a life sentence by the Emperor (in 1918 Adler was freed and elected to Parliament; he died as an exile in the United States in 1960).

In 1917 Einstein sent to the German Society of Air Service (*Luftverkehrsgesellshaft*) a project on a more effective wing profile for planes; experiments on it were performed as a consequence, with no convincing results.

In 1918 and after the war Einstein also acted as a consultant for, and then collaborator with, Hermann Anschütz-Kaempfe, concerning the latter's invention of a gyrocompass, a device of essential importance in the U-boat war. It is hard to believe that he failed to perceive the obvious inconsistency between this work and his pacifist and internationalist stand ([19, p. 185], [13, pp. 399–402]).

At the end of the First World War, Einstein, who had not signed the notorious manifesto of the German intellectuals (stating among other things that "German culture and German militarism are identical"), found himself in an ideal position as an 'ambassador' for the reconciliation of the scientific communities of previously belligerent countries. This role he heartily embraced, among other things by joining in 1922 the Committee on Intellectual Cooperation of the League of Nations, which had been established in 1919.

In 1916 he published the first complete account of general relativity. It had been preceded by nine years of thinking, publishing and polemics on the possibility of extending the principle of relativity to accelerated systems. Only three years before, in Zürich, Planck had warned him against working on such a topic: "[...] you will not succeed; and even if you succeed, no one will believe you" [24, p. 239].

The exposition of relativity, *On the Special and General Theory of Relativity (A Popular Account)*, which is probably his best known piece of work, first appeared in 1917 [7], and had in the following decades several editions and countless translations.

Einstein's apotheosis occurred in 1919, after the Eddington–Crommelin expeditions to detect the deflection of light predicted by general relativity. Einstein suddenly became a world-famous celebrity. As many authors have pointed out, this can be partly explained by taking into account the historical context: a team of British astronomers confirming the theory of a German physicist by an adventurous enterprise was bound to capture the imagination of a public opinion tired of war and longing for international cooperation. The official endorsement during the session of the Royal Society of November 6 accom-

---

[4]See [14, pp. 104–105] for an interesting account of this visit.

plished Einstein's "canonization" [24, p. 303]. Mathematician and philosopher Alfred N. Whitehead remembered:

> It was my good fortune to be present at the meeting when the Astronomer Royal for England announced that the photographic plates of the famous eclipse, as measured by his colleagues in Greenwich Observatory, had verified the prediction of Einstein that rays of light are bent as they pass in the neighbourhood of the sun. The whole atmosphere of tense interest was exactly that of the Greek drama. we were the chorus commenting on the decree of destiny as disclosed in the development of a supreme incident. There was a dramatic quality in the very staging – the traditional ceremonial, and in the background the picture of Newton to remind us that the greatest of scientific generalizations was now, after more than two centuries, to receive its first modification. (Cit. in [34, p. 44].)

From now on Einstein was to personify the power of pure intellect in penetrating the deepest mysteries of nature.

Thus Einstein started to receive invitations from all over the world, which as a rule he accepted: so he visited countries including Holland, the US, Great Britain, Italy, France, Japan, Palestine, Spain, Argentina, Brazil, Uruguay, Belgium, Norway and Denmark. He met famous people in the literary and artistic world (Thomas Mann, Rabindranath Tagore, Luigi Pirandello, Bernard Shaw, Henri Bergson, Charles Chaplin, among others) as well as in politics (Winston Churchill, Queen Elizabeth of Belgium, Franklin D. Roosevelt), and in 1933 had a famous exchange ("Why war?") with Sigmund Freud, whom he had met in Berlin six years earlier ("Einstein understands as much about psychology as I do about physics", Freud had complacently commented to a friend after the meeting [5, p. 157]).

In 1921 he delivered four lectures in Princeton, New Jersey, USA, which were collected in a booklet of about 120 pages entitled *The Meaning of Relativity*. This slim volume, with his updated editions, can be regarded as his own authoritative survey of relativity; the last edition, the 5th, was published in 1955, with a new version of Appendix II ("Relativistic Theory of the Non-Symmetric Field", first published in 1950 as "Generalized Theory of Gravitation").

He was retroactively awarded the (1921) Nobel prize in 1922, "for his services to theoretical physics and especially for his discovery of the law of the photoelectric effect". (The 1922 prize went to Niels Bohr, "for his services in the investigation of the structure of atoms and of the radiation emanating from them".) Relativity was not mentioned, because within the Nobel committee doubts had been raised as to the tenability of the general theory. When the prize was decreed, Einstein was in Japan, so he received it at home in Germany from the hands of the Swedish Ambassador. Somewhat defiantly his Nobel lecture, which he pronounced in Göteborg in 1923, was entirely devoted to relativity.

In Japan he had delivered (in German) an important talk on the origins of the theory of relativity, which survives only in a Japanese rendering (December 14, 1922) [1].

During his life he received many other prizes (the gold medal of the Royal Astronomical Society in 1926, the Planck medal in 1929, the Franklin medal in 1935, etc.) and several honorary academic degrees.

The fame of Einstein and his theory was accompanied by the flourishing of a new branch of popular scientific literature, aiming at 'making relativity understandable to everybody', and also by sustained attacks from both colleagues and outsiders, starting in 1920. In the same year, his engagement in writing on a wide variety of subjects sharply increased (an useful and ample, but not exhaustive, collection of his general contributions is *Ideas and Opinions* [9], published in 1954; see also *Out of my later years*, of 1950).

In 1920, a crowded meeting was convened by a new association in Berlin in order to criticize the theory of relativity. Einstein, who attended the meeting, wrote an article in the *Berliner Tageblatt*,[5] where he surmised that the criticism levelled at his theory and at him personally was connected with his distaste for German nationalism and with his being "a Jew with liberal international convictions". At the end of the year, at the conference of the Society of German Scientists and Physicians at Bad Nauheim, there was a debate between the famous experimental physicist Philipp Lenard (himself a Nobel laureate, in 1905) and Einstein on relativity, foreshadowing the later involvement of Lenard as apologist of a supposedly pure tradition of "German physics" during the Nazi regime [18].[6]

Although he had left all forms of organized religion in his adolescence, in the 1920s he began to contribute to the Zionist movement (although he never joined the Zionist organisation), touring America with Chaim Weiszmann in 1921 to collect money for the Hebrew University in Jerusalem, and writing articles on the 'Jewish problem'. This did not prevent him from criticizing in later years the militaristic bent of the Israeli government, and emphasizing the necessity of cooperation between Arabs and Jews.

While both fame and, to a lesser extent, criticism were growing on his work, Einstein voiced his disagreement towards the 'complementarity' interpretation of quantum mechanics, which had soon developed into an orthodoxy. Two famous public discussions between Einstein and Niels Bohr, the main supporter and advertiser of 'complementarity', took place in 1927 and 1930 at the 5th and 6th Solvay Conferences.

In 1931 Einstein wrote to the Italian authorities protesting against the enforcement on university teachers of the loyalty oath by the Fascist regime.[7] In 1933 he resigned from the Prussian Academy of Sciences and renounced his German citizenship, as he "did not wish to live in a country where the individual does not enjoy equality before the law, and freedom of speech and teaching". The Academy accepted the resignation, pointing out that Einstein's public statements "were bound to be exploited and abused by the enemies not merely of the present German government but of the whole German people".[8] As a consequence the Bavarian Academy of Sciences also sought his resignation [9, p. 227–231].

In 1933, Einstein's works were among those burned in the book bonfires organized by the Nazis throughout Germany, together with those of such different anti-fascist writers as Thomas and Heinrich Mann, Freud, Proust, Hemingway, H.G. Wells, Gide, Upton Sinclair, etc. At the time Einstein was staying in Le Coq sur Mer in Belgium; the Belgian king ordered that a protection scheme be set up for him.

Einstein visited the United States three times: in 1921, 1930–1 and 1932, the year in which he left Germany for good. He came back to Europe (Belgium) for a time, but from October 1933, apart from a short trip to Bermuda, he never again left the States. In

---

[5]"[...] A distinctly weak piece of writing, out of style with anything else he ever allowed to be printed under his name", according to one biographer [24, p. 316].

[6]The participants in this instructive debate also included Born and Gustav Mie, and it was focused on Einstein's theory and its relationship to the classical ether.

[7]The text of the letter and its consequences are presented in [26].

[8]This is of course the same paralogism by which criticism of, for example, the American government is today often re-labelled as "anti-Americanism".

1935 he applied for permanent residency, and five years later became a US citizen, while retaining Swiss citizenship. In 1932 he had accepted a professorship at the Institute for Advanced Study in Princeton, and there he stayed after his return to US for the rest of his life, at 112 Mercer Street, with his loyal secretary Helene Dukas, his second wife (who died in 1936), her daughter Margot and his slightly younger sister Maja (who died in 1951).

At the Institute he was an attraction and a symbol, but he had little power, and in particular his requests were not given a high priority – for instance, as regards his collaborators, like Leopold Infeld, or the scientists, like Schrödinger or Born, whom he would have been happy to see there as temporary or permanent members. He told Infeld: "My fame begins outside Princeton. My word counts for nothing in Fine Hall". Evidence confirming this is that in 1937 he could not obtain a one-year extension of Infeld's fellowship, although he had ensured the committee that he and Infeld were doing important work together [17, pp. 301ff].

In the United States he was investigated for many years by the FBI, which filed a huge amount of 'politically sensitive' information on him; an official summary of this file was prepared, and "it still amounted to 1,160 pages, a weird mixture of fact and fantasy, of lies, rumors, and ravings" [5, p. 403].

He defended in 1940 Bertrand Russell when the latter was forbidden to lecture at a New York college on the grounds of "immorality", spoke against discrimination of blacks, made public his hostility to the McCarthy anti-Communist witch-hunt, and protested the death sentence of the Rosenbergs.[9]

In 1939 he signed a letter to Roosevelt, written with Leo Szilard, where he described the possibility that German scientists were trying to build a new bomb exploiting the chain-reaction started in a big mass of uranium, and he asked the US government to promote experimental work in that field. Though he later denied having worked for the military, in fact he acted from June 1943 as a consultant on the topic of explosions for the US Navy's Bureau of Ordnance, for a daily fee of $25. He also advised his son Hans Albert to do likewise, and the latter complied by joining the Army Corps of Engineers [16, p. 245].

Harry Truman took over the US presidency after Roosevelt's death in April 1945. The two atomic bombs were dropped on 6 August (Hiroshima) and 9 August (Nagasaki), three months after Germany's unconditional surrender. Einstein did not make any statements to the press until a year later, when in a short article on the front page of the *New York Times* ("Einstein deplores Use of Atom Bomb", August 19, 1946) it was reported that

> Prof. Albert Einstein [. . .] said that he was sure that President Roosevelt would have forbidden the atomic bombing of Hiroshima had he been alive and that it was probably carried out to end the Pacific war before Russia could participate.[10]

---

[9]Cf. the letter of 16 May 1953, to a New York teacher "who had refused to testify before a Congressional Committee", reproduced in [9, pp. 36–37].

[10]Cf. the conclusion reached as early as 1945 by the members of the US Strategic Bombing Survey: "[. . .] Certainly prior to 31 December 1945, and in all probability prior to 1 November 1945, Japan would have surrendered even if the atomic bombs had not been dropped, even if Russia had not entered the war and even if no invasion had been planned or contemplated" (cit. in [2, p. 545]; in this important book virtually all the available evidence is analysed to give an answer to the question of precisely why the bombs were used).

In 1949 one of the volumes of the series edited by P.A. Schilpp, the *Library of Living Philosophers*,[11] appeared under the title *Albert Einstein Philosopher–Scientist* [28], containing not only 25 papers by famous philosophers and physicists and a bibliography of his writings, but also (in agreement with the series requirements) the nearest Einstein ever produced to an intellectual autobiography (simply titled *Autobiographisches* ["Autobiographical Notes"]), and an enlightening reply to a selection of the contributors. This volume is especially important, as it provides evidence of both the historical proportions that Einstein's contribution had reached by wide agreement among scholars in the 1940s, and the ideological isolation in which he spent the last three decades of his life, searching for a unified theory of gravitation and electromagnetism and refusing to accept the orthodox interpretation of quantum mechanics. According to Infeld, with whom he published in 1938 a famous popular book, *The Evolution of Physics*:

> It was distressing to see Einstein's isolation and aloofness from the main stream of physics. On several occasions this man, probably the greatest physicist in the world, said to me in Princeton: "Physicists consider me an old fool, but I am convinced that the future development of physics will depart from the present road" [16, p. 259].

In his last years he devoted himself relentlessly to the elaboration of a unified field theory and to some fundamental problems of general relativity, working with several collaborators (Walther Mayer, Peter Bergmann, Infeld, Banesh Hoffmann, Ernst Straus, Thomas Kemeny, etc.).

Einstein had a strongly built body and was 1.76 m tall. He was an inveterate pipe smoker, and had suffered a stomach disorder since his youth. He was not afraid of making dangerous physical efforts – for example, when sailing. In 1928 he had collapsed in Switzerland when carrying heavy luggage, and his doctor had diagnosed an "acute dilation of the heart" [5, p. 167]. For several months this ailment left him weak. In 1948 a laparotomy had discovered an aneurysm in the abdominal aorta, which, a year and a half later, was found to have grown. When seven years later it burst, doctors advised him to have surgery using recently developed techniques. Notwithstanding the pressures of the people assisting him, Einstein rejected the proposal, saying: "I do not believe in artificially prolonging life"; he commented afterwards to his secretary, "I can die without the help of the doctors". He died on April 18, 1955, at the Princeton Hospital. His autopsy revealed that surgery would not have been successful [5, p. 426; 16, p. 262].

His last political statement was his signature, received by Bertrand Russell a few days later, to what has come to be known as the Russell–Einstein appeal, issued on July 9 and submitting the following resolution:

> In view of the fact that in any future world war nuclear weapons will certainly be employed, and that such weapons threaten the continued existence of mankind, we urge the governments of the world to realize, and to acknowledge publicly, that their purpose cannot be furthered by a world war, and we urge them, consequently, to find peaceful means for the settlement of all matters of dispute between them.

His last scientific statement (April 4) is the lengthy "Prefazione" to an Italian edited book celebrating the fiftieth anniversary of special relativity. After dealing with his uni-

---

[11] In 1944 Einstein had contributed a philosophical essay to the volume of that series dedicated to Bertrand Russell [9, pp. 19–26].

fication programme, he concluded by emphasizing "how far, in my opinion, we are from having a somewhat reliable conceptual basis for physics".[12]

The publishing of *The Collected Papers of Albert Einstein* series, with annotated original texts and companion volumes providing English translations,[13] began in 1987.

For its issue of December 31, 1999 *Time* magazine elected Einstein as "Person of the Century", from a final three-person shortlist comprising also Mohandas 'Mahatma' Gandhi and US President Roosevelt. This choice shows the extent Einstein has come to be, in the collective imagination and through the myth-making by the media, much more than a great scientist, or a by-word for 'genius'; in a sense it might be said of him what Auden said of Freud:

> To us he is no more a person
> Now but a whole climate of opinion
> Under whom we conduct our differing lives.[14]

Of course, this has little or no connection with a widespread understanding of Einstein's place in the history of physics and human thought.

II

Some psychiatrists and others have expressed the view that Einstein had autistic traits, or Asperger's syndrome (characterized by obsessive interests and difficulty in forming social relationships and communicating) [21]. In fact, the anecdotal evidence for this is rather weak, even if examined with an unwarranted confidence in what the neurobiological study of Einstein's brain may be taken to indicate [35,12]. When in Berlin he complained about what his biographer, Philipp Frank, calls "the cold, somewhat mechanical manner of the Prussians and their imitators":

> "These cool blond people make me feel uneasy; they have no psychological comprehension of others. Everything must be explained to them very explicitly" [14, p. 113].

The story of Einstein's daughter, whom he rejected at birth, shows an unexpected side to Einstein's personality (cf. [16,5,23]). It is not known how long his daughter lived, whether she died very early or even survived him; in the latter case one can only speculate why her father never recognized her, even after his marriage to her mother Mileva. What is certain is that in the following decades he never mentioned her. The whole episode sheds an unpleasant light on how the young Einstein, for all his nonconformity, shared in, or at least submitted to, the bourgeois prejudices of his milieu; it also says something, probably, of his family's influence on him. A further lesson is that Einstein's biographers should not assume he was as outspoken on personal affairs as he was on impersonal matters.

* * *

Another fact to which the letters between Albert and Mileva bear witness is that she must have acted as a sounding board (more or less like Besso), at least during his early speculations on physics, up to and including special relativity. For instance, he

---

[12]"[…] wie weit wir nach meiner Meinung davon entfernt sind, eine irgendwie verlässliche begriffliche Basis für die Physik" [8]. This important document escaped the attention of [24] and most other biographers.

[13]Starting with the sixth volume (*The Berlin Years: Writings, 1914–1917*), published in 1997, only a selection of the documents is translated (for instance, [18] is not translated in [11]).

[14]W.H. Auden, "In Memory of Sigmund Freud", *Collected Shorter Poems*, London, 1950.

asked her to check formulas, stated to her his doubts about the relevance of ether, and mentioned "our work on the relative motion". Needless to say, the healthy attention to misconduct in scientific research which has been growing in the last two decades has prompted much guesswork about Mileva's actual, unrecognized contribution to Albert's developing research.

The issue is not likely to be settled to everybody's satisfaction. It is certain that Mileva never claimed to deserve recognition for her husband's early work. Nonetheless it is intriguing to consider that, when they separated, Albert promised to give her *the whole amount* of the Nobel Prize. Moreover, if one takes into account the many subtle ways in which new ideas come to light in discussions with other people, it is hard to draw a line between collaboration and 'mere' emotional support.

Having said that, it is known that Einstein enjoyed explaining his last insights even to people completely ignorant of physics, like his sister Maja: according to one of his collaborators, "He did not feel that he understood something until he himself had understood it in these simple and basic terms" [16, p. 249]. Though Mileva was a physics student herself, and had also studied for a few months at Heidelberg with Lenard, it is possible that her role in the conception of relativity and other theories was not much more decisive than this.

* * *

Though he was susceptible to feminine charm, Einstein's general opinion of women from an intellectual point of view was not flattering, nor was he cautious in voicing it. In the 1920s he said to a female student: "Very few women are creative. I should not have sent a daughter of mine to study physics. I'm glad my wife doesn't know any science; my first wife did" [16, p. 158]. In 1934, in a letter to a "young girl" which he later published, he wrote with unintended and yet conspicuous rudeness: "[Your manuscript] is so typically feminine, by which I mean derivative and steeped in personal resentment" [9, p. 61]. Max Born's wife, a writer, repeatedly heard him endorsing a very narrow view of woman's place in society [4, n. 23, 82].

At this point it will come as no surprise, perhaps, that Einstein's relationships with both his wives were not successful, by his own admission. In a satirical reply to "the women of America" – or more exactly to an American association of 'patriotic' women protesting against his visit to the United States – he said that he was "low-down enough to reject every sort of war, except the unavoidable war with one's own wife" [9, p. 8]. And in a letter to the Besso family in 1955, written for Michele's death, he stated:

> What I most admired in him as a human being is the fact that he managed to live for many years not only in peace but also in lasting harmony with a woman – an undertaking in which I twice failed rather disgracefully [24, p. 302].

* * *

His disregard for formal dress and haircut is proverbial; he liked to wear a leather jacket, did not use socks, and in the last decade did not care to take baths frequently [27, p. 79n].

He liked classical tradition in music, particularly Bach, Mozart and Haydn. He played violin for most of his life, also in public, at fund-raising group performances. In his last years he gave up the violin, but still enjoyed improvising on the piano every day.

His preferred outdoor pastime was sailing on his little boat; he could not swim, but apparently carried no life belts in the boat [5, pp. 262, 264].

As a professor he encouraged students to come into his office and interrupt him, if they needed his help. There are stories testifying to his charm as a teacher, but no particular ability; as a matter of fact he seems to have welcomed academic positions with no teaching duties.[15] In ordinary interactions he was usually available to people who wished to discuss things with him. Philosopher Alfred J. Ayer, who met him at Oxford in the early 1930s, remarked that "he talked to young and unknown men, on topics connected with his own subject, as though he could learn something from them" [3, p. 145].

In Princeton his best friend was the famous logician Kurt Gödel. He is reported to have said that "to have the privilege to walk home with Gödel" was the main reason he came to the Institute in the last years. By the way, even Gödel did not think that quantum mechanics could be the last word in microphysics; it must be added that he was also sceptical of the unification programme pursued by Einstein [33, pp. 57–58].

* * *

Einstein repeatedly criticized the "cult of individuals" as "always [...] unjustified" (1921) and pleaded that no man be "idolized" (1930), though realizing that, ironically, his own case was a major example of the social phenomenon he condemned [9, pp. 4, 10]; indeed, even cultivated people were not immune to Einstein's cult.[16]

However, notwithstanding several statements to the contrary (by himself and by several friends and biographers), it is clear that he enjoyed the popular perception of him as a celebrity (although he once said: "I appear to myself as a swindler because of the great publicity about me without any real reason" [17, p. 290]). When journalists started to be interested in his opinions on all kinds of subjects, he readily complied with their requests, giving short shrift to the qualms and, indeed, outrage that friends such as the Borns felt and showed him for this behaviour [4, n. 20–27]. Of course, he was not the first scientist to be the object of admiration far exceeding the circle of his peers, but he may have been the first scientist in history to accept to actively play the role of popular hero (this circumstance may have been instrumental in arousing criticism, both sound and unsound, on his theories). He probably put on the balance that such choice enabled him to make himself widely and usefully heard on issues of importance in politics, education, and so on, which were dear to his heart. He was indeed a rare specimen, for his and our time, of a scientist *engagé* who was also an all-round intellectual.

III

On his university experience as a student Einstein wrote:

> [...] after I passed the final examination [at the ETH], I found the consideration of any scientific problems distasteful to me for an entire year. In justice I must add, moreover, that in

---

[15]"Indeed, he occupied a regular teaching position for fewer than six years" [27, p. 65]. On Einstein as a professor, see [14, pp. 89–91].

[16]Some stories are quite disconcerting. For instance, the editor of *The American Scholar* and the dean of Princeton University met Einstein in 1935 on the way to a tea party given in his honour. The former said that Einstein's smiling at him "constituted the most religious experience of my life"; the latter commented after the party, in Schopenhauerian terms, "Such moments [i.e. meeting Einstein] tear a rent in ordinary perceptions, cut a hole in the fabric of things, through which we see new visions of reality" [5, p. 283].

Switzerland we had to suffer far less under such coercion, which smothers every truly scientific impulse, than is the case in any other locality. [...] It is, in fact, nothing short of a miracle that the modern methods of instruction have not yet entirely strangled the holy curiosity of inquiry; for this delicate little plant, aside from stimulation, stands mainly in need of freedom; without this it goes to wreck and ruin without fail. It is a very grave mistake to think that the enjoyment of seeing and searching can be promoted by means of coercion and a sense of duty. To the contrary, I believe that it would be possible to rob even a healthy beast of prey of its voraciousness, if it were possible, with the aid of a whip, to force the beast to devour continuously, even when not hungry, especially if the food, handed out under such coercion, were to be selected accordingly [28, pp. 17–19].

Comparing his job at the Patent Office with academic life, he said that "he found the atmosphere more friendly, more human, less marred by intrigue than at the universities, and he had plenty of time for scientific work" [17, p. 286].

On education generally, he pointed out that: "Overemphasis on the competitive system and premature specialization on the ground of immediate usefulness kill the spirit on which all cultural life depends, specialized knowledge included" [9, p. 72]. The function of schools, even technical schools, is not to produce technically skilled people:

The school should always have as its aim that the young man leave it as a harmonious personality, not as a specialist. This in my opinion is true in a certain sense even for technical schools, whose students will devote themselves to a quite definite profession. The development of general ability for independent thinking and judgment should always be placed foremost, not the acquisition of special knowledge [9, p. 69].

* * *

To his collaborator Leopold Infeld he said: "I am really more of a philosopher than a physicist" [17, p. 258]. He wrote that "Epistemology without contact with science becomes an empty scheme. Science without epistemology is – insofar as it is thinkable at all – primitive and muddled" [28, p. 684]. In 1919 he wrote a short article for a newspaper, "Induction and deduction in physics", which reads like an outline of Karl Popper's doctrine as expounded 15 years later in *Logik der Forschung*: induction has a very limited place in science, a scientist starts with conjectures obtained by means of an "intuitive grasp" of a group of facts, and then derives its consequences by deductive means; only the falsity of a theory can be proven, not its truth:

For one never knows if future experience will contradict its conclusion; and furthermore there are always other conceptual systems imaginable which might coordinate the very same facts. When two theories are available ad both are compatible with the given arsenal of facts, then there are no other criteria to prefer one over the other besides the intuitive eye of the researcher. In this manner one can understand why sagacious scientists, cognizant of both – theories and facts – can still be passionately adherents of opposing theories [11, p. 109].

His general view of the actual development of physics is well outlined in these concluding statements from an important essay of 1936:

Evolution is proceeding in the direction of increasing simplicity of the logical basis. In order further to approach this goal, we must resign to the fact that the logical basis departs more and more from the facts of experience, and that the path of our thought from the fundamental basis to those derived propositions, which correlate with sense experiences, becomes continually harder and longer [9, p. 355].

His attitude to scientific methodology was eclectic: a scientist

> must appear to the systematic epistemologist as a type of unscrupulous opportunist: he appears
> as a *realist* insofar as he seeks to describe the world independent of the acts of perception;
> as *idealist* insofar as he looks upon the concepts and theories as the free inventions of the
> human spirit (not logically derivable from what is empirically given); as *positivist* insofar
> as he considers his concepts and theories justified *only* to the extent they furnish a logical
> representation of relations among sensory experiences. He may even appear as *Platonist* or
> *Pythagorean* insofar as he considers the viewpoint of logical simplicity as an indispensable
> and effective tool of his research [28, p. 684].

His opinion of the value of mathematics in physics greatly increased after he created
general relativity, although he still wrote in 1917 to a famous mathematician, Felix Klein,
that "the formal points of view [. . .] may be valuable when an *already found* truth needs
to be formulated in a final form, but fail almost always as heuristic aids" [24, p. 325]. In
the 1920s he made the following remark to a physician friend, concerning a certain problem: "I'm afraid I'm wrong again. I can't put my theory into words. I can only formulate
it mathematically, and that's suspicious" [27, p. 28]. He would say: "No scientist thinks
in formulae" [17, p. 312]. At the end of his life he complained that physicists considered
him a mathematician, and mathematicians considered him a physicist [29].

He advised not to rely on what scientists themselves report about their work, because
of their tendency to take the "creations of thought" as items in the furniture of the world:

> If you want to find out anything from the theoretical physicists about the methods they use,
> I advise you to stick closely to one principle: don't listen to their words, fix your attention on
> their deeds. To him who is a discoverer in this field, the products of his imagination appear so
> necessary and natural that he regards them, and would like to have them regarded by others,
> not as creations of thought but as given realities [9, p. 296].

In the same vein, to the historian of science I.B. Cohen he said in 1955 that "he
thought the worst person to document any ideas about how discoveries are made is the
discoverer" [5, p. 423].

* * *

All scientists pay at least lip-service to empirical evidence as judge of scientific
theories, but their epistemological outlook can be usefully classified in terms of how
much weight they place on contrary experimental results. To see Einstein's position in
this spectrum the right approach is to follow his own advice: to fix one's attention on his
deeds. Here are a few which are relevant, and from which the readers can try to make
their judgement.

In 1907, after Walter Kaufmann's cathode ray experiments had produced evidence
"contradicting the Lorentz–Einstein" postulate, and confirming the theories of Max
Abraham and Alfred Bucherer, established scientists such as Lorentz and Poincaré commented on these results in worried terms. Instead, the young theoretical physicist wrote
that

> "these theories should be ascribed a rather small probability because their basic postulates
> concerning the mass of the moving electron are not made plausible by theoretical systems
> which encompass wider complexes of phenomena" [24, p. 159].

In 1914, when an expedition was on its way to observe the deflection of light during
a solar eclipse, he wrote to Michele Besso: "I do not doubt any more the correctness

of the whole system, whether the observation of the solar eclipse succeeds or not". At that time general relativity had not reached its last stage, and Einstein's prediction of the deflection was *half* the correct value (in fact there was no difference between the 1914 prediction and the one obtained by applying the Newtonian attraction law to light conceived as a flow of corpuscles!).

In 1917 he stated:

> If the displacement of spectral lines towards the red by the gravitational potential does not exist, then the general theory of relativity will be untenable [7, p. 135].

In the 1920s Dayton Miller found a positive 'aether' effect in an interferometric experiment, of the Michelson–Morley kind. When he was informed of that he said, "The Lord is subtle, but He is not malicious". He later explained that this meant: "Nature hides her secret because of her essential loftiness, but not by means of ruse" [24, p. vi]. However, a freer translation was given in a letter to Besso: "I have not for a moment taken them [Miller's experiments] seriously" [30, 25.XII.1925].

In 1950, in presenting his generalized theory of gravitation for the readers of *Scientific American*, he wrote:

> The skeptic will say: "It may well be true that this system of equations is reasonable from a logical standpoint. But this does not prove that it corresponds to nature". You are right, dear skeptic. Experience alone can decide on truth [9, p. 393].

Two years later, on hearing from Born that their common acquaintance, the astronomer Erwin Freundlich, had presented data which contradicted the both relativistic formulas for the deflection of light rays and for the gravitational redshift, he replied:

> Freundlich [...] does not move me in the slightest. Even if the deflection of light, the perihelial movement or line shift were unknown, the gravitation equations would still be convincing because they avoid the inertial system (the phantom which affects everything but is not itself affected). It is really strange that human beings are normally deaf to the strongest arguments while they are inclined to overestimate measuring accuracies [4, p. 192].

* * *

He was not loath to discuss issues with fringe scientists and self-taught laymen. An assistant said that "[...] Einstein had nothing against talking to a crank. This amused him" [5, p. 325]. A more perceptive comment is due to one of his biographers, the physicist Philipp Frank, who knew him well:

> Many professors, even outstanding ones, are so immersed in their own ideas that it is difficult for them to comprehend ideas that deviate from the traditional, or are merely expressed in a way differing from that commonly used in scientific books. This difficulty frequently manifests itself in hatred or contempt for amateurs since the professors are often actually incapable of refuting the ingenious objections made by dilettantes to scientific theories. As a result they give the impression of incompetence and the falsehood of 'academic science'. Einstein, on the other hand, did not regard the differences between the layman and the professional as being very great. He liked to deal with every objection and had none of the reluctance that makes such work so difficult for others; [...] [14, pp. 117–118].

Talking to a friend he admitted that there could be "human emanations of which we are ignorant. Remember how everyone mocked the existence of electrical currents and invisible waves. The science of the human being is still in its infancy" [32, p. 165]. In fact, in 1929 Einstein wrote a positive preface to a book describing experiments on telepathy [15, p. 176]. In 1946, in a letter to a psychoanalyst and parapsychologist, he commented thus on tests performed to find evidence of paranormal vision or communication:

> It seems to me, at any rate, that we have no right, from a physical point of view, to deny a priori the possibility of telepathy. For that sort of denial the foundations of our science are too unsure and too incomplete. [. . .] But I find it suspicious that "clairvoyance" [tests] yield the same probabilities as "telepathy", and that the distance of the subject from the cards or from the "sender" has no influence on the result. This is, a priori, improbable to the highest degree, consequently the result is doubtful [15, pp. 155–156].

This attitude is consistent with an aphorism he wrote in 1953: "Whoever undertakes to set himself up as a judge in the field of Truth and Knowledge is shipwrecked by the laughter of the gods" [9, p. 30].

* * *

As to scientific research as a social activity, he thought that true science is hampered by being organized the way corporations are:

> I do not believe that a great era of atomic science is to be assured by organizing science, in the way large corporations are organized. One can organize to apply a discovery already made, but not to make one. Only a free individual can make a discovery [9, p. 133].
>
> It is well known, on the other hand, that centralization – that is, the elimination of independent groups – leads to one-sidedness and barrenness in science and art because centralization checks and even suppresses any rivalry of opinions and research trends [9, p. 211].

Einstein was very far-sighted as regards, to use the phrase of one of his collaborators, "the corrupting influence that the need to be successful has on the scientist":

> He frequently discussed this both in print and in conversation. He suggested that it would be a very nice profession for a scientist to be a lighthouse-keeper, for it would not be very demanding intellectually and would leave plenty of time to think about other matters. (P.G. Bergmann, in [34, p. 73]; cf. also [14, p. 110].)

He had no esteem for that kind of research industry producing intimidatingly framed and formulated solutions of more or less irrelevant and routine questions – an industry which, today surely no less than a hundred years ago, is responsible for the bulk of scientific literature. Of one colleague, "a fairly well-known physicist", he said, "He strikes me as a man who looks for the thinnest spot in a board and then bores as many holes as possible through it" [14, p. 117]. Of "a well-known American physicist" he once said that "he 'couldn't really understand how anybody could know so much and understand so little' " [34, p. 52].

As to the character of the creative scientist, he outlined it when Besso's sister asked him why Michele had not made any great discovery:

> "[. . .] this is a very good sign. Michele is a humanist, a universal spirit, too interested in too many things to become a monomaniac. Only a monomaniac gets what we commonly refer to as results".

Later he commented to Besso himself: "A butterfly is not a mole; but that is not something any butterfly should regret" [16, p. 261].

As to what led him and a few others (but by no means all scientists) into the "temple of science", he wrote that he agreed with Schopenhauer "that one of the strongest motives that leads men to art and science is escape from everyday life with its painful crudity and hopeless dreariness, from the fetters of one's own ever shifting desires". Most of the scientists belonging to this class he described as "somewhat odd, uncommunicative, solitary fellows" [9, p. 245].

* * *

He did not believe in free will; as he wrote in 1931:

> I do not at all believe in human freedom in the philosophical sense. Everybody acts not only under external compulsion but also in accordance with inner necessity. Schopenhauer's saying, " A man can do what he wants, but not want what he wants", has been a very real inspiration to me since my youth; it has been a continual consolation in the face of life's hardships, my own and others', and an unfailing well-spring of tolerance. This realization mercifully mitigates the easily paralyzing sense of responsibility and prevents us from taking ourselves and other people all too seriously; it is conducive to a view of life which, in particular, gives humour its due [9, p. 9].

This view is linked to the "sense of universal causation" of which "the scientist is possessed", and to the notion of "cosmic religion":

> The future, to him [i.e. the scientist], is every whit as necessary and determined as the past. There is nothing divine about morality; it is a purely human affair. His religious feeling takes the form of rapturous amazement at the harmony of natural law, which reveals an intelligence of such superiority that, compared with it, all the systematic thinking and acting of human beings is an utterly insignificant reflection [9, p. 43].

In 1929 he wrote that "the conviction [...] of the rationality or intelligibility of the world", which "lies behind all scientific work of a higher order", is "akin to religious feeling":

> This firm belief, a belief bound up with deep feeling, in a superior mind that reveals itself in the world of experience, represents my conception of God. In common parlance this may be described as "pantheistic" (Spinoza) [9, p. 286].

From "the wonderful writings of Schopenhauer" he also learnt that Buddhism "contains a much stronger element" of the "cosmic religious feeling" than is to be found in the Bible, where it appears "in many of the Psalms of David and in some of the Prophets" [9, p. 41]. (In his study in Berlin on the walls Schopenhauer's portrait was hanging together with those of Faraday and Maxwell.)[17]

About the Bible he said that it was "in part beautiful, in part wicked", and that "to take it as eternal truth seems to me also superstition which would have vanished a long time ago would its conservation not be in the interest of the privileged class" [5, p. 334]. In his *Autobiographisches* he remembered his rebellion against religion at the age of 12, during which he was convinced, from reading popular scientific books, that "the stories of Bible could not be true" and that "youth is intentionally being deceived by the state through lies; it was a crushing impression" [28, p. 5].

He viewed Judaism as a cultural tradition rather than a religion:

> Judaism is not a creed: the Jewish God is simply a negation of superstition, an imaginary result of its elimination. It is also an attempt to base the moral law on fear, a regrettable and

---

[17]However, in a letter to his son Eduard in 1926 he wrote: "I like what you write about Schopenhauer. I too find the splendid style is worth far more than the actual content" [16, p. 230].

discreditable attempt. Yet it seems to me that the strong moral tradition of the Jewish nation has to a large extent shaken itself free from the fear [9, p. 203].

In particular the values of the "Jewish tradition" with which he identified were "the pursuit of knowledge for its own sake, an almost fanatical love of justice and the desire for personal independence" [9, p. 202].

* * *

Concerning social justice, he would "regard class distinctions as unjustified and, in the last resort, based on force" [9, p. 9]. His view of the general political trends in Western democracies was as prescient as it was gloomy. In 1939 he wrote that, notwithstanding the inventions of many "inventive minds", which could "facilitate our lives considerably",

> the production and distribution of commodities is entirely unorganized so that everybody must live in fear of being eliminated from the economic cycle, in this way suffering for the want of everything [9, p. 19].

He viewed the increasing concentration of financial resources and control on the mass media by special interest groups as one of the curses of the age:

> Private capital tends to become concentrated in few hands [. . .] The result [. . .] is an oligarchy of private capital the enormous power of which cannot be effectively checked even by a democratically organized political society. This is true since the members of legislative bodies are selected by political parties, largely financed or otherwise influenced by private capitalists who, for all practical purposes, separate the electorate from the legislature. The consequence is that the representatives of the people do not in fact sufficiently protect the interests of the underprivileged sections of the population. Moreover, under existing conditions, private capitalists inevitably control, directly or indirectly, the main sources of information (press, radio, education). It is thus extremely difficult, and indeed in most cases quite impossible, for the individual citizen to come to objective conclusions and to make intelligent use of his political rights [9, pp. 171–172].

This was written in 1949. A Princeton librarian and close acquaintance, Johanna Fantova, wrote in her recently discovered diary, referring to conversations she had with him in his last years, that he defined himself politically a "revolutionary" and a "fire-belching Vesuvius"; moreover,

> [Einstein] expressed himself very decisively about many developments in world politics, felt partially responsible for the creation of the atom bomb, and this responsibility oppressed him greatly [29].

However, in 1952, summing up for a Japanese magazine his attitude towards the creation of the atomic bomb, he said:

> My part in producing the atomic bomb consisted in a single act: I signed a letter to President Roosevelt, pressing the need for experiments on a large scale in order to explore the possibilities for the production of an atomic bomb.
>
> I was fully aware of the terrible danger to mankind in case this attempt succeeded. But the likelihood that the Germans were working on the same problem with a chance of success forced me to this step. I could do nothing else although I have always been a convinced pacifist. To my mind, to kill in war is not a whit better than to commit ordinary murder [9, p. 181].

Einstein's pacifism was one of the most prominent features of his political value system, although it was not immune to doubts and ambiguities. He consistently criticized "the military system, which I abhor":

> That a man can take pleasure in marching in fours to the strains of a band is enough to make me despise him. He has only been given his big brain by mistake; unprotected spinal marrow was all he needed. [...] How vile and despicable seems war to me! I would rather be hacked in pieces than take part in such an abominable business [9, p. 11].

This was written in 1931; in the same year, at the California Institute of Technology he delivered a speech where he said: "I am not only a pacifist but a militant pacifist. I am willing to fight for peace. Nothing will end war unless the peoples themselves refuse to go to war". In 1932 he insisted at the Disarmament Conference that

> The introduction of compulsory military service is therefore, to my mind, the prime cause of the moral decay of the white race, which seriously threatens not merely the survival of our civilization but our very existence [9, p. 107].

He had started voicing his pacifist stand in 1914, and repeatedly encouraged conscientious objectors [9, pp. 117, 120]. However, in 1933, answering an invitation to speak in favour of two Belgian conscientious objectors, he stated that had he been a Belgian citizen, under the impending danger of his country being invaded by Germany, he would have "cheerfully" served in the army; and in 1949, when a member of War Resisters International asked whether he agreed that Israel should legally recognize conscientious objection, he answered yes, "but with an important reservation: he would not presume to advise people who had overcome what seemed to be insurmountable obstacles to save their nation" [5, p. 374]. As we have seen, he worked with the military during the Second World War, but even during the First World War he involved himself in research of military relevance and did not publicly criticize scientists such as Walther Nernst and Fritz Haber, who were involved in the development of poisonous gases and powders; indeed, he stayed on very friendly terms with both.[18] Still in 1952, nevertheless, he praised Gandhi as "the greatest political genius of our time".[19]

*   *   *

After the war he never forgave the Germans for the treatment inflicted on the Jews; in 1944 he had written:

> The Germans as an entire people are responsible for these mass murders and must be punished as a people if there is justice in the world and if the consciousness of collective responsibility in the nations is not to perish from the earth entirely. Behind the Nazi party stands the German people, who elected Hitler after he had in his book and in his speeches made his shameful intentions clear beyond the possibility of misunderstanding [5, p. 233].

---

[18]See the sympathetic 1942 obituary for Nernst in [6], where no reference to the late scientist's military work is made. One should not think that 'at that time' the involvement of scientists with warfare was taken as an indifferent matter by everybody. In 1915, because of her husband's callousness and, indeed, enthusiasm in his military work, Clara Immerwahr Haber (who herself owned a Ph.D. in Chemistry) committed suicide.

[19]Einstein's endorsement of Gandhi's non-violence programme is sometimes presented as evidence of his political "naïveté" (cf. for instance also [5, p. 301]). The charge of naiveté is usually based not on any explicit historical evidence and argument, but on misunderstanding of non-violent resistance and unsubstantiated counterfactual guessing (cf. Hook's criticism as cited in [5, p. 397]: "Gandhi could have been successful only with the British or people with the same high human values").

When a few years later philosopher Sydney Hook challenged him on the notion of collective guilt ("How can you? There were two millions Germans, non-Jews, who were in Hitler's concentration camps"), he lamely "said something like he didn't mean those Germans", but he did not retract [5, p. 388].[20]

As to Zionism, his view of the state where Jews had to settle was quite different from what came up eventually; in 1938 he wrote:

> I should much rather see reasonable agreement with the Arabs on the basis of living together in peace than the creation of a Jewish state. Apart from practical consideration, my awareness of the essential nature of Judaism resists the idea of a Jewish state with borders, an army, and a measure of temporal power no matter how modest [9, p. 207].

His respect and support for the Israeli settlers never blinded him as to the nature of Israeli governments. For instance, in 1948 he signed a letter together with other prominent Jewish intellectuals, such as Hannah Arendt, Zelig Harris, and Hook, in which the favourable reception of M. Begin, leader of the Israeli "Freedom Party", by "several Americans of national repute" during his visit to the United States, was severely criticized. Begin's party was qualified as "closely akin in its organization, methods, political philosophy and social appeal to the Nazi and Fascist parties", and its criminal acts and terrorist methods were condemned. Moreover, the "top leadership of American Zionism" was criticized for having "refused to campaign against Begin's efforts, or even to expose to its own constituents the dangers to Israel from support to Begin" [10].

Though grateful to the United States for the hospitality it had granted him, Einstein saw the general attitude of its leadership towards the rest of the world as being laden with the "military mentality [which] raises 'naked power' as a goal in itself". He wrote in 1947:

> I must frankly confess that the foreign policy of the United States since the termination of the hostilities has reminded me, sometimes irresistibly, of the attitude of the Germany under Kaiser Wilhelm II, and I know that, independent of me, this analogy has most painfully occurred to others as well [9, p. 145].

Here, as in so many other topics, what Einstein had to say was as insightful as it was outspoken. He always expressed himself in terms that were understandable and thought-provoking – after his landmark work in physics, he certainly did not need to be pompous and obscure in order to acquire the reputation of being profound (by the way, his technical papers are also written simply and "honest in structure")[21]. Although he was not a systematic thinker, his statements connect to form a consistent world-view, and they well deserve the attention and study of both lay people and historians of ideas. In particular, his lucid political reflections are well worth pondering for our and future generations.

References

[1] Abiko S. 2000: "Einstein's Kyoto address: 'How I created the theory of relativity' ", *Historical Studies in the Physical Sciences*, **31**: 1–35.

[2] Alperovitz G. 1995: *The Decision to Use the Atomic Bomb*, New York, Vintage.

[3] Ayer A.J. 1977: *Part of My Life*, Oxford, etc., Oxford University Press.

---

[20] A wise appraisal of Einstein's opinion is given by Born in [4, n. 103–104].

[21] See Einstein's criticism of Hilbert in 1916 as quoted in [24, p. 261n].

[4]   Born M. (ed.) 1969: *Born–Einstein Letters*, London, Born.
[5]   Brian D. 1996: *Einstein. A Life*, New York, Wiley.
[6]   Einstein A. 1950: *Out of My Later Years*, New York, Philosophical Library.
[7]   Einstein A. 1954: *Relativity*, London, Routledge 2001.
[8]   Einstein A. 1955: "Prefazione di Albert Einstein", pp. xv–xvii of M. Pantaleo (ed.), *Cinquant'anni di relatività*, Firenze, Giunti 1955.
[9]   Einstein A. 1994: *Ideas and Opinions* [1954], New York, The Modern Library.
[10] Einstein A. *et al.* 1948: "To the Editors of *New York Times*", *New York Times*, Dec. 4.
[11] Engel A., Shucking E. (ed.) 2002: *The Collected Papers of Albert Einstein, vol. 7, The Berlin Years: Writings, 1918–1921, English Translation*, Princeton-Oxford, Princeton University Press.
[12] Fitzgerald M. 2000: "Einstein: brain and behaviour", *Journal of Autism and Developmental Disorders*, **30**: 620–621.
[13] Fölsing A. 1998: *Albert Einstein. A Biography* [1993], transl. from the German, Penguin Books.
[14] Frank P. 1953: *Einstein: His Life and Times*, New York, Da Capo.
[15] Gardner M. 1989: *Science. Good, Bad and Bogus*, Buffalo (NY), Prometheus Books.
[16] Highfield R., Carter P. 1993: *The Private Lives of Albert Einstein*, London, Faber and Faber.
[17] Infeld L. 1980: *Quest* [1941, 1965], New York, Chelsea.
[18] Lenard P., Einstein A. *et al.* 1920. "Allgemeine Diskussion über Relativitätstheorie", *Physik. Zeitschr.*, **21**, pp. 666–668.
[19] Martin C.-N. 1979: *Einstein*, Paris, Hachette.
[20] Merleau-Ponty J. 1993: *Einstein*, Paris, Flammarion.
[21] Muir H. 2003: "Einstein and Newton showed signs of autism", *New Scientist*, April 30.
[22] Mulligan J.F. 1992: "Doctoral oral examination of Heinrich Kayser, Berlin, 1879", *Am. J. Phys.*, **60**: 38–43.
[23] Overbye D. 2000: *Einstein in Love. A Scientific Romance*, Viking.
[24] Pais A. 1982: *"Subtle Is the Lord…". The Science and Life of Albert Einstein*, New York, Oxford University Press.
[25] Pais A. 1994: *Einstein Lived Here*, New York, Oxford University Press.
[26] Polverino L. 1991: "Albert Einstein e il giuramento fascista del 1931", *Rivista Storica Italiana*, **103** (I): 268–280.
[27] Pyenson L. 1985: *The Young Einstein*, Bristol–Boston, Adam Hilger.
[28] Schilpp P.A. (ed.) 1949: *Albert Einstein: Philosopher–Scientist*, Evanston (Ill.), Library of Living Philosophers.
[29] Schultz S. 2004: "Newly discovered diary chronicles Einstein's last years", *Princeton – Weekly Bulletin*, April 26.
[30] Speziali P. (ed.) 1972: *Albert Einstein–Michele Besso, Correspondance 1903–1955*, Paris, Hermann (Italian translation: Napoli, Guida, 1995).
[31] Sugimoto K. 1987: *Albert Einstein, die kommentierte Bilddokumentation*, Gräfelung vor München (FDR), Verlag Moos & Partner.
[32] Vallentin A. 1958: *Albert Einstein*, Club des Libraires de France.
[33] Wang H. 1996: *A Logical Journey. From Gödel to Philosophy*, London (UK) and Cambridge (Mass.), The MIT Press.
[34] Whitrow G.J. (ed.) 1967: *Einstein. The Man and His Achievements*, Dover.
[35] Witelson S.F., Kigar D.L., Harvey T. 1999: "The exceptional brain of Albert Einstein", *The Lancet*, **353**: 2149–2153.

Physics Before and After Einstein
M. Mamone Capria (Ed.)
IOS Press, 2005
© 2005 The authors

Chapter 2

Mechanics and Electromagnetism in the Late Nineteenth Century: The Dynamics of Maxwell's Ether

Roberto de Andrade Martins

*Group of History and Theory of Science, Physics Institute, State University
of Campinas (Unicamp), Brazil*

1. Introduction

The most important effects predicted by the theory of special relativity had already been obtained before Einstein's work. This fact is well known by historians of science, but not widely known among scientists and laymen.

In popular accounts of the theory of relativity, Einstein is usually depicted as the only person responsible for that theory. The physicists who had witnessed the rise of the theory knew otherwise, and when one consults early books on relativity (such as those written by Max von Laue, Wolfgang Pauli and other authors) it is possible to find that they acknowledged the central contributions of several other physicists to the theory. However, after a few decades the Einstein myth had established itself. The situation changed in 1953, when Sir Edmund Whittaker published the second volume of his book *A History of the Theories of Ether and Electricity*, which contained a chapter titled "Relativity theory of Poincaré and Lorentz" [120]. In that work, Whittaker minimized Einstein's contribution to the special theory and ascribed all relevant steps to Lorentz, Poincaré, Larmor and other physicists. This book triggered the publication of several papers discussing the relative contributions of Einstein and the other physicists who contributed to the building of special relativity.[1] Nowadays, historians of physics acknowledge that many other researchers before Einstein made very important contributions to relativity, although they have not reached a consensus about the relative importance of their works.

This chapter will present the development of those ideas, in the late nineteenth century and during the first years of the twentieth century – that is, before Einstein's first publication on the subject, in 1905. It will be shown that Maxwell's electromagnetism, together with the ether concept, which played a central role in his theory, led to a new dynamics where mass increased with velocity, and energy changes were accompanied by mass changes.[2]

---

[1] See, for instance, [15,100,30–32,98].

[2] Hirosige [46] studied the role of electromagnetism and ether in the rise of relativity theory.

In order to condense the subject into one chapter of suitable size, it was necessary to select specific topics to describe. It has been impossible to include here a general overview of the history of electromagnetism.[3] Therefore, many relevant subjects – such as Weber's electrodynamics – have had to be left out. Only Maxwell's field approach and some of its consequences will be dealt with here, giving special emphasis to the developments that led to results closely related to relativistic dynamics.

Another simplification adopted here is the use of modern notation and terminology. The mathematical notation used in electromagnetism in the late nineteenth century was seldom the one that is used today [107]. Besides, the period under discussion was a time when several systems of units were used by different authors. A faithful rendering of the historical contributions should use the original words, notation and systems of units employed by each author. Unfortunately, that would increase the size and complexity of the present work beyond acceptable limits. The approach I have thus adopted in this chapter is to use contemporary notation and the international (MKS) system of units throughout. I apologize for this anachronism.

2. Faraday and the Lines of Force

The starting point of this story will be the concept of ether developed by Faraday and Maxwell. Michael Faraday's (1791–1867) experimental work on electromagnetism led him to accept that electric and magnetic forces are not direct forces at a distance, but forces carried by a physical medium, possessing mechanical properties.[4] Faraday usually described the electric and magnetic effects using the concept of "lines of force". According to Faraday, electromagnetic forces are carried by lines of force stretched between electric charges.[5]

In the prevailing teaching of physics (especially at a basic level) we still make use of lines of force. However, the contemporary use is not equivalent to Faraday's concept, because we do not conceive of those lines as substantial physical realities. Michael Faraday accepted that the electromagnetic field has physical reality and quantitative properties, even in the absence of matter. The same idea was later used by Maxwell, in his electromagnetic theory.[6]

Faraday ascribed the effect of electrostatic induction, for instance, to "inductive lines of force", not to a direct action at a distance [23, § 1164]. This inductive effect seemed to propagate from point to point, through the medium. Metallic shields can impede the action of this effect, and accordingly Faraday thought that the lines of force reached to surface of the grounded metal barrier and could not pass through it. On the other hand, inductive effects can be noticed behind the borders of a metallic plate connected to the ground, hence the inductive lines of force seemed to bend around the metallic border [23, § 1221].

---

[3] An account of the history of electromagnetism in the nineteenth century can be found in [20].

[4] The evolution of Faraday's ideas can be found in [122, Chap. 10].

[5] The interpretation of Faraday's ideas is not straightforward, and they changed over time (see [81]). Here, his concept of lines of force is presented taking into account his later ideas.

[6] Hesse [45] presents a historical and philosophical discussion of the tension between the field approach and the action-at-a-distance approach.

Why did Faraday introduce this conception? The main reason seems to be the possibility of intervening in the electric and magnetic interaction. Gravitational effects behave as direct actions at a distance, in a straight line, without suffering any change when matter is interposed. Electric and magnetic forces, on the other hand, seem to propagate along curved lines, and are influenced by interposed matter. Therefore, they behave as influences that gradually propagate in space – a space filled with some substance – from one point to the next. Since 1832 Faraday believed that actions took some time to pass from one point to another, and he thought that it would be possible to measure their corresponding speed [6, p. 107].

To account for the bending, Faraday supposed that the lines of force repel each other, and for that reason they separate from one another and bend towards regions where there are no other lines [23, §§ 1224–5, 1231]. Therefore, Faraday's inductive lines of force had two main properties: they produced induction and electrostatic attraction along the direction of the line; and they produced a mutual lateral repulsion [23, § 1297].

Faraday used a similar model for the lines of magnetic force – indeed, it seems that the study of magnetism was instrumental in shaping his belief in lines of force [34]. The effects of attraction between magnets or parallel electric currents were accounted for by the tendency of the lines to contract along their lengths [23, §§ 3266–7, 3280, 3294]. The repulsion of two similar magnetic poles, or two parallel opposite electric currents, was explained by the tendency of the magnetic lines of force to repel each other transversely [23, §§ 3266–8, 3295]. The effects observed in ferromagnetic, diamagnetic and paramagnetic bodies were also explained by those properties of magnetic lines of force, under the assumption that they tend to concentrate more or less in different substances [23, § 3298].

Faraday clearly stated that the lines of magnetic and electric forces are real [23, §§ 3263, 3269], and discussed their physical nature. He suggested that they could correspond to a vibration of the ether, or a state of stress of the ether, or some other static or dynamical state of the medium [23, § 3263]. It seems that in most cases he favoured the idea of tension. However, in 1845, when he discovered the magneto-optical effect [108], where a beam of polarized light is rotated when travelling along the direction of a magnetic field, he convinced himself that the magnetic lines of force have a dynamic nature – those lines turn around their length [23, §§ 2162–75; 122, pp. 386–391].

Using these ideas, Faraday emphasized the importance of the intermediate field, to the detriment of the idea of action at a distance. The electric charges, currents and magnets produce lines of force, but the lines of force are responsible for all distant interaction.

3. Maxwell and the Energy of the Electromagnetic Field

James Clerk Maxwell (1831–1879) adopted Faraday's views, adding a mathematical treatment to the qualitative model. In his *Treatise on Electricity and Magnetism,* Maxwell took up the idea of lines of force, and stated that he would adopt a theory where electric action is a phenomenon of tension of the medium, or tension along the lines of force [76, §§ 47–8].

Because of this approach, Maxwell characterized the electromagnetic phenomena by physical magnitudes distributed in space, instead of forces concentrated into points. The idea of an electric and magnetic field distributed in space, even in the absence of matter, used up to the current day, is a vestige of Maxwell's ether.

According to Faraday's views, the intermediary lines of forces produced forces, but the mechanical properties were always ascribed to matter, not to ether. In Maxwell's theory, on the other hand, the medium – the ether – is endowed with momentum, potential energy and kinetic energy (see [103,104]). Before Maxwell, the electrostatic energy of a system of charges, in a space devoid of matter, was calculated taking into account the value of the electric potential *at places where each charge was*:

$$W_e = \frac{1}{2} \sum (e_i V_i). \tag{1}$$

Starting from this equation (§ 84), and following a mathematical analysis that had already been developed by William Thomson (1824–1907) [118], Maxwell proved that electrostatic energy can also be computed from the knowledge of the electric field in the whole space, without any explicit mention of the charges that generate the field or react to the field. Indeed, the electrostatic energy can be computed from:

$$W_e = \frac{1}{2} \int \varepsilon E^2 \mathrm{d}V \tag{2}$$

where the integral is computed over all the space surrounding the charges [76, § 99a]. The integrated quantity ρ_e can be interpreted as the density of the electric field energy:

$$\rho_e = \frac{\mathrm{d}W_e}{\mathrm{d}V} = \frac{\varepsilon E^2}{2}. \tag{3}$$

According to Eq. (2), all regions of space where there is an electric field contribute to the total energy of the system, even when the field is generated by a set of charges contained in a small region. This energy scattered in space is, according to Maxwell, essentially a form of elastic potential energy [76, §§ 630, 638], connected to the ether tension [76, § 110]. This is a mathematical rendering of Faraday's lines of force [76, § 109].

In the case of a spherical body with a superficial charge q and radius a, in the vacuum, the total electrostatic energy around the charge can be computed by integration over all the space:

$$W_e = \frac{q^2}{8\varepsilon_0 \pi a}. \tag{4}$$

The same result was obtained by the old approach, of course. Maxwell presented a detailed quantitative analysis of those tensions in order to compute electrostatic actions taking into account only *local* actions, that is, eliminating actions at a distance. The mathematical development of the theory led him to associate nine components of electrostatic tension to each point of the electromagnetic medium, reduced to six by imposing the absence of rotational effects. In the absence of matter they reduce to only three stresses [76, § 106]: a longitudinal attractive tension (along the electrical field) and a transverse pressure (perpendicular to the direction of the electrical field). The import of those stresses is the same, both in the longitudinal and the transverse directions, when the sign is omitted. Their value p is equal to the density of electrostatic energy. In the presence of a material medium, Maxwell obtained [76, § 631] an equation equivalent to this:

$$\rho_e = \frac{\mathrm{d}W_e}{\mathrm{d}V} = -\frac{1}{2}\vec{E}\cdot\vec{D} = \frac{\varepsilon}{2}E^2. \tag{5}$$

Maxwell presented an analogous treatment for the magnetic field. However, in this case he took as the fundamental sources of the field the electric currents (not the magnetic poles), and regarding the currents as the motion of electric charges, he described the magnetic energy as "electrokinetic energy" [76, § 634]. He also computed the magnetic energy corresponding to a set of electric currents and proved that this energy can be computed by integration of a function of the magnetic field in all space. The density of magnetic energy – which, according to Maxwell, is essentially kinetic energy [76, §§ 630, 636, 638] – is given by:

$$\rho_m = \frac{\mathrm{d}W_m}{\mathrm{d}V} = -\frac{1}{2}\vec{B}\cdot\vec{H} = \frac{\mu}{2}H^2. \tag{6}$$

Maxwell states that this magnetic energy "exists in the form of some kind of motion of the matter in every portion of space" [76, § 636]. Maxwell accepted Faraday's magneto-optical effect as evidence that there is a rotational motion of the ether around the lines of magnetic force [76, §§ 821, 831].

Lagrange's method in mechanics was widely used in the nineteenth century. Maxwell described this method [76, §§ 553–567] and then showed that it could be applied to electromagnetism [76, §§ 568–584]. He did not explicitly present the Lagrangean function describing the electromagnetic field. However, since he accepted that the electromagnetic field has both potential energy W and kinetic energy T, it is possible to write down a Lagrangean function $L = T - W$ describing this field.

The possibility of describing electromagnetism according to the Lagrangean formalism was interpreted by Maxwell and his contemporaries as strong evidence that it was possible to provide a mechanical theory for electromagnetism. Several authors (including Maxwell himself) had proposed detailed mechanical models of the ether. However, all such models had limitations, and for that reason Maxwell, in his *Treatise on Electricity and Magnetism*, chose to avoid any specific model. Instead, he adopted a general approach, stating that all electromagnetic phenomena could be understood as effects transmitted through the underlying ether, and calculating the general mechanical properties ascribed to that medium, without proposing detailed mechanical models for the ether [89, pp. iv–viii].

4. The Ether Stresses according to Maxwell

The magnetic interactions, according to Maxwell, are also associated with ether stresses [76, §§ 641–3]. The description of those stresses is highly complex, in the general case (in the presence of matter), but in a vacuum they reduce to a longitudinal attraction and a transverse pressure, both with the same absolute value p_m, equal to the density of magnetic energy ρ_m:

$$\rho_m = p_m = \frac{\mathrm{d}W_m}{\mathrm{d}V} = \frac{\mu H^2}{2}. \tag{7}$$

At several places, Maxwell emphasized that he was following Faraday's ideas. It was also following Faraday's suggestion that Maxwell developed the electromagnetic theory of light [76, § 782]. At this point, Maxwell made use of the theory of ether stresses to compute the pressure produced by an electromagnetic wave [76, § 792]. Suppose that a polarized plane electromagnetic wave propagates in the z direction. The electric and magnetic fields are perpendicular to one another and transverse to the propagation of the wave. Let us suppose that the electric field is in the x direction and the magnetic field in the y direction. If the total energy density of the wave is ρ, on average half of this energy will be electric and the other half magnetic. Therefore, the mean electric stresses will be:

$$T_{ex} = -\frac{\rho}{2}, \tag{8}$$

$$T_{ey} = T_{ez} = +\frac{\rho}{2}. \tag{9}$$

The mean magnetic stresses will be:

$$T_{my} = -\frac{\rho}{2}, \tag{10}$$

$$T_{mx} = T_{mz} = +\frac{\rho}{2}. \tag{11}$$

Adding those stresses, Maxwell obtained the resultant effect:

$$T_x = T_y = 0, \tag{12}$$

$$T_z = \rho. \tag{13}$$

Therefore, there is no resultant pressure in the x and y directions (perpendicular to the direction of wave propagation), and there is a pressure in the direction of wave propagation (z) equal to the total energy density of the wave.

Maxwell's *Treatise on Electricity and Magnetism* was published in 1873, containing the theoretical prediction that light should produce pressure. Independently, in 1876, Adolfo Bartoli (1851–1896) deduced a formula for the pressure of radiation, using a purely thermodynamic argument [14]. He showed that, if radiation did not produce a pressure, it would be possible to violate the second law of thermodynamics. Bartoli's argument was generalized by Ludwig Boltzmann (1844–1906) and Prince B. Galitzine (c. 1840–1916), some years later [10,28]. The coincidence between the equations obtained by the thermodynamic argument and Maxwell's theory of ether stresses led to strong confidence in those results.

The later confirmation of this effect in 1901 by Pyotr Lebedew (1866–1912), Ernest Fox Nichols (1869–1924) and Gordon Ferrie Hull (1870–1956), was no surprise [59,83,84].[7] As a matter of fact, this result is not a specific property of electromagnetic waves. Any wave carrying energy (such as sound or water waves) will also produce pressure [94].

In Maxwell's mature theory, therefore, the ether had several mechanical properties. It amassed energy, it produced forces, it could produce pressure. To ascribe motion and momentum to the ether was the next step.

---

[7]Worrall [124] describes the history of the search for the pressure of light.

5. The Momentum of the Magnetic Field

Following Faraday once more, Maxwell remarked that the phenomenon of self-induction exhibited properties similar to inertia [76, §§ 546–7]. When an electric current flows through a circuit, and the electromotive force which produces that current is interrupted, the current does not immediately stop; it has a tendency to continue for some time. The resistance of the current to stop depends on both the value of the current and the geometric configuration of the wires through which it is flowing, which can be described by a quantitative property – the self-inductance of the circuit. If this effect depended only on the value of the current and the length of the wire, it would be possible to ascribe it to the inertia of the particles that carry the electric charge (note that, at the time, the electrons were unknown). However, when the same wire, carrying the same electric current, is bent or transformed into a coil, its self-inductance changes [76, §§ 548–9]. Therefore, the self-inductance cannot be interpreted as a property of the electricity carriers. How then can it be interpreted?

Maxwell provided several interpretations, all of them involving the dynamics of ether. On one hand, it was possible to account for the effect taking into account the energy of the magnetic field around the wires [76, § 552] – the energy depends on the geometric configuration of the circuit. On the other hand, Maxwell associated the magnetic field with ether motion; and therefore, the self-inductance could be associated with some dynamic effect around the wires, which Maxwell called "electrokinetic momentum" [76, §§ 578, 585]. Maxwell proved that it was possible to evaluate this momentum by integrating the vector potential around the circuit [76, § 590]. The vector potential could be interpreted as a momentum associated to the ether.

The path that led Maxwell to this concept of the vector potential is very interesting (see [11]). Faraday had interpreted the electromagnetic induction as due to changes of "the electrotonic state" – an obscure concept he introduced in the 1830s. Inspired by a paper published in 1847 by William Thomson, in 1856 Maxwell was able to represent the induced electromotive force at each point of a circuit in very simple form [73], which can be represented in modern notation as:

$$\vec{E} = \frac{d\vec{A}}{dt}.$$ (14)

In this paper, Maxwell interpreted the vector $\vec{A}$ as the "electro-tonic function", or "electro-tonic intensity". In the same paper, he clearly established the relation between this function and the magnetic induction $\vec{B}$. In modern notation,

$$\vec{B} = \vec{\nabla} \times \vec{A}.$$ (15)

Several years later, Maxwell gave the same vector another name: "electromagnetic momentum" [75]. The new name was probably inspired by Eq. (14) because, when we multiply both sides by an electric charge, the left side of the equation becomes a force, and according to Newton's second law the right side should be the time derivative of the momentum. Afterwards, in his *Treatise*, Maxwell introduced a third name for the same quantity: "vector potential" – the name we use nowadays [76, § 405].

Note that here Maxwell is ascribing a new dynamical property to the ether: momentum. In the presence of a magnetic field – or, more exactly, in the presence of the vector potential – a charge will have a dynamical momentum, *even if it is at rest*.

Maxwell's electrodynamics is essentially macroscopical, in the sense of dealing with electric charge and currents as continuous entities. However, the atomic theory of matter, associated with the study of electrolysis by Faraday, led to the idea that electricity is also atomistic. This provided motivation for the study of a theory involving the motion of charged particles.

6. The Energy of a Moving Charge: J.J. Thomson

In 1881, J.J. (Joseph John) Thomson (1856–1940) studied the electric and magnetic fields associated with an electrically charged particle moving in the vacuum [112].[8] His approach was valid only in the case of velocities much smaller than that of light. He supposed that the electric field was not changed by the motion of the charge. However, the motion of the charge is equivalent to an electric current, and it therefore generates a magnetic field. In modern notation, the magnetic field $\vec{H}$ can be computed (for low speeds) using the formula:

$$\vec{H} = \vec{v} \times \vec{E}, \tag{16}$$

where $\vec{v}$ is the velocity of the particle and $\vec{E}$ is the electric field created by the moving charge.

If the electric field did not change with the motion of the charge, its energy was also independent of the speed. However, the magnetic field generated by the motion of the charge was associated with an energy density proportional to the square of the magnetic field, according to Maxwell's theory. As this magnetic field is proportional to the speed, the additional magnetic energy associated with the moving charge will be proportional to the square of the speed. A detailed computation (not shown here) provides the following result, for a spherical charge:

$$W_m = \frac{q^2 v^2}{12\varepsilon_0 \pi a c^2}, \tag{17}$$

where a is the radius of the particle, v is its speed, and q is its electrical charge (supposed to be distributed on its surface).

This magnetic energy could be regarded as kinetic energy in two different senses. First, because Maxwell and his followers regarded magnetic energy as kinetic energy of the ether. Second, because in this case, the magnetic energy associated with a moving charge was proportional to the square of the speed, exactly like the usual mechanical formula for the kinetic energy K of a particle:

$$K = \frac{mv^2}{2}. \tag{18}$$

Equation (17) can be rewritten to become similar to Eq. (18):

$$W_m = \frac{1}{2} \left(\frac{q^2}{6\varepsilon_0 \pi a c^2} \right) v^2. \tag{19}$$

---

[8]Thomson's paper contained a small mistake which was corrected some months later by George Francis FitzGerald (1851–1901). See [24].

Thus the expression between parentheses could be regarded as an "electromagnetic mass":

$$m_e = \frac{q^2}{6\varepsilon_0\pi ac^2}.\tag{20}$$

Comparison with Eq. (4) shows that the electromagnetic mass m_e is proportional to the electrostatic energy W_e of the charge. Indeed, they have the following relation:

$$m_e = \frac{4}{3}\frac{W_e}{c^2}.\tag{21}$$

This formula (not presented by Thomson, but easily obtained from his result) is a special case of the mass–energy relation $m = E/c^2$. The numerical factor $4/3$ will be discussed later.

Thomson supposed that the charged sphere had a mechanical mass M, as well as its electromagnetic mass. Therefore, its total kinetic energy would be:

$$K = \frac{1}{2}\left(M + \frac{q^2}{6\varepsilon_0\pi ac^2}\right)v^2.\tag{22}$$

Note that the electromagnetic mass is not inside the particle; it is outside it, spread all over the ether, and occupying an infinite volume.

In 1895, Joseph Larmor suggested that matter could simply be a collection of electrical particles, and that, in that case, all inertia would be of electromagnetic origin.[9]

The ether is now full of mechanical properties: force, pressure, potential energy, kinetic energy and electromagnetic mass. Electromagnetic mass is not the mass of the ether itself. It is the mass associated with a change in the ether – the magnetic field produced by a moving mass.

As we have seen, this concept of electromagnetic mass was derived by comparison with the formula of kinetic energy. This is a specific concept that could be called "kinetic mass", or "capacity of electromagnetic kinetic energy", if we adopt the term later proposed by Henri Poincaré (1854–1912) and Paul Langevin (1872–1946).[10] We shall soon see that there are other relevant mass concepts.

One should also bear in mind that the electromagnetic mass obtained by Thomson and Fitzgerald is a first approximation, valid only for small speeds, and that they were aware of this limitation.

7. Electromagnetic Mass of Fast Moving Charges

In 1889, Oliver Heaviside (1850–1925) obtained exact equations for the field around a moving charge, using an elegant operational approach [38]. However, this new mathematical technique was not regarded as reliable at that time. For that reason, J.J. Thomson deduced again the same results [113,114], confirming Heaviside's results. The electro-

---

[9]On Larmor's contribution to the theory of relativity, see [55,18].

[10]It is possible to define inertial mass in several ways, and the various definitions lead to different equations, in the case of the theory of relativity (and in the case of the electromagnetic theory described here). See [57].

static field of a moving point charge was radial, as that of a charge at rest. However, the field intensity depended on the direction, and was given by a complex formula:

$$E = \frac{q(1 - v^2/c^2)}{4\varepsilon_0 \pi r^2 (1 - (v^2 \operatorname{sen}^2 \theta/c^2))^{3/2}}, \tag{23}$$

where θ is the angle of the given direction with the direction of motion. The magnetic field turned in circles around the direction of the motion of the charge, and its intensity is given by:

$$B = \frac{v}{c} E = \frac{q v \operatorname{sen} \theta (1 - v^2/c^2)}{4\varepsilon_0 \pi c r^2 (1 - (v^2 \operatorname{sen}^2 \theta/c^2))^{3/2}}. \tag{24}$$

Heaviside proved that the field produced by a moving charged sphere of radius a was not equal to that of a point charge – it was equivalent to that of a short charged wire with a uniform charge density and length equal to $2av/c$ [40, § 164, p. 269]. However, in 1896, George Frederic Charles Searle (1864–1954) proved that the field of a point charge was equivalent to that of a revolution ellipsoid with the smaller axis along the direction of the motion, with the ratio of the smaller to the larger axis equal to $\sqrt{1 - v^2/c^2}$ [101]. He called this flattened sphere "Heaviside's ellipsoid".

In the same paper, Searle computed the energy associated to the field of a charged sphere. The result showed that the additional electromagnetic energy was not proportional to the square of the speed, in high speeds. He obtained the result:

$$W = \frac{q^2}{8\varepsilon_0 \pi a} \left(\frac{c}{v} \log \frac{c+v}{c-v} - 2 \right). \tag{24}$$

For small speeds, this formula reduces to the classical Eq. (17) for kinetic energy, proportional to the square of the speed. Taking into account this result, the electromagnetic mass could not be independent of the speed of the charge: the electromagnetic mass should increase with the speed.

The above formula for the additional energy of the electromagnetic field of a charged particle in uniform motion showed that this energy increased without limit as the speed approached the speed of light. For that reason, Thomson and Searle concluded that it is impossible to accelerate a charged particle to a speed equal to or greater than c [102].

Heaviside, however, did not agree with this conclusion [40, vol. 2, Appendix G, p. 533]. He proved that there were two situations when the energy would remain finite, even for speeds equal to c. The first case was that of a pair of particles with equal charges, but opposite signs. In that case, the two particles could move together and their field would remain finite, even when $v = c$. The second exception was that of a charge particle that underwent strong acceleration. The above formula is valid only for charges in uniform motion. If the charge starts from rest and attains a speed $v > c$ after time Δt, only the field inside a sphere of radius $c\Delta t$ will change, because changes in the electromagnetic field propagate with the speed of light. Outside the sphere, the total energy is finite (as in the case of a particle at rest), and inside the sphere the energy is also finite, because the density of energy is finite.

However, four years later, Heaviside proved that an electric charge undergoing strong acceleration and attaining a speed greater than c would emit electromagnetic radiation that would react upon the particle, reducing its speed to less than c [40, vol. 3, §§ 498, 500, 511, pp. 120, 125, 165].

8. Energy Flux and Momentum in Electromagnetic Fields

The concept of electromagnetic mass was born from the study of the magnetic energy associated with a moving charge. Another approach was the study of the electromagnetic *momentum* associated with a moving charge. It was shown above that Maxwell ascribed a momentum $q\vec{A}$ to a charge q moving at a place where the vector potential is $\vec{A}$. This, however, is a momentum independent of the velocity of the charge, and therefore is not due to its motion. There is another kind of electromagnetic momentum, associated with the motion of the charge, that will be explained below.

Let us first introduce the concept of the flux of electromagnetic energy. This concept was proposed independently by John Henry Poynting (1852–1914) in 1884, and Heaviside in 1885 [93,37,39]. The main result they obtained can be expressed in modern notation as:

$$\vec{S} = \vec{E} \times \vec{H}. \tag{25}$$

That is, the flux of electromagnetic energy $\vec{S}$ (today called "Poynting's vector") depends on both the electric and magnetic field, and is perpendicular to both. In the case of an electromagnetic wave, for instance, both electric and magnetic fields are perpendicular to the direction of propagation of the wave; the flux of energy, of course, has the direction of propagation of the wave.

If the electric and magnetic fields are parallel, there is no flux of energy. If they are not parallel, there is a flux, even if the fields do not vary. For instance, if a charged capacitor is put between the poles of a magnet, with the electric and magnetic fields in perpendicular directions, there will be an energy flux flowing inside the capacitor, perpendicular to both fields. This is an unexpected result.

It is not possible to *prove* that the flux of energy is given by Poynting's equation. There are other formulae compatible with the basic equations of electromagnetism, as proved by Ritz 20 years later [95]. However, most authors understood at that time (as they do today) that the Poynting vector is the simplest and most acceptable equation for the flux of electromagnetic energy.

The simplest interpretation of the energy flux, in the late nineteenth century, was that it was somehow related to the motion of the ether due to the electric and magnetic fields carrying the electromagnetic energy. This suggested that, if the ether has mechanical properties, it should also have a momentum whenever the Poynting vector is not null.

J.J. Thomson introduced the concept of an electromagnetic momentum associated with the electromagnetic field in 1893 [115, Chap. 1; 117]. According to him, wherever there is an energy flux $\vec{S}$ there is also a density of electromagnetic momentum $\vec{g}$ proportional to the Poynting vector:

$$\vec{g} = \frac{\vec{S}}{c^2}. \tag{26}$$

Today, this formula is interpreted as a description of the momentum density of the electromagnetic field in empty space. At that time, electromagnetic fields were regarded as special conditions of the ether, and accordingly Thomson's proposal should be understood as the description of a new physical property of the ether. The rationale for introducing a momentum associated with the ether was very simple and convincing. The ether can produce forces upon charged particles. Therefore, the particles acted on by the ether undergo momentum changes. If the ether did not have a momentum, this would violate the law of momentum conservation.

In 1894, Hermann von Helmholtz (1821–1894) suggested that the ether probably moves with a very large velocity in strong electromagnetic fields [41]. Three years later, W.S. Henderson and J. Henry attempted to observe the motion of the ether in electromagnetic fields, using an optical interferometer [42]. This experiment was different to and independent of the Michelson–Morley experiment, since it did not attempt to detect the motion of the Earth through the ether, but attempted to put the ether in motion, in the neighbourhood of the Earth, by applying a strong electromagnetic field and measuring an eventual change in the velocity of the light in that region. No effect was observed, although the experiment seemed able to detect a variation in the speed of light of about 10 m/s.

In theories where there is instantaneous direct action at a distance between two particles, this problem does not occur, because the total momentum of the system of particles is constant. However, in field theories – such as Maxwell's theory – the interaction between two particles is not instantaneous: each particle interacts directly only with the field – that is, with the ether – and the effect is propagated from one point to a distant one at the speed of light. Therefore, in field theories, the field itself must have a momentum, otherwise Newton's third law is violated.

9. Maxwell and the Motion of the Earth through the Ether

When Maxwell developed the theory of electromagnetic waves, he proved that the speed of those waves should have a speed equal to (or close to) the measured speed of light, and concluded that light was an electromagnetic wave.[11] When in 1887 Heinrich Hertz (1857–1894) produced short wavelength electromagnetic waves in the laboratory and showed that they had indeed the speed of light, Maxwell's theory was strongly confirmed [43]. The reduction of light and optics to electromagnetism required the unification of two ethers: the ether that had been hypothesized by the defenders of the wave theory of light (such as Fresnel), and the electromagnetic ether.

The two main theories of the light ether had been proposed by Augustin Fresnel (1788–1827) and George Gabriel Stokes (1819–1903), in the early nineteenth century. Fresnel put forward the theory of an ether at rest in all places of the universe [27]. Bodies would move through the ether (or, conversely, the ether was able to flow through all material bodies). Stokes suggested that the ether could move, as a viscous fluid, attaching itself to the surface of material bodies [109].

Fresnel's theory was highly successful. It explained the aberration of star light, explained the null result of experiments carried out by François Arago (1786–1853) and other researchers who tried to detect the motion of the Earth relative to this stationary ether, and correctly predicted the velocity of light in a moving liquid [82]. This was confirmed in 1851 by Hippolyte Fizeau (1819–1896), using interference phenomena to measure changes of speed of light in moving water [25].

---

[11] Before the development of his electromagnetic theory, Maxwell had already been led to believe that light was an electromagnetic wave. One of the arguments that led to this conclusion was dimensional analysis. See [17].

In order to account for the polarization of light, the luminous ether should be able to transmit *transverse* waves, such as those produced in an elastic solid (which are different from the *longitudinal* waves transmitted by gases, such as sound in the air). It was difficult to envisage how those properties could be combined with those of the electromagnetic ether, because magnetic energy was regarded as kinetic energy of the ether.[12] Maxwell did not address this problem. However, he accepted the idea of a stationary ether and wondered about the possibility of detecting the motion of the Earth relative to this ether [54].

At the time when Maxwell devoted his attention to this problem, there were several known optical experiments that had been unable to detect the motion of the Earth though the ether. Fresnel had proved theoretically that this motion cannot be detected by measuring the deflection of light by a prism, but Maxwell attempted this kind of experiment – and got no positive results. Stokes proved that light experiments using reflection and refraction could not detect any influence of the motion of the Earth through the ether, in Fresnel's theory, to the first order of v/c [110]. Assuming that the speed of the solar system through the ether was similar to the orbital speed of the Earth (about 30 km/s), one had $v/c = 0.0001$ (or 10^{-4}); therefore, no effect of this order of magnitude could be expected to occur in optical experiments.

However, there had been two experiments that had reported positive results in the search for the influence of the motion of the Earth upon optical experiments.

One of them was carried out by Fizeau in 1859, and amounted to a small rotation of the plane of polarization of light when it traversed a pile of glass plates. The rotation was different depending on if the light was travelling in the direction of the motion of the Earth or in the opposite direction [26]. The other experiment had been carried out by Anders Jonas Ångström (1814–1874) in 1865. He measured the deflection of light by a diffraction grating, and reported that this deflection was different when light travelled in the same direction as the Earth and in the opposite direction. In both cases, the effects were very small, and the experiment was difficult to reproduce. Eleuthère Élie Nicolas Mascart (1837–1908) repeated both experiments, and reported no positive effect [72].

Maxwell was aware of Fizeau's and Ångström's experiments (however, he was not aware of Mascart's experiments), but thought that they were not decisive. He suggested two new experiments [54]: one based on the study of the eclipses of Jupiter's satellites; the other one was a proposal for experiments measuring the time taken by light to travel to and fro between the source, a mirror, and back to the source. In the first case, he expected a first-order effect, that is, about 10^{-4}. In the second case, he expected a second-order effect, that is, about 10^{-8} – a very small effect indeed. He thought that it was impossible, with current techniques, to measure such a small effect.

Maxwell consulted an astronomer, David Peck Todd (1855–1939), who told him that the available data on Jupiter's satellites were not precise enough to test the predicted effect. However, Maxwell's ideas stimulated a young experimental physicist, Albert Abra-

---

[12]Some contemporary authors emphasize that the concept of the ether, in the late nineteenth century, was contradictory and absurd. There were, indeed, serious difficulties, but there was no proof that the concept of the ether was contradictory, or that it was impossible to surmount those difficulties. On the other hand, those critics of the ether should be asked whether physicists confronted with contradictions should always reject their theories, and should instead devote some time to considering wave-particle duality, and wondering whether the usual interpretation of quantum mechanics does not pose more serious difficulties than the old ether theory.

ham Michelson (1852–1931), who decided to follow the second path suggested by him. This led to the famous interferometer experiment, a few years later.

The story of the Michelson–Morley experiment is well known [105]. Michelson attempted to measure the second-order effect of the motion of the Earth relative to the ether, assuming Fresnel's theory. His first optical interferometer was built in Berlin, and the measurements were made in April 1881. He was unable to observe the predicted regular shift of the interference fringes, and concluded that the hypothesis of the stationary ether (Fresnel's theory) was wrong. He insinuated that Stokes' theory could be the correct one. However, Michelson's first experiment was not reliable. The instrument was not sufficiently stable, and its sensitivity was too low. Besides that, Michelson's mathematical analysis was wrong, and was soon criticized by Lorentz (1886), Potier and other physicists.

Fresnel's theory had been strikingly confirmed by Fizeau's 1851 measurements of the speed of light in moving water. In 1886 Michelson, with the help of Edward W. Morley (1838–1923), repeated and confirmed Fizeau's experiment. Urged by Lord Rayleigh, Michelson and Morley decided to repeat the 1881 interferometer experiment. In 1887, with improved apparatus, they obtained a significant null outcome (irregular displacements, much smaller than the predicted theoretical displacement).[13] The result disagreed with Fresnel's theory, but neither Michelson and Morley nor most of the physicists of that time concluded that the ether did not exist. The experiment simply seemed a new constraint to be taken into account in constructing ether theories.

10. Lorentz, Poincaré and the Impossibility of Detecting Motion Relative to the Ether

Hendrik Antoon Lorentz (1853–1928) was the main physicist who attempted to reduce optics to electromagnetism, following Maxwell's theory. He adopted Fresnel's stationary ether, and studied the properties of light using electromagnetism together with an atomic model of matter [77,78]. To account for reflection and refraction, he assumed that matter contained charged particles ("ions") that could respond to electromagnetic waves.

Fresnel's theory had accounted for the null results of optical attempts to measure the speed of the Earth through the ether (to the first order of v/c). Lorentz decided to tailor Maxwell's theory in such a way that it would also predict the same electromagnetic phenomena (to the first order of v/c), independently of the motion of the system through the ether. Other authors, such as Joseph Larmor, in England, and Henri Poincaré also followed a similar path [18].

In a striking series of papers, those authors gradually built the electrodynamics of moving bodies. The main results were the set of field transformation equations, allowing the calculation of electric and magnetic fields relative to different reference systems; and the space and time transformation equations that, together with the field transformation equations, preserve Maxwell's equations in reference systems moving relative to the ether.

Those equations were not found immediately. Lorentz first attempted to find only first-order equations – that is, the electromagnetic counterpart to Fresnel's optical the-

---

[13]Later experiments, by Morley and Dayton C. Miller (1866–1941), produced small *positive* results [111].

ory [98]. However, when Michelson and Morley were unable to detect a second-order optical effect, the situation changed. Under the pressure of this experimental outcome, several authors began to suspect that the negative results had a deeper meaning. Henri Poincaré suggested that all future attempts, of any kind, would also be unable to detect the motion of the Earth relative to the ether, and that this result should be taken into account in all future theories. This is the essence of the principle of relativity [100]. After struggling for several years, at last Lorentz developed the transformations of the electromagnetic field and the space and time transformations that preserved Maxwell's equation *exactly* in reference systems moving through the ether.[14] Those developments are described in Chapter 4.

11. Lorentz, Poincaré and the Momentum of the Ether

Let us now return to the problem of the dynamics of the ether.

In his first versions of the electrodynamics of moving bodies, up to 1895, Lorentz did not introduce the concept of electromagnetic momentum, because he accepted that the ether was always at rest [66]. Therefore, in his theory, the ether acted upon charged particles, but the particles did not act upon the ether, and therefore there was a violation of the conservation of momentum. In 1900, in a paper where he discussed this problem, Poincaré proved that, in order to maintain the principle of conservation of momentum, it was necessary to ascribe a density of momentum to the electromagnetic field, by following Eq. (26), that is, J.J. Thomson's formula [87].[15] According to Poincaré, it was necessary to accept that the ether (or something associated with the ether) is equivalent to a fluid endowed with mass, able to produce forces and to be put in motion, and capable of carrying momentum.

At first Lorentz did not accept Poincaré's analysis. However, the concept of an electromagnetic momentum was soon accepted by most theorists, and the ether acquired an additional dynamical property. In particular, Poincaré associated a momentum $p = E/c$ to a directed pulse of electromagnetic radiation. This momentum is another way of explaining the pressure produced by radiation – or, conversely, it can be deduced from the formula for radiation pressure. In the same paper, Poincaré also showed that the theorem of centre of mass would hold for a charged particle plus the ether if a mass $m = E/c^2$ was associated with the energy E of the electromagnetic field. He did not apply this relation to material bodies, however.[16]

Poincaré also introduced the concept of an angular momentum associated with electromagnetic fields where the Poynting has a non-vanishing rotational. The only relevant application seems to be the angular momentum of circularly polarized light – a property that was confirmed by experiment many years later [9].

---

[14]There were some mistakes in Lorentz's 1904 paper, such as his inadequate transformation of current density, which were corrected by Poincaré in 1905. Only Poincaré's version of Lorentz's theory was completely covariant.

[15]On Poincaré's contribution one might consult [15,30] and [29].

[16]Once Poincaré established this result, it would be very easy, employing the principle of relativity (which Poincaré had already proposed) to prove that the relation $m = E/c^2$ also applies to material bodies. See [47,61].

12. The Discovery of the Electron

The successful atomic theory of matter led Lorentz, in 1892, to propose an atomic theory of electricity. He suggested that neutral atoms (or molecules) were composed of one larger charged particle connected to a smaller particle of opposite charge. The smaller particle (called an "ion" by Lorentz) was responsible for several properties of matter, including refraction and emission of light.

In 1896 Pieter Zeeman (1865–1943) discovered the splitting of spectral lines of sodium when the source was placed in a strong magnetic field. Zeeman's effect was immediately explained by Lorentz using his theory. He was able to compute the order of magnitude of the ratio of charge to the mass of the hypothetical ions, and showed that they should have a negative charge.

At the same time, following an independent line of investigation, J.J. Thomson had measured the charge-to-mass ratio e/m of cathode rays [116] and obtained a result similar to Lorentz's value for the ion. Independently of Thomson, Walter Kaufmann also studied the deflection of cathode rays and obtained similar results [48]. The agreement between the two parallel developments led to the rapid acceptance of the existence of the "electron" (a name proposed by G. Johnstone Stoney) as an universal component of matter.

The theoretical studies previously described had shown that one should ascribe an electromagnetic mass to a moving charge. Instead of a generic charged particle, the next research attempted to improve the analysis of the electromagnetic mass of the electron.

In 1898 and 1900, Phillip Lenard (1862–1947) measured e/m for very fast electrons, with speeds up to $c/3$ [60]. Although his measurements were regarded as inconclusive, they seemed to reveal an increase of mass with speed. The "purely mechanical" mass of a particle was supposed to be independent of its speed, while the electromagnetic mass was a function of its speed.

In 1900, Wilhelm Wien (1864–1928) conjectured (as Oliver Lodge had already proposed) that perhaps all mass was of electromagnetic origin, and in that case the fundamental mechanical properties of matter would be reduced to electromagnetism [121]. Maxwell had shown that electromagnetism could be described according to Lagrange's approach, and this was understood as proof that it was possible to provide a mechanical foundation for electromagnetism. Conversely, around 1900, this was interpreted as proof that mechanics could be reduced to electromagnetism. Of course, this reduction was only possible if all mass was of electromagnetic origin.

It became imperative to improve Lenard's measurements, and to establish the theory of electromagnetic mass on solid grounds, in order to compare the theory to a experiment, to decide whether the electron had a "purely mechanical" mass, or whether its mass was solely electromagnetic.

13. Kaufmann's Measurement of the Mass of Fast Electrons

In 1901 Walter Kaufmann (1871–1947) published the result of measurements of e/m for very fast electrons (beta radiation emitted by a radium compound), with speeds between 0.8–$0.9c$ [49]; see also [16]. He noticed a very clear increase of e/m (and, therefore, an increase of the mass of the electron) in the range of velocities he studied. The quantitative analysis presented by Kaufmann assumed that the electrons had a "real mass" m_0 (that is, a purely mechanical mass, independent of its charge) and an "apparent mass" μ (that is, its electromagnetic mass). He calculated the "apparent mass" from Searle's formula of the energy of a charged spherical particle:

$$W = \frac{q^2}{8\varepsilon_0 \pi R}\left(\frac{c}{v}\log\frac{c+v}{c-v} - 2\right). \tag{27}$$

However, the electromagnetic mass cannot be directly obtained from this formula by a simple comparison with the classical equation for the kinetic mass of a particle. Therefore, Kaufmann used a new approach. The kinetic energy of a particle is equal to the total work produced by the force that accelerates it. The change dW of the kinetic energy is therefore equal to the work $F\,dx$ produced by the force F when the particle undergoes a displacement dx. But $dx = v\,dt$, therefore $F = (1/v)(dW/dt)$.

Now, defining mass as the ratio between force and acceleration, $m = F/a$, and since $a = dv/dt$, we obtain:

$$m = \frac{F}{a} = \frac{dW/(v\,dt)}{dv/dt} = \frac{dW}{v\,dv}. \tag{28}$$

Applying this formula to Searle's equation, Kaufmann obtained the "apparent" electromagnetic mass:

$$\mu = \frac{q^2}{8\varepsilon_0 \pi R v^2}\left[-\frac{c}{v}\log\left(\frac{c+v}{c-v}\right) + \frac{2}{1-v^2/c^2}\right]. \tag{29}$$

Comparison between the experimental data and this formula showed that they did not match. Kaufmann was therefore led to assume that only *part* of the electron mass was electromagnetic, and the other part was mechanical (or "real"). A good fit with the experimental data was obtained assuming that about one-third of the total mass was "apparent" (or electromagnetic) in the low speed limit.

14. The Momentum of an Electron

Kaufmann's theoretical analysis was unsound. His derivation of the mass of the electron from energy considerations could only be valid if he were studying the *longitudinal acceleration* of electrons – that is, their change of speed due to some force. All equations he used had a scalar form (that is, they did not take into account the *direction* of motion) and could only make sense if the speed changed, because of the relations used in the derivation shown above. However, in Kaufmann's experiments he only measured the deflection of electrons when submitted to forces perpendicular to their velocities. The comparison between theory and experiment required a different analysis, taking into account the momentum of the electrons.

In January 1902, a few months after the publication of Kaufmann's measurements, Max Abraham (1875–1922) published a new theoretical analysis of the dynamics of moving electrons [1].[17] He conceived the electron as a rigid sphere, with its charge ei-

---

[17]See also [2–4]. On Abraham's contributions, see [32,16].

ther spread on its surface, or homogenously distributed in its volume. Starting from the Thomson relation between density of momentum and energy flux (Poynting vector), and integrating for the whole space, Abraham was able to compute the momentum associated with a moving electron.[18] After lengthy calculations, he obtained, in the case of the superficial charge:

$$G = \frac{e^2}{8\varepsilon_0\pi Rc}\left[\left(\frac{1+\beta}{2\beta^2}\right)\ln\left(\frac{1+\beta}{1-\beta}-1\right)\right],$$ (30)

where $\beta = v/c$. This momentum is a vector, and its direction is the same as that of the velocity of the electron. If we represent by $\hat{v}$ an unit vector parallel to the velocity of the electron, then we have:

$$\vec{G} = G\hat{v} = G\vec{v}/c\beta.$$ (31)

Therefore, Eq. (30) can also be written as:

$$\vec{G} = \frac{e^2}{8\varepsilon_0\pi Rc^2}\left[\left(\frac{1+\beta}{2\beta^3}\right)\ln\left(\frac{1+\beta}{1-\beta}-1\right)\right]\vec{v}.$$ (32)

Abraham used Newton's second law in the form $\vec{F} = d\vec{p}/dt$. When the electron is submitted to an external force, its momentum can change in both magnitude and direction. When the force acting upon the electron is parallel to its initial velocity, only the magnitude of the momentum will change. In that case (longitudinal acceleration), we have:

$$\vec{F}_{//} = \frac{d\vec{G}}{dt} = \frac{dG}{dt}\hat{v} = \frac{dG}{dv}\frac{dv}{dt}\hat{v} = \frac{dG}{dv}\vec{a}_{//}.$$ (33)

In this case (longitudinal force), $(dv/dt)\hat{v}$ is the longitudinal acceleration $\vec{a}_{//}$, and dG/dv can be interpreted as the electron mass. Abraham called it the "longitudinal mass".

If the force is perpendicular to the velocity of the electron (for example, for a magnetic force acting upon a moving charge), the speed and the value of the momentum will not change, but the motion will suffer a deflection and the path will be circular. In that case, we have:

$$\vec{F}_\perp = \frac{d\vec{G}}{dt} = G\frac{d\hat{v}}{dt} = \frac{G}{v}\frac{d\vec{v}}{dt} = \frac{G}{v}\vec{a}_\perp.$$ (34)

The transverse acceleration $\vec{a}_\perp$ is the centripetal acceleration of the electron's circular motion, and G/v can be interpreted as the electron mass. Abraham called it the "transverse mass". From Eq. (32), we obtain for the longitudinal mass:

$$m_{//} = \frac{e^2}{8\varepsilon_0\pi Rc^2}\frac{1}{\beta^2}\left[-\frac{1}{\beta}\ln\left(\frac{1+\beta}{1-\beta}\right)+\frac{2}{1-\beta^2}\right]$$ (35)

---

[18]In 1901, Lorentz had computed the momentum of an electron, but he used an approximation that was only valid for low speeds.

that is, Kaufmann's formula (29). And, for the transverse mass:

$$m_\perp = \frac{e^2}{8\varepsilon_0 \pi Rc^2} \frac{1}{\beta^2}\left[\left(\frac{1+\beta}{2\beta}\right)\ln\left(\frac{1+\beta}{1-\beta}-1\right)\right].$$ (36)

Both equations reduce to $e^2/(6\varepsilon_0 \pi Rc^2)$ for speeds much smaller than c.

Using this analysis, Abraham showed that Kaufmann's experimental data should be compared to the transverse mass, because he measured the deflection of electrons. However, Kaufmann had used an equation equivalent to Abraham's longitudinal mass. Therefore, his analysis was wrong.

Kaufmann recognized his mistake, and in 1902 he published a new paper, with additional measurements analysed using Abraham's theory [50]. He concluded that there was good agreement between the experimental data and the transverse mass formula, and concluded that the entire mass of the electron was purely electromagnetic.

15. The Mass of a Box Full of Light

The experimental confirmation of the pressure of light in 1901 led to new theoretical work. In 1904, Max Abraham computed the pressure produced by radiation upon a *moving* surface, when the beam of light reaches the surface of a mirror in any angle [5]. Starting from Abraham's results, Friedrich Hasenöhrl (1874–1916) studied the dynamics of a box full of radiation [36].

Imagine a cubic box with perfectly reflecting internal surfaces, full of light. When the box is at rest, the radiation produces equal forces upon all those surfaces. Now, suppose that the box is accelerated, in such a way that one of its surfaces moves in the x direction. It is possible to prove that, when the radiation inside the box strikes this surface, the pressure will be smaller, and when it strikes the opposite surface, the pressure will be greater, than in the case when the box is at rest (or in uniform motion). Therefore, the radiation inside the box will produce a *resultant force* against the motion of the box. So, to accelerate a box full of light requires a greater force than to accelerate the same box without light. In other words, the radiation increases the inertia of the box. In the case when the radiation inside the box is isotropic, there is a very simple relation between its total energy E and its contribution m to the inertia of the box:[19]

$$m = \frac{4E}{3c^2}.$$ (37)

Note that here, as in the theory of the electron, there appears a numerical factor $4/3$. This is not a mistake. The relation between those equations and the famous $E = mc^2$ will be made clear later.[20]

Hasenöhrl also computed the change of the radiation energy as the box was accelerated. He proved that the total radiation energy would be a function of the speed of the box. Therefore, when the box is accelerated, part of the work done by the external forces is transformed into the extra radiation energy. Since the inertia of the radiation is

---

[19]Hasenöhrl arrived at a different result, in 1904, but his integration mistake was corrected by Max Abraham in the same year and acknowledged by Hasenöhrl in 1905.

[20]Fadner [21] describes several contributions concerning the establishment of $E = mc^2$.

proportional to its energy, and since this energy increases with the speed of the box, the inertia of the box will increase with its speed. Of course, if the internal temperature of the box were increased, the radiation energy would augment, and the inertia of the box would also increase. Therefore, Hasenöhrl stated that the mass of a body depends on its kinetic energy and temperature.

16. Lorentz's Electron

In 1892, Lorentz studied the null result of the Michelson–Morley experiment and came to the conclusion that it could be explained if all bodies moving through the ether underwent a longitudinal contraction,[21] according to this equation:

$$L = L_0\sqrt{1 - v^2/c^2}. \tag{38}$$

Afterwards, Lorentz assumed that this contraction should apply to the microscopic components of material bodies, including the electron. Therefore, he was led to develop a new theory of the electron, similar to that proposed by Abraham, but using a contracted electron, instead of a spherical one [69]. The principle of calculation is the same as Abraham's: Lorentz assumed that the momentum of the electron is the momentum of the electromagnetic field around it. Only the geometrical limits of integration are different.

The results were published in 1904. For the momentum of the electron, Lorentz obtained this formula:

$$G = \frac{e^2}{6\varepsilon_0 \pi Rc^2} \frac{1}{\sqrt{1 - v^2/c^2}} v. \tag{39}$$

Following Abraham's approach, he computed the longitudinal and transverse masses of the electron as follows:

$$m_{//} = \frac{e^2}{6\varepsilon_0 \pi Rc^2} \frac{1}{(1 - v^2/c^2)^{3/2}}, \tag{40}$$

$$m_\perp = \frac{e^2}{6\varepsilon_0 \pi Rc^2} \frac{1}{(1 - v^2/c^2)^{1/2}}. \tag{41}$$

Lorentz's equations look quite different from Abraham's (and are much simpler). However, for low speeds they provide similar results. If we develop Abraham's formula for the transverse mass of the electron in series, we obtain:

$$m_\perp = \frac{e^2}{8\varepsilon_0 \pi Rc^2} \frac{1}{\beta^2} \left[\left(\frac{1+\beta}{2\beta} \right) \ln \left(\frac{1+\beta}{1-\beta} - 1 \right) \right]$$

$$\cong \frac{e^2}{6\varepsilon_0 \pi Rc^2} \left(1 + \frac{2}{5}\beta^2 + \cdots \right). \tag{42}$$

---

[21] In his 1892 paper, Lorentz published an equation valid to second-order approximation: $L = L_0(1 - v^2/2c^2)$. Lorentz accepted that both the length and the width of the interferometer could change when the apparatus moved through the ether. The null result could be accounted for if the ratio between them obeyed Eq. (38).

In the case of Lorentz's formula, the development in series gives:

$$m_\perp = \frac{e^2}{6\varepsilon_0 \pi Rc^2} \frac{1}{(1-v^2/c^2)^{1/2}} \cong \frac{e^2}{6\varepsilon_0 \pi Rc^2} \left(1 + \frac{1}{2}\beta^2 + \cdots\right). \qquad (43)$$

Comparing Eqs (42) and (43), we see that, for low speeds (that is, when $v/c = \beta$ is much smaller than 1) Abraham's and Lorentz's formulas provide similar results.

The comparison between the theories and the experimental results was indirect. Lorentz analysed Kaufmann's experiment using his own theory, and concluded that the data was compatible both with his theory and with Abraham's.

17. Bucherer's Electron

Besides Abraham's and Lorentz's models of the electron, there were other possibilities, of course. In the same year when Lorentz published his results, Alfred Heinrich Bucherer (1863–1927) proposed another theory [12,13]. He assumed that the electron contracted due to its motion, as Lorentz had assumed, but he supposed that the *volume* of the electron remained constant. Therefore, the longitudinal radius of the contracted electron would become $L = R(1 - v^2/c^2)^{1/3}$ and its transverse radius becomes $L' = R(1 - v^2/c^2)^{-1/6}$, where R is the radius of the electron at rest.

Following the same basic theoretical ideas as Abraham and Lorentz, Bucherer obtained the following values for the transverse and longitudinal masses of the electron:

$$m_{//} = \frac{e^2}{6\varepsilon_0 \pi Rc^2} \frac{1}{(1-v^2/c^2)^{4/3}}, \qquad (44)$$

$$m_\perp = \frac{e^2}{6\varepsilon_0 \pi Rc^2} \frac{1}{(1-v^2/c^2)^{1/3}}. \qquad (45)$$

Independently of Bucherer, the same theory was proposed by Paul Langevin [56].

Other alternatives could also be envisaged. The charge of the electron could be spread over its surface or across its volume; the distribution of its charge could change, due to its motion, etc.

It seemed possible to test (by deflection experiments) which formula provided the best fit for the experimental data. It was also possible to provide theoretical arguments for choosing between the options.

In the next years, Kaufmann published new experimental data and compared his measurements to the three available theories of the electron [52,53]. He concluded that Abraham's formula provided the best fit. Max Planck (1858–1947), however, criticized Kaufmann's analysis [85], and concluded that his measurements were compatible with both Abraham's and Lorentz's equations – and that Lorentz's equation provided the best fit. At that time the situation was not clear [8,16]. Only ten years later, new experiments were able to confirm Lorentz's formula and rule out all other models.

18. The Contribution of Stresses to Mass

The experimental evidence was not sufficiently clear to provide a choice between the several models of the electron. Poincaré, however, provided a significant theoretical argument for Lorentz's theory.

In a paper written in 1905 and published the following year, Poincaré provided a review of Lorentz's theory and an analysis of the three theories of the electron [92]. He showed that Lorentz's theory should be supplemented by the assumption of a non-electromagnetic force holding the electron together. This force could be described as a negative pressure, of unknown origin. It was necessary to take into account this stress when computing the energy and momentum of the electron, and therefore the dynamics of the electron could not be derived solely from electromagnetism. However, after introducing this complement to Lorentz's theory, he proved that only Lorentz's theory was compatible with the principle of relativity. Conversely, if Abraham's or Bucherer's theory of the electron was valid, it should be possible to check, by measuring the dynamical properties of the electron, whether the Earth is at rest or in motion relative to the ether.

In Lorentz's theory, the electromagnetic mass of the low-speed electron should be equal to $m_0 = e^2/(6\varepsilon_0\pi Rc^2)$. The same result holds in the theories of Abraham and Bucherer. Now, the electrostatic energy of an electron at rest is $W_0 = e^2/(8\varepsilon_0\pi R)$. Therefore, we can write the following relation between mass and energy of low-speed electrons:

$$m_0 = \frac{4W_0}{3c^2}. \tag{46}$$

For any contemporary reader, this relation seems very odd, because we are used to Einstein's equation $m = E/c^2$, without the numerical factor 4/3. This difference was not due to any mistake made by Lorentz and the other theoreticians. It is an unavoidable consequence of electromagnetic theory.

However, there was one small blunder. Poincaré proved that electromagnetism could not provide a complete analysis of the electron. It was necessary to introduce other non-electromagnetic forces both to explain the stability of the electron and to produce a completely coherent dynamics (see [15]). Therefore, Poincaré was led to introduce a non-electromagnetic force that could be described as a negative pressure, counteracting the repulsive force produced by the charge at the surface of the electron. Taking this force into account, it is necessary to include a non-electromagnetic component in the equations of energy and momentum. This correction leads to a new relation between the total inertial mass m_0' and the total energy W_0'

$$m_0' = \frac{W_0'}{c^2} \tag{47}$$

which is compatible with the well known result of Einstein's theory.[22]

Most results of relativity theory were embodied in a paper Poincaré wrote in 1905, but this was published only the following year, in Italy [92].[23]

---

[22]Poincaré's approach is not accepted by all authors. F. Rohrlich and other physicists have criticized the introduction of Poincaré's stress and attempted to establish compatibility between electrodynamics and special relativity by a different route. It seems, however, that Poincaré was correct, and that we have two different viable approaches: Poincaré's and Rohrlich's. See [35].

[23]In 1905 Poincaré published a short note in the *Comptes Rendus* of the French Academy of Science, anticipating (without proof) some of the results of the larger paper. A lot of ink has already been used concerning the similarities and differences between Poincaré's contribution and Einstein's work. See Schwartz [99] (who translated a large part of Poincaré's 1906 paper); [79,62,29].

An argument similar to Poincaré's can be applied to the case of the box full of light. The mass associated to the light obeys a relation such as Eq. (46). However, this radiation produces pressure upon the walls of the box that contains it. The box is stretched by this pressure, and this tension must be taken into account when computing the mass of the whole system. For the complete system (box plus radiation), the relation between mass and energy obeys Eq. (47). This result was established by Max Planck, in 1907, after the publication of Einstein's theory ([86] see also [33]). In this fundamental paper, Planck proved that $m = E/c^2$ is not a general law. Indeed, it is valid for *closed* systems. However, any system submitted to an external pressure[24] will obey a different law: its mass will be proportional to its enthalpy $H = E + PV$, that is, $m = H/c^2$.

19. Conclusion

The development of Maxwell's electromagnetism, in his hands and in those of his followers, led to a new world view. On one hand, they provided a dynamical theory of the ether, showing that it was possible to ascribe forces, pressures, energy, momentum and mass to the electromagnetic field. The general dynamical relations were then applied to the electron, which was regarded as the fundamental constituent of matter, and the dynamics of the electron was derived from electromagnetism. Confrontation with experiment confirmed the theoretical prediction that the mass of the electron increased with its speed. This was an outstanding confirmation of the theory.

Since the dynamical properties of the electron could be derived from electromagnetism, and as the electron was regarded as the fundamental constituent of matter, many physicists thought that matter should be regarded as an electromagnetic phenomenon, and that all laws of matter (including mechanics) would be soon regarded as consequences of electromagnetism. This reduction of physics to electromagnetism was however shown to be impossible by Poincaré, who proved that it was necessary to introduce non-electromagnetic forces in the theory of the electron.

Notice that the development of the mass–velocity and the mass–energy relations depended on many distinct contributions by several different physicist. This is the rule, not the exception, in the history of science. To attribute a complex theory, such as relativity, to a single person, is a complete distortion of history.

Physicists usually praise Maxwell for his four equations (which he never wrote), and pardon him his belief in the ether (which was central to his work). We have seen, however, that the despised ether concept led to a series of dynamical studies that generated some of the most important results of the theory of relativity.

From Maxwell to Lorentz and Poincaré, the belief in the ether as the fundamental substratum of electromagnetic phenomena guided the study of its dynamical properties. Without that belief, the developments described in this chapter could not have occurred. Therefore, belief in the ether and the study of its properties was the fundamental step in the unfolding of relativistic dynamics.

Of course, confirmation of several consequences of the ether theory are not a proof that ether exists – just as the confirmation of several consequences of any theory (including Einstein's relativity or quantum mechanics) never prove that the theory is correct.

---

[24]This effect is relevant only when the body is submitted to a pressure or stress – independently of being accelerated – and when its volume is not neglected.

The theory that reached its peak at the hands of Lorentz and Poincaré was not Einstein's theory. Their world-view was different. They accepted the ether, although they also accepted that it was impossible to detect motion relative to this medium. Their epistemological approach was also different to Einstein's. However, the main predictions of Einstein's theory were already there, in the papers written before his first article. It seems impossible to distinguish, by any experiment, Lorentz's and Poincaré's theory from Einstein's special relativity. The empirical content of those theories is identical.[25,26]

References

[1] Abraham M. 1902: "Dynamik des Elektrons". *Königliche Gesellschaft der Wissenschaften zu Göttingen. Mathematisch-Physikalische Klasse. Nachrichten*, pp. 20–41.

[2] Abraham M. 1903: "Prinzipien der Dynamik des Elektrons". *Annalen der Physik* [series 4] **10**, pp. 105–179.

[3] Abraham M. 1903: "Prinzipien der Dynamik des Elektrons". *Physikalische Zeitschrift* **4**, pp. 57–63.

[4] Abraham M. 1904: "Die Grundhypothesen der Elektronentheorie". *Physikalische Zeitschrift* **5**, pp. 576–579.

[5] Abraham M. 1904: "Zur Theorie der Strahlung und des Strahlungsdruck". *Annalen der Physik* [series 4] **14**, pp. 236–287.

[6] Agassi J. 1971: *Faraday as a natural philosopher*. Chicago: University of Chicago Press.

[7] Barnett S.J. 1910: "On the question of the motion of the ether in a steady electromagnetic field". *Physical Review* **31**, pp. 662–665.

[8] Battimelli G. 1981: "The electromagnetic mass of the electron: A case study of a non-crucial experiment". *Fundamenta Scientiae* **2**, pp. 137–150.

[9] Beth R.A. 1936: "Mechanical detection and measurement of the angular momentum of light". *Physical Review* **50**, pp. 115–125.

[10] Boltzmann L. 1884: "Ueber eine von Hrn. Bartoli entdeckte Beziehung der Wärmestrahlung zum zweite Hauptsätze". *Annalen der Physik und Chemie* [series 3] **22**, pp. 31–72.

[11] Bork A.M. 1967: "Maxwell and the vector potential". *Isis* **58**, pp. 210–222.

[12] Bucherer A.H. 1904: *Mathematische Einführung in die Elektronentheorie*. Leipzig: Teubner.

[13] Bucherer A.H. 1905: "Das deformierte Elektron und die Theorie des Elektromagnetismus". *Physikalische Zeitschrift* **6**, pp. 833–834.

[14] Carazza B. and Kragh H. 1989: "Adolfo Bartoli and the problem of radiant heat". *Annals of Science* **46**, pp. 183–194.

[15] Cujav C. 1968: "Henri Poincaré's mathematical contributions to relativity and the Poincaré stresses". *American Journal of Physics* **36**, pp. 1102–1113.

[16] Cushing J.T. 1981: "Electromagnetic mass, relativity, and the Kaufmann experiments". *American Journal of Physics* **49**, pp. 1133–1149.

---

[25]There have been several claims, old and new, that Einstein's relativity theory and Lorentz's and Poincaré's ether theory can be experimentally contrasted in specific situations. Chalmers Sherwin [106] proposed and performed one such experiment, involving a fast rotating structure, and he maintained that it was incompatible with Lorentz's theory. I am not aware of any published criticism of Sherwin's experiment, but I would like to point out that some previously proposed experiments involved a series of mistakes that have been analysed by Torr and Kolen [119], by Rodrigues and Tiomno [96,97] and other authors.

[26]Editor's note: Lorentz's theory was valid only for "systems moving with any velocity less than that of light", as the title of his 1904 paper reads; in the text he pointed out that his theory did not have universal validity: "The only restriction as regards the velocity will be that it be less than that of light". This means that Lorentz did not rule out faster-than-light velocities, for signals or even for reference systems, but he only dealt with ordinary, 'subluminal' systems. On the other hand, in the final section of his 1905 paper Einstein explicitly says that, "Velocities greater than that of light have – as in our previous results – no possibility of existence". It seems to me that two theories such that one accepts, while the other denies, that faster-than-light systems and particles exist, cannot be considered to have "identical empirical content".

[17] D'Agostino S. 1986: "Maxwell's dimensional approach to the velocity of light". *Centaurus* **29**, pp. 178–204.

[18] Darrigol O. 1994: "The electron theories of Larmor and Lorentz: a comparative study". *Historical Studies in the Physical and Biological Sciences* **24**, pp. 265–336.

[19] Darrigol O. 1996: "The electrodynamic origins of relativity theory". *Historical Studies in the Physical and Biological Sciences*, **26**, pp. 241–312.

[20] Darrigol O. 2000: *Electrodynamics from Ampère to Einstein*. New York: Oxford University Press.

[21] Fadner W.L. 1988: "Did Einstein really discover $E = mc^2$?" *American Journal of Physics* **56**, pp. 114–122.

[22] Faraday M. 1852: "On the physical character of the lines of magnetic force". *Philosophical Magazine* [series 4] **3**, pp. 401–428.

[23] Faraday M. 1952: "Experimental researches in electricity". In: HUTCHINS, Robert Maynard (ed.) *Great books of the western world*. Chicago: Encyclopaedia Britannica, Inc.

[24] Fitzgerald G.F. 1881: "Note on Mr. J.J. Thomson's investigation of the electromagnetic action of a moving electrified sphere". *Proceedings of the Royal Dublin Society* **3**, 250–254. Reprinted in: *Philosophical Magazine* [series 5] **3**, 302–305, pp. 1892.

[25] Fizeau H. 1851: "Sur les hypothèses relatives à l'éther lumineux, et sur une expérience que paraît démontrer que le mouvement des corps change la vitesse avec laquelle la lumière se propage dans leur intérieur". *Comptes Rendus de l'Académie des Sciences de Paris* **33**, pp. 349–355.

[26] Fizeau H. 1859: "Sur un méthode propre à rechercher si l'azimut de polarisation du rayon réfracté est influencé par le mouvement du corps réfringent". *Comptes Rendus de l'Académie des Sciences de Paris* **49**, pp. 717–723.

[27] Fresnel A. 1818: "Lettre d'Augustin Fresnel à François Arago sur l'influence du mouvement terrestre dans quelques phénomènes d'optique". *Annales de Chimie et Physique* **9**, pp. 57–66.

[28] Galitzine B. 1892: "Über strahlende Energie". *Annalen der Physik und Chemie* [series 3] **47**, pp. 479–495.

[29] Giannetto E. 1999: "The rise of special relativity: Henri Poincaré's works before Einstein". Pp. 181–216, in: Tucci P. (ed.). *Atti del Diciottesimo Congresso Nazionale di Storia della Fisica e dell'Astronomia*. Milano: Universita' di Milano.

[30] Goldberg S. 1967: "Henri Poincaré and Einstein's theory of relativity". *American Journal of Physics* **35**, pp. 934–944.

[31] Goldberg S. 1969: "The Lorentz theory of electrons and Einstein's theory of relativity". *American Journal of Physics* **37**, pp. 498–513.

[32] Goldberg S. 1970: "The Abraham theory of the electron: the symbiosis of experiment and theory". *Archive for History of Exact Sciences* **7**, pp. 7–25.

[33] Goldberg S. 1976: "Max Planck's philosophy of nature and his elaboration of the special theory of relativity". *Historical Studies in the Physical Sciences* **7**, pp. 125–160.

[34] Gooding D. 1981: "Final steps to the field theory: Faraday's study of magnetic phenomena, 1845–1850". *Historical Studies in the Physical Sciences* **11**, pp. 231–75.

[35] Griffiths D.J. and Owen R.E. 1983: "Mass renormalization in classical electrodynamics". *American Journal of Physics* **51**, pp. 1120–1126.

[36] Hasenöhrl, F. 1904–1905: "Zur Theorie der Strahlung in bewegten Körpern". *Annalen der Physik* [series 4] **15**, 344–370, 1904; **16**, pp. 589–592, 1905.

[37] Heaviside O. 1885: "Electromagnetic induction and its propagation". *Electrician* **14**, pp. 178–181, 306–310.

[38] Heaviside O. 1889: "On the electromagnetic effects due to the motion of electrification through a dielectric". *Philosophical Magazine* [series 5] **27**, pp. 324–339.

[39] Heaviside O. 1893: "On the forces, stresses, and the fluxes of energy in the electromagnetic field". *Philosophical Transactions of the Royal Society of London* **183A**, pp. 423–484.

[40] Heaviside O. 1971: *Electromagnetic theory*. New York: Chelsea Publishing Company, 1971. 3 vols.

[41] Helmholtz H. von 1893: "Folgerungen aus Maxwell's Theorie über die Bewegung des reinen Äthers". *Sitzungsberichte der Akademie der Wissenschaften zu Berlin*, 649–656. Reprinted in *Annalen der Physik und Chemie* [series 3] **53**, pp. 135–143, 1894.

[42] Henderson W.S. and Henry J. 1897: "Experiments on the motion of the aether in an electromagnetic field". *Philosophical Magazine* [series 5] **44**, pp. 20–26.

[43] Hertz H. 1887: "Über sehr schnelle elektrische Schwingungen". *Annalen der Physik und Chemie* **21**, pp. 421–448.

[44] Hertz P. 1904: "Kann sich ein Elektron mit Lichtgeschwindigkeit bewegten?". *Physikalische Zeitschrift* **5**, pp. 109–113.

[45] Hesse M. 1961: *Forces and fields. The concept of action at a distance in the history of physics.* New York: Philosophical Library.

[46] Hirosige T. 1976: "The ether problem, the mechanistic world view, and the origins of the theory of relativity". *Historical Studies in the Physical Sciences* **7**, pp. 3–82.

[47] Ives H.E. 1952: "Derivation of the mass-energy relation". *Journal of the Optical Society of America* **42**, pp. 520–543.

[48] Kaufmann W. 1897: "Die magnetische Ablenkbarkeit der Kathodenstrhlen und ihre Abhängigkeit vom Entladungspotential". *Annalen der Physik und Chemie* [series 3] **61**, pp. 544–552; **62**, pp. 596–598.

[49] Kaufmann W. 1901: "Die magnetische und elektrische Ablenkbarkeit der Becquerel Strahlen und die scheinbare Masse der Elektronen". *Königliche Gesellschaft der Wissenschaften zu Göttingen. Mathematisch-Physikalische Klasse. Nachrichten* **2**, pp. 143–155.

[50] Kaufmann W. 1902: "Über die elektromagnetische Masse des Elektrons". *Königliche Gesellschaft der Wissenschaften zu Göttingen. Mathematisch-Physikalische Klasse. Nachrichten* **3**, pp. 291–296. Reprinted in *Physikalische Zeitschrift* **4**, pp. 54–57.

[51] Kaufmann W. 1903: "Über die elektromagnetische Masse der Elektronen". *Königliche Gesellschaft der Wissenschaften zu Göttingen. Mathematisch-Physikalische Klasse. Nachrichten* **4**, pp. 90–103, 148.

[52] Kaufmann W. 1905: "Über die Konstitution des Elektrons". *Königlich Preussische Akademie der Wissenschaften* (Berlin). *Sitzungsberichte*, pp. 949–956.

[53] Kaufmann W. 1907: "Bemerkungen zu Herrn Plancks: 'Nachtrag zu der Besprechung der Kaufmannschen Ablenkungsmessungen' ", *Verh. D. Phys. Ges.* **9**, pp. 667–673.

[54] Keswani G.H. and Kilmister C.W. 1983: "Intimations of relativity: Relativity before Einstein". *British Journal for the Philosophy of Science* **34**, pp. 343–354.

[55] Kittel C. 1974: "Larmor and the prehistory of Lorentz transformations". *American Journal of Physics* **42**, pp. 726–729.

[56] Langevin P. 1905: "La physique des électrons". *Revue Générale des Sciences Pures et Appliquées* **16**, pp. 257–276.

[57] Langevin P. 1913: "L'inertie de l'énergie et ses conséquences". *Journal de Physique Théorique et Appliquée* [series 4] **3**, pp. 553–591.

[58] Larmor J. 1895: "A dynamical theory fo the electric and luminiferous mediu. Part II: theory of electrons". *Philosophical Transactions of the Royal Society of London A* **86**, pp. 695–743.

[59] Lebedew P. 1901: "Untersuchungen über die Druckskräfte des Lichtes". *Annalen der Physik* [series 4] **6**, pp. 433–458.

[60] Lenard P. 1898: "Über die electrostatischen Eigenschaften der Kathodenstrahlen". *Annalen der Physik und Chemie* [series 3] **64**, pp. 279–289.

[61] Lewis G.N. 1908: "A revision of the fundamental laws of matter and energy". *Philosophical Magazine* [series 4] **16**, pp. 705–717.

[62] Logunov A.A. 1996: "On the article by Henri Poincaré: "On the dynamics of the electron" ". *Hadronic Journal* **19**(2), pp. 109–183.

[63] Lorentz H.A. 1886: "De l'influence du mouvement de la terre sur les phénomènes lumineux". *Archives Néerlandaises des Sciences Exactes et Naturelles* **21**, pp. 103–176. Reprinted in Lorentz 1934–39, Vol. 4, pp. 153–214.

[64] Lorentz H.A. 1892: "La théorie électromagnétique de Maxwell et son application aux corps mouvants". *Archives Néerlandaises des Sciences Exactes et Naturelles* **25**, pp. 363–552. Reprinted in Lorentz 1934–39, Vol. 2, pp. 164–343.

[65] Lorentz H.A. 1892: "The relative motion of the Earth and the ether". [De relatieve beweging van der aarde en den aether]. *Verslagen Koninklijke Akademie van Wetenschappen te Amsterdam* **1**, pp. 74–79. Reprinted in Lorentz 1934–39, Vol. 4, pp. 219–223.

[66] Lorentz H.A. 1895: *Versuch einer Theorie der electrischen und optischen Erscheinungen in bewegten Körpern.* Leiden: Brill, 1895. Reprinted in Lorentz 1934–39, Vol. 5, pp. 1–138.

[67] Lorentz H.A. 1899: "Simplified theory of electrical and optical phenomena in moving bodies". *Proceedings of the Section of Sciences, Koninklijke Akademie van Wetenschappen te Amsterdam* **1**, pp. 427–442.

[68] Lorentz H.A. 1901: "Über die scheinbare Masse der Ionen". *Physikalische Zeitschrift* **2**, pp. 78–80.

[69] Lorentz H.A. 1904: "Electromagnetic phenomena in systems moving with any velocity less than that of light". *Proceedings of the section of sciences, Koninklijke Akademie van Wetenschappen te Amsterdam* **6**, pp. 809–836. Reprinted in Lorentz 1934–39, Vol. 5, pp. 172–197, and (without the final section 14) in Lorentz *et al.* 1952.

[70] Lorentz H.A. 1934–1939: *Collected Papers*. 9 Vols. The Hague: Nijhoff.

[71] Lorentz H.A., Einstein A., Minkowski H. and Weyl H. 1952: *The principle of relativity*. New York: Dover.

[72] Mascart E. 1874: "Sur les modifications qu'éprouve la lumière par suite du mouvement de la source et du mouvement de l'observateur". *Annales de l'École Normale* **1**, pp. 157–214, 1872; **3**, pp. 363–420.

[73] Maxwell J.C. 1856: "On Faraday's lines of force". *Transactions of the Cambridge Philosophical Society* **10**, pp. 27–83.

[74] Maxwell J.C. 1861–1862: "On physical lines of force". *Philosophical Magazine* [series 4] **21**, pp. 161–175, 281–291, 1861; **22**, pp. 12–24, 85–95, 1862.

[75] Maxwell J.C. 1865: "A dynamical theory of the electromagnetic field". *Philosophical Transactions of the Royal Society of London* **155**, pp. 459–512.

[76] Maxwell J.C. 1954: *Treatise on electricity and magnetism*. 3rd ed. New York: Dover.

[77] Mccormmach R. 1970: "Einstein, Lorentz, and the electron theory". *Historical Studies in the Physical Sciences* **2**, pp. 41–87.

[78] Mccormmach R. 1970: "H.A. Lorentz and the electromagnetic view of nature". *Isis* **61**, pp. 459–497.

[79] Miller A.I. 1973: "A study of Henri Poincaré's "Sur la dynamique de l'électron" ". *Archive for History of Exact Sciences* **10**, pp. 207–328.

[80] Miller A.I. 1981: *Albert Einstein's special theory of relativity: Emergence (1905) and early interpretation (1905–1911)*. Reading: Addison-Wesley.

[81] Nercessian N.J. 1989: "Faraday's field concept". Pp. 174–187, in: Gooding D. and James F.A.J.L. *Faraday rediscovered. Essays on the life and work of Michael Faraday, 1791–1867*. New York: American Institute of Physics.

[82] Newburgh R. 1974: "Fresnel drag and the principle of relativity". *Isis* **65**, pp. 379–386.

[83] Nichols E.F. and Hull G.F. 1901: "A preliminary communication on the pressure of heat and light radiation". *Physical Review* **13**, pp. 307–320.

[84] Nichols E.F. and Hull G.F. 1903: "The pressure due to radiation". *Astrophysical Journal* **17**, pp. 315–351.

[85] Planck M. 1906: "Die Kaufmannschen Messungen der Ablenkbarkeit der β-Strahlen in ihrer Bedeutung für die Dynamik der Elektronen". *Physikalische Zeitschrift* **7**, pp. 753–761.

[86] Planck M. 1907: "Zur Dynamik bewegter Systeme". *Königlich Preussische Akademie der Wissenschaften (Berlin). Sitzungsberichte* **13**, pp. 542–570. Reprinted in: *Annalen der Physik* **26**, pp. 1–34, 1908.

[87] Poincaré H. 1900: "Sur les rapports de la physique expérimentale et de la physique mathématique". Vol. 1, pp. 1–29 in: *Rapports presentés au Congrès international de Physique réuni à Paris en 1900*. Paris: Gauthier–Villars.

[88] Poincaré H. 1900: "La théorie de Lorentz et le principe der réaction". Pp. 252–278 in: Bosscha, Johannes (ed.). *Recueil de travaux offerts par les auteurs à H.A. Lorentz, professeur de physique à l'université de Leiden, à l'occasion du 25me anniversaire de son doctorat de 11 décembre 1900*. The Hague: Martinus Nijhoff, 1900. Reprinted in *Oeuvres de Henri Poincaré*, Vol. 9, pp. 464–488.

[89] Poincaré H. 1901: *Eléctricité et optique*. Paris: Gauthier-Villars.

[90] Poincaré H. 1904: "L'état actuel et l'avenir de la physique mathématique". *Bulletin des Sciences Mathématiques* **28**, pp. 302–324.

[91] Poincaré H. 1905: "Sur la dynamique d'électron". *Comptes Rendus de l'Académie des Sciences* **140**, pp. 1504–1508.

[92] Poincaré H. 1906: "Sur la dynamique de l'électron". *Rendiconti del Circolo Matematico di Palermo* **21**, pp. 129–176. Reprinted in: *Oeuvres de Henri Poincaré*, Vol. 9, pp. 489–550. Paris: Gauthier-Villars, 1954.

[93] Poynting J.H. 1884: "On the transfer of energy in the electromagnetic field". *Philosophical Transactions of the Royal Society of London A* **175**, pp. 343–361.

[94] Poynting J.H. 1905: "Radiation pressure". *Proceedings of the Physical Society* **19**, pp. 475–490.

[95] Ritz W. 1908: "Recherches critiques sur l'éléctrodynamique générale". *Annales de Chimie et Physique* [series 8] **13**, pp. 145–275.

[96] Rodrigues Jr. W.A. and Tiomno J. 1984: "Einstein's special relativity versus Lorentz's aether theory". *Revista Brasileira de Física* **14** (suppl.), pp. 450–465.

[97] Rodrigues Jr. W.A. and Tiomno J. 1985: "On experiments to detect possible failures of relativity theory". *Foundations of Physics* **15**, pp. 945–961.

[98] Schaffner K.F. 1969: "The Lorentz electron theory of relativity". *American Journal of Physics* **37**, pp. 498–513.

[99] Schwartz H.M. 1971–1972: "Poincaré's *Rendiconti* paper on relativity". *American Journal of Physics* **39**, pp. 1287–1294, 1971; **40**, pp. 862–872, 1282–1287, 1972.

[100] Scribner Jr. C. 1964: "Henri Poincaré and the principle of relativity". *American Journal of Physics* **32**, pp. 672–678.

[101] Searle G.F.C. 1896: "Problems in electric convection". *Philosophical Transactions of the Royal Society of London A* **187**, pp. 675–713.

[102] Searle G.F.C. 1897: "On the motion of an electrified ellipsoid". *Philosophical Magazine* [series 5] **44**, pp. 329–341.

[103] Siegel D.M. 1975: "Completeness as a goal in Maxwell's electromagnetic theory". *Isis* **66**, pp. 361–368.

[104] Siegel D.M. 1981: "Thomson, Maxwell, and the universal ether in Victorian physics". In: Cantor and Hodge (eds.), *Conceptions of ether studies in the history of ether theories 1740–1900*. Cambridge: Cambridge University Press.

[105] Shankland R.S. 1964: "Michelson–Morley experiment". *American Journal of Physics* **32**, pp. 16–35.

[106] Sherwin C.W. 1987: "New experimental test of Lorentz's theory of relativity". *Physical Review A* **35**, pp. 3650–3654.

[107] Silva C.C. 2002: *Da força ao tensor: evolução do conceito físico e da representação matemática do campo eletromagnético*. Campinas: Universidade Estadual de Campinas.

[108] Spencer B. 1970: "On the varieties of nineteenth-century magneto-optical discovery". *Isis* **61**, pp. 34–51.

[109] Stokes G.G. 1845: "On the aberration of light". *Philosophical Magazine* [series 3] **27**, pp. 9–15.

[110] Stokes G.G. 1846: "On Fresnel's theory of the aberration of light". *Philosophical Magazine* [series 3] **28**, pp. 76–81.

[111] Swenson L.S. 1970: "The Michelson–Morley–Miller experiments before and after 1905". *Journal of the History of Astronomy* **1**, pp. 56–78.

[112] Thomson J.J. 1881: "On the electric and magnetic effect produced by the motion of electrified bodies". *Philosophical Magazine* [series 5] **11**, pp. 229–249.

[113] Thomson J.J. 1889: "On the magnetic effects produced by motion in the electric field". *Philosophical Magazine* [series 5] **28**, pp. 1–14.

[114] Thomson J.J. 1891: "On the illustration of the properties of the electric field by means of tubes of electrostatic induction". *Philosophical Magazine* [series 5] **31**, pp. 149–171.

[115] Thomson J.J. 1893: *Recent researches in electricity and magnetism*. Oxford: Clarendon Press.

[116] Thomson J.J. 1897: "Cathode rays". *Philosophical Magazine* [series 5] **44**, pp. 293–316.

[117] Thomson J.J. 1904: "On momentum in the electric field". *Philosophical Magazine* [series 6] **8**, pp. 331–356.

[118] Thomson W. 1853: "On the mechanical values of distribution of electricity, magnetism, and galvanism". *Proceedings of the Glasgow Philosophical Society* **3**, pp. 281–285.

[119] Torr D.G. and Kolen P. 1982: "Misconceptions in recent papers on special relativity and absolute space theories". *Foundations of Physics* **12**, pp. 265–284.

[120] Whittaker E.T. 1973: *A history of the theories of aether and eletricity*. New York: Humanities Press, 2 Vols.

[121] Wien W. 1900: "Über die Möglichkeit einer elektromagnetischen Begründung der Mechanik". Pp. 96–107, in: BOSSCHA, Johannes (ed.). *Recueil de travaux offerts par les auteurs à H.A. Lorentz, professeur de physique à l'université de Leiden, à l'occasion du 25me anniversaire de son doctorat de 11 décembre 1900*. The Hague: Martinus Nijhoff, 1900. Reprinted in *Annalen der Physik* [series 4] **5**, pp. 501–513, 1901.

[122] Williams L.O. 1965: *Michael Faraday*. New York: Basic Books.

[123] Wise N.T. 1981: "The Maxwell literature and British dynamical theory". *Historical Studies in the Physical Science* **13**, pp. 175–205.

[124] Worrall J. 1982: "The pressure of light: the strange case of the vacillating 'crucial experiment' ". *Studies in the History and Philosophy of Science* **13**, pp. 133–171.

Physics Before and After Einstein
M. Mamone Capria (Ed.)
IOS Press, 2005
© 2005 The authors

Chapter 3

Mechanistic Science, Thermodynamics, and Industry at the End of the Nineteenth Century

Angelo Baracca

Department of Physics, University of Florence, Italy

1. Introduction

The physical theories developed during the first decades of the twentieth century have introduced a radical new way of considering both natural phenomena and the structure and tasks of scientific explanation. Albert Einstein can be considered as the father of this revolution in physics, having introduced some of the most innovative concepts and theories: although, at the end, he strongly opposed the extreme formal drift of the "orthodox" formulation of quantum mechanics by the Copenhagen school.

The level of abstraction and mathematical treatment reached by the latter theory makes it not only extremely difficult for lay people to understand, but also problematic to teach to students in scientific fields that do not require deep mathematical bases and strong abstraction capacities. This last difficulty implies a serious contradiction with respect to the growing diffusion and pace of technological advancement, since the quantum theory is a century old, is a fundamental tool in this field, and should therefore have a deeper diffusion.

It does not help, in this respect, to point out, as most physicists do, that the structure of quantum mechanics is one that corresponds to the properties of the atomic world. We do not question the success of the quantum theory, but we hold that in the path that led to its formulation, specific choices were made, related to the social and cultural environment and currents, and to the spread, growth and evolution of the Industrial Revolution in Western Europe [25]. Scientific activity obviously has its own specificity, but scientists studying the laws of nature are not insulated from the world outside, they are instead men of their time, involved with its social problems, mentality, cultural paradigms and philosophical currents. A rational reconstruction of this environment and these choices could then help in the difficult task of making the final structure and concepts of quantum mechanics more acceptable.

This chapter presents a historical analysis, discussing how the twentieth-century revolution in physics was the result of a long and complex evolution, from an initial empirical and phenomenological formulation based on experimental evidence and data, towards increasingly complex and formal structures based on models [1]: mathematical models in physics, models of molecules and reactions in chemistry, other hypothetical structures in other branches. This evolution overturned the relationship between science and technology, and transformed the first into a powerful productive force: the drift towards formalisation made science a flexible tool, applicable to every field, while production was growing into a complex system of increasingly specialised branches. At the same time, the organisation of scientific research changed from one of individual scientists investigating on the basis of their personal resources and interests, to a complex structure financed by the State or corporations; its material means and instruments became increasingly convoluted, reaching the dimensions of huge laboratories with an international status. In the course of such a process, the understanding of, and control over, natural phenomena have deepened astonishingly, just as in the field of technical applications and innovation. This evolution was neither linear nor completely understandable on a purely "internal" basis, in the sense that the changes and choices involved are hardly reducible to purely scientific requirements.

2. Limitations of the Early Phenomenological Scientific Approach

The first phase of industrialisation, starting with the eighteenth-century Industrial Revolution in England, was essentially based on empirical inventions. Technical innovations (like the steam engine, or many elementary chemical processes) could not rely on previous scientific knowledge and instead derived from individual inventiveness and creativity. During the first half of the nineteenth century, scientific development remained substantially dependent on technical innovation, and functioned as a tool to provide understanding and sounder bases for the underlying principles, in order to allow further improvement. Scientific knowledge undoubtedly became increasingly systematic, as the specific scientific branches acquired greater autonomy and rigorous foundations. The prevailing scientific attitude was however still based on a prescription of strictly conforming to objective and well ascertained experimental facts and data, avoiding any recourse to concepts or entities that were not directly measurable.

This prescription was rarely fully implemented in practice, due to the unavoidable presence of common sense, or ideological concepts, like the "hydraulic analogy" adopted by Sadi Carnot in 1824, or the explanation of interactions in terms of forces acting at a distance. However, it was a wholesome reaction to the indiscriminate use of metaphysical speculations in the past, although it severely limited the possibility of unfolding or foreseeing really new properties or phenomena, and could hardly lead to completely new technological devices. One of the most explicit examples was the general rejection of the atomic model, although Dalton had already formulated the atomic concepts at the beginning of the nineteenth century, and Avogadro (1811) and Ampère (1814) had introduced the molecular hypothesis. Such a rejection delayed, for instance, the development of thermodynamics and of the energetic concepts (which had been introduced by John Smeaton on an empirical basis in the second half of the eighteenth century): thermodynamics was born mainly as the science of the steam engine, and its further development did not overcome this horizon until the second half of the nineteenth century. Chemistry, on the other hand, although its quantitative settlement had definitely overcome the limits of alchemy, maintained an empirical and phenomenological attitude, which was to prove a substantial obstacle to the understanding of the nature of chemical substances and of the mechanisms of chemical processes. This is particularly evident in the case of organic chemistry: once the idea of a non-physical vital force was overcome, Gerhardt proposed (in 1844) a classification based on "structural types", "without the need of turning to hypotheses, but strictly keeping inside the limits of experience" [18]. Gerhardt's classification turned out to be substantially misleading, due to the absence of models of molecular structures.

3. The Mid-Century New Breath (1850–1870): From Empiricism to Theoretical Models

This scientific attitude changed around mid-century, when the approach of early nineteenth-century science showed its inadequacy in the face of the changes of the social and cultural situation and of the productive structure.

Joule had already referred to models of matter and atoms in his fundamental investigations on the "mechanical theory" of heat (1842–48), although he limited himself to qualitative considerations in order to support the new concept. The following decade saw a true upsetting of the previous methodological foundations, with the introduction of models both, and almost simultaneously, in physics and in chemistry, in order to reach a deeper understanding of the processes that were studied, and to get new results: these models were formulated in quantitative, mathematical terms.

On the social and economic front, in the second half of the century the middle class established its power in central Europe; and in the first two decades of this period the new social situation began to influence every aspect of life – social, cultural and practical. The defeat of the old aristocratic class and the establishment of a capitalistic economy and an industrial system posed the need, before the possibility, for a big step forward in every field. In the previous century, the accomplishment of an industrial system had been limited to Great Britain, which had developed an impressive industrial power; early industrialisation began to develop in France after the 1789 revolution, but was almost stopped by the Restoration. When the contracting middle class reached the conditions for developing its economic and entrepreneurial activity in the countries of central Europe (essentially the German-speaking area), the problem of competing with the overwhelming British industrial and economic power posed problems and challenges that were overcome through innovations in every field: practical, cultural and ideological. Protectionism began to decline, and an international area of free trade was established. This free enterprise was encouraged and enhanced by the setting up of new forms of credit and by new developments in banking systems.

These changes were accompanied by considerable technical innovation. The technological model of the first Industrial Revolution proved to be increasingly restrictive for the new requirements: a number of bottlenecks had to be overcome by means of a technological jump. In the two decades after 1850 there was a spectacular increase in the rate of inventions and innovations, although they still remained essentially independent of scientific advances.

Let us limit ourselves here to some relevant examples. New techniques were introduced for steel production (Bessemer, Siemens, Gilchrist-Thomas). Unlike British industry – which had already developed in the course of the previous century into a massive, rather rigid structure, that proved to be quite difficult to reorganise – the German steel industry was essentially developed on the basis of these new processes: at the beginning of the twentieth century the average German steelworks was about four times the size of a British one, while German production overtook British production.

Something similar happened in soda production when Solvay introduced a new and much more efficient method of synthesis. British industry, based on the old Leblanc process, held a monopoly on world soda production, but it was not flexible enough to respond quickly to the new process, trying instead to improve the old process as far as possible. The emerging German industry, on the other hand, used the new process, and outstripped British production within a few decades, becoming the main world producer. Chemical production in general advanced rapidly in Germany, in particular organic chemistry, and the new-born dye industry. Another important instance of innovation was the invention of the internal combustion engine.

Along with these processes, however, an awareness grew in leading technological fields that leaving the process of innovation to almost haphazard activity or to the ingenuity of inventors was inadequate. Some kind of guide to technical and industrial innovation was now needed, and such a guide could only be provided by scientific research, if it could overcome the substantially empirical approach that strongly limited the possibility of new results or discovering new processes. It is worth noting here that this change did not result from a conscious decision: instead it was a response to a new spirit of inquiry and investigation into natural phenomena that broke with the old, traditional methods, and reflected the participation of science and technology in the new society and economy. It was a process of general maturation which reflected itself in all aspects of the activity of members of the emerging class.

This new attitude induced a deep methodological renovation in the investigation and explanation of natural phenomena that turned the scientific practice of the previous half-century upside down. Thus models based on non-observable entities began to be adopted as useful tools to lead to the prediction of new properties or the discovery of new phenomena or empirical facts. To this end, however, models could no longer be used in a speculative form, but had to be formulated and developed in mathematical terms in order to be tested rigorously. A new theoretical physics was thus born, based on a hypothetical-deductive approach. The new predictions, reached on the basis of these models, might turn out to be right or wrong when tested against experiments: in the first case, the model was to be considered as substantially correct, in the second case it had to be rejected, improved or changed. In any case, insights and advances were made in the understanding and practical control of processes or phenomena, or new phenomena were predicted. One of the most astonishing examples, as we will see in more detail, was Maxwell's prediction of electromagnetic waves, made possible by the mathematical representation of the electromagnetic field in terms of fluids.

It is emblematic that this change occurred almost simultaneously in physics and chemistry during the 1850s, starting with the adoption of the atomic molecular theory.

a) Physics. In physics, this started with an explanation of thermodynamic properties in terms of the *atomic structure of matter*. Krönig in 1856 and Clausius in 1857 derived the expression of pressure in an ideal gas through a mathematical treatment of the average effect of the elastic shocks of gas atoms on the walls of a container, thus providing a justification of the fundamental equation of state of the ideal gas. It is worth noting that similar calculations had been previously performed by Herapath in 1820 and by Waterson in 1843, but their papers had not been accepted for publication. In the following years, Maxwell and Boltzmann formulated, in a more compact and rigorous mathematical form, the kinetic model of gases, deducing in a systematic way its consequences (Appendix, *1*). They found unexpected connections between properties that appeared independent on an empirical basis (e.g. the relationship between transport coefficients, based on the concept of the mean free path of atoms), and laws that could not have been established by pure experimentation (such as the independence of viscosity on pressure). Boltzmann reached a general formulation of kinetic theory (*Boltzmann equation*, 1872) [6], providing an interpretation of entropy and the irreversible character of thermodynamic processes in a gas in terms of collisions between atoms. In particular, he introduced the fundamental distinction between the "microscopic", or "dynamic", state of a gas (determined by the exact positions and velocities of all atoms) and its "macroscopic", or "thermodynamic", state (determined by a restricted number of macroscopic magnitudes, defined as averages over atoms). As we will see in more detail, Botzmann believed he had found a mechanical explanation for the thermodynamic properties of gases.

At the same time, the use of mathematical models based on *fluids* produced no less interesting results. Stokes developed the mathematical theory of physical optics, identifying light with waves propagating in a highly elastic fluid, the "optical ether". An analogous treatment was introduced by Lord Kelvin, and was fully implemented by Maxwell for electric and magnetic phenomena, as a development of Faraday's early qualitative approach in terms of contact actions, in contrast with the traditional approach which used forces acting at a distance, introduced by Newton, but adopted mainly by the French school [34].

Maxwell explicitly theorised the resort to "physical analogies", namely "the partial similarity between the laws of one science and those of another one, which allow that one of them illustrate the other" [28]: like the analogy of a gas with a system of elastic spheres, or of the electromagnetic field with a mechanical fluid, called the "electromagnetic ether". Mathematical theory developed on these grounds, identifying electric and magnetic actions with the states of pressure, stress or torque inside the hypothetical fluid, and came to extremely important results. First, Maxwell formulated the general laws of the electromagnetic field (*Maxwell equations*) [27], fulfilling the unification between electric and magnetic phenomena. Second, he predicted the existence of electromagnetic waves, on the basis of the physical properties of the ether which, in order to reproduce experimental properties, had to be a highly elastic fluid. Finally, he predicted the electromagnetic nature of light, on the basis of the identity of the properties of his "electromagnetic ether" with those of Stokes' "optical ether". The existence of electromagnetic waves was experimentally confirmed by Hertz in 1888.

These results strikingly confirmed the superiority of the new theoretical approach based on models, since it would have been difficult, if not impossible, to reach such conclusions through a purely empirical approach: electromagnetic waves constitute one of the discoveries that have transformed and renewed technology, production and social relations.

It is worth noting that these advances in theoretical physics were accompanied by progress in experimental physics, with a widening of fields of research, an improvement in equipment and a growing accuracy in experimental procedures.

b) Chemistry. The changes in chemistry proceeded along similar lines. In 1859, at an international meeting held in Karlsruhe, Stanislao Cannizzaro reproposed Avogadro's hypothesis of the distinction between atoms and molecules [7]: contrary to what had happened half a century before, now the atomic–molecular model of matter was not only immediately accepted in the German-speaking countries (in France, for instance, this theory still met with opposition, showing the scientific and technological lag accumulated by this country after the splendour of the first half of the century), but it became the basis of a kind of "molecular engineering" that promoted chemical technology and production, giving Germany a clear superiority in this field. Models of the internal structure of complex molecules were systematically developed, connecting the macroscopic properties to such structures, or to specific atomic groups: this allowed the design of, on the one hand, new molecules with specific chemical properties, starting from known atomic groups, and, on the other hand, more efficient industrial processes of synthesis. One of the most astounding results was Kekulé's hexagonal model of the benzene molecule [23] (1865; Appendix, 7), which became the basis for the modern classification of organic compounds. Although these chemical models were not mathematical in the strictest sense, the new level of abstraction and formal reasoning are evident.

Another fundamental advance made possible by the new conception was the concept of chemical equilibrium. Previously, only the simplest processes up to the exhaustion of reactants had been developed, but the development of organic compounds involved more complex reactions that often do not even occur in normal thermodynamic conditions and require exceptional pressure and temperature values. The concept of chemical equilibrium was introduced for the first time in 1864 by Guldberg and Waage, by considering a combination of activities of reactants and reaction products: it is worth emphasising that their treatment, although formulated in rather obscure terms, presented a strong analogy with Boltzmann's formulation of gas theory (see Appendix, 6).

To sum up, the main feature of the new scientific approach introduced after 1850 was that all the properties of the phenomena, and their characteristic parameters and functions were mathematically or formally deduced on the basis of models of their microscopic structure and interactions between the elementary, unobservable components of the system (atoms and molecules in matter, fluids in wave and electromagnetic phenomena). This new approach allowed a big jump in the understanding and treatment of physical and chemical systems, strengthening the belief that a mechanical interpretation of all natural processes could be provided.

The use of hypotheses and models also became commonplace in other scientific disciplines, according to the level they had attained, and allowed important advances. For instance, in the biological theory of evolution (Darwin, 1859) a great advance was made possible by the conscious use of hypothetical, but rigorous (however qualitative) considerations:

> I have always considered the doctrine of natural selection as an hypothesis that, if it should
> explain wide orders of facts, would merit to be considered a theory worthy of acceptance [9].

4. Triumph and Contradictions of Mechanism

The developments connected with the new method obtained the most remarkable scientific results in the last decades of the nineteenth century. These advances opened new perspectives in science and technology, leading to the overturning of the previous relationship of dependence of the former on the latter. This took place at the turn of the century, when a new phase of industrialisation, centred in Germany, took off. science became an effective guide for technical innovation and productive development, i.e. a real productive force.

It is important to note however, that the routes followed by physics and chemistry diverged: while physics went on developing and refining the theories introduced during the 1850s and 1860s, chemistry, as we will discuss in the next section, took a different approach, which anticipated the revolution in physics of the beginning of the twentieth century. Let us begin with physics.

The impressive advances in physics led to an intensive construction of fundamental theories which gave the impression that almost all the fundamental phenomena of nature had been essentially understood and explained. Field and matter were the two basic aspects (a dichotomy whose criticism was the basis of Einstein's 1905 paper on "light quanta"). On one hand, Maxwell's electromagnetic theory, based on the electromagnetic ether, fulfilled the unification of the electric and magnetic fields, providing a basis for the treatment of every aspect of these phenomena (Appendix, *8*). On the other hand, Boltzmann's kinetic theory provided a general basis for calculation of the thermodynamic properties of gases, and an explanation of irreversibility and the second principle: it seemed therefore to open the way to a generalisation including every aspect of matter.

This impressive theoretical construction was, however, undermined by an intrinsic contradiction. It appeared in fact as the triumph of mechanics, providing a mechanical explanation of all known natural phenomena. But this underlying mechanical philosophy severely limited its potentialities, and led at the same time to deep and unexpected paradoxes. The end of the nineteenth century thus marked a time of triumph and of crisis for mechanistic philosophy.

The full development of the kinetic theory of gases, accomplished by Boltzmann [8], seemed to fulfil the ideal of a complete reduction of thermodynamics to mechanics, successfully explaining and calculating all the properties of gases in terms of the interactions and collisions of their constituent atoms and molecules. Boltzmann maintained such an attitude in spite of the further advance he himself had made in 1877 with the introduction of the probabilistic interpretation [4].

A first criticism had been made the previous year by Loschmidt (although with a positive intention), with the remark that mechanical processes are reversible, while the Boltzmann equation predicted irreversible behaviour, corresponding to the second principle of thermodynamics (*reversibility paradox*). In order to overcome this criticism, Boltzmann introduced a new fundamental concept, showing that entropy can be given a probabilistic interpretation, and the final state of the thermodynamic evolution, i.e. the equilibrium state of the gas, can be obtained as the most probable state (see Appendix, *3*). However, it seems that Boltzmann did not fully perceive the fundamental overturn of the relationship between mechanics and thermodynamics implied by the introduction of probability, in spite of the further, and fundamental, developments he brought to this concept in subsequent years.

But the contradictions and paradoxical consequences raised by the kinetic theory did not end here. In the last decades of the nineteenth century, kinetic theory was subject to a hail of criticisms and attacks that led Boltzmann to declare in 1896: "I am conscious to be only an individual feebly fighting against the current of time" [5]. In fact, the year before Zermelo had pointed out a second paradox of the kinetic theory, known as the *recurrence paradox.* In order to appreciate the relevance of this criticism, it must be recalled that Zermelo was an assistant of Planck, and Boltzmann was aware that behind this criticism lay the negative attitude of Planck himself to the kinetic approach and the philosophy behind it. The paradox derived from a theorem proven by Poincaré for a (bounded) mechanical system, according to which such a system has to return to a state arbitrarily near the initial one over a sufficiently long period of time: this behaviour seemed to imply that entropy could not go on growing until it becomes a constant, but that sooner or later it has to return to a value close to its initial value. Boltzmann, upset and depressed by this criticism, retorted that his previous considerations had not been understood, reaffirmed that the evolution of the system towards equilibrium was a probabilistic process, and evaluated that the "recurrence time" for a macroscopic system was tremendously long, much more than the life of the universe. In spite of the relevance of such considerations and calculations, he seemed to miss the fundamental point, which was that Poincaré's theorem concerned the "microscopic state" of the system, but had nothing to do with the "macroscopic" one, since the latter is defined in terms of an ensemble of microscopic states, compatible with the macroscopic one. Such a consideration would have opened the way to a wider perspective, in which dynamic properties such as recurrence could be viewed as manifestations of thermodynamic fluctuations, but one had to wait for Einstein in order to attain a full awareness of this.

There were even more problems raised by the kinetic theory, such as those relating to the specific heats of gases, whose values were correctly predicted, on the basis of the theorem of equipartition of energy (Appendix, *1*), by taking into account only the translational and rotational degrees of freedom of atoms and molecules. But an internal structure of these components was slowly being discovered, and it seemed a fatal contradiction that their inclusion in the kinetic formalism would have led to unacceptable values for specific heats. Such a contradiction was to be solved only when quantum theory showed that the internal degrees of freedom are "frozen" (i.e. they cannot be excited by thermal motions) at ordinary temperatures (see Appendix, *2*).

It is however important to remark here that the criticisms of and attacks on kinetic theory did not derive only from such contradictions, but were rooted in a more general philosophical and methodological attitude. The cultural milieus in central Europe were dominated by currents of thought that rejected the new insights into natural phenomena and the very legitimacy of using models based on non-observable entities. They critically re-examined the positivistic philosophy, supporting a scientific approach restricted to observable phenomena and data. The most profound, authoritative and influential current was critical empiricism, formulated by Ernst Mach and Avenarius, who denied the reality of anything beyond direct empirical evidence, which was in turn reduced to sensations. This viewpoint – although it also produced important results, such as studies on the connections between sensations and perceptions, and the development of psychophysics – led to the rejection of the reality of atoms, considered as "economical" tools, and to idealistic positions denying matter itself. Nevertheless the anti-mechanistic polemic reached a deep level of critical awareness; so much so that Einstein acknowledged his debt towards Mach in retrenching the role of mechanics [11]. Conclusions not much different from Mach's were reached by a different, much rougher theory named "Energetics", formulated by the chemist Ostwald, who pretended to reduce every phenomenon to manifestations of energy. Planck himself strongly criticised these conceptions, but recalled in his memoirs that it was practically impossible to be heard against Mach's and Ostwald's authority. Boltzmann accused Mach's philosophy of reaching sterile solipsistic conclusions, and had direct and harsh disputes with Ostwald, who denied the possibility of providing a mechanical explanation of irreversibility.

One can see how the theme of mechanical interpretation was at the core of the cultural and scientific debate at the end of the nineteenth century, and was really the weak side of the most advanced physical theories.

In a similar context, one may also discuss the difficulties that arose in electromagnetic theory, as it was formulated on the basis of the electromagnetic ether or, more precisely, a certain kind of "mechanical" ether, conceived as a classical fluid. The most generally known was the paradox known as the "ether wind" (Appendix, 9). It may be described as conceiving of the ether as a kind of absolute frame of reference: an interpretation corresponding to the formulation of Newtonian mechanics, the one that Einstein was to criticise in his formulation of the theory of special relativity in 1905. At the end of the nineteenth century it seemed obvious (besides being mathematically demonstrated using the Galilean transformation laws for uniform translations) that the behaviour of the electromagnetic phenomena was to change with the motion of the experimental apparatus (or the Earth) with respect to the ether, just as we experience wind when moving through the atmosphere. In these years a series of experiments tried to measure such effects, culminating in the negative results of the Michelson and Morley experiments (from 1881 up to 1904). Here again the contradiction was not inherent in Maxwell's electromagnetic theory, but in its mechanical interpretation, as Einstein showed in 1905. In fact Maxwell's theory met with much opposition, in spite of its successes and aplications to electromagnetic devices. After all, one common feature of the spectacular advance in physics at the end of the nineteenth century was its mechanical and reductionist foundation, which is essential in order to understand the basis of the twentieth-century revolution in physics, and the formal and abstract turn it introduced. At the end of the nineteenth century the mechanical philosophy succeeded in the construction of a massive building that seemed to demonstrate the triumph of mechanics: the latter provided the common basis for the increasing proliferation and specialisation of branches of natural sciences. Such developments brought about an increase in mathematical complexity: models and theories however, based on clear mechanical concepts, grew into highly formalised systems of differential and integral equations (such as Maxwell's equations for the electromagnetic field and the Boltzmann equation for a rarefied gas).

With respect to the contradictions raised by end-of-century reductionist physics, it is important to emphasise that the latter contained possibilities for their solution – however involved they may appear – which would have constituted different routes from the one that eventually prevailed. These routes were interrupted by the start of the twentieth-century revolution in physics, but it is important to state that the evolution of science implies choices that are not merely of a scientific nature, but involve more general factors.

In the realm of thermodynamics Boltzmann, as we have seen, reacted to the criticisms introducing new fundamental developments of kinetic theory, namely probabilistic considerations, that enormously enriched the mechanical foundation of the theory. His considerations on the "recurrence time" and its dependence on how close should the initial state recur, could have been kinetically interpreted in terms of fluctuations, with frequencies inverse to their amplitudes: irreversibility is a probabilistic trend, and the recurrence of the microscopic molecular configuration of initial state is not impossible, only extremely improbable (in the same sense as a monkey randomly typing would compose, although it would take an incredibly long time, the sequence of letters of the *Divina Commedia*). In order for a chair to rise spontaneously, all the molecules in the air should move upward together, an extremely improbable, but not impossible event: if it were impossible, Brownian motion would not exist. Many of Boltzmann's points of view are in agreement with modern developments in the dynamics of complex systems.

Boltzmann was a dramatic figure, since he introduced new approaches and concepts that have remained cornerstones in natural science, but he remained substantially locked into a reductionist position that prevented him from fully utilising the novelty of the new hypotheses at the beginning of the twentieth century, even those directly connected with his own proposals. He was deeply upset by the attacks on his work, and this seems to have been one of the reasons that led him to commit suicide in 1906.

For electromagnetism and the problem of the ether, the situation developed even further. Lorentz worked out the "electron theory" [26], in which matter was conceived as composed of elementary electric charges of a corpuscular nature (This an hypothesis that preceded the discovery of the electron by J.J. Thomson in 1897; Appendix, 9). This theory in a sense complemented Maxwell's theory of the electromagnetic field, by unifying it with the theory of matter: the basic equations of electron theory are in fact Maxwell's equations together with Lorentz's equation for the motion of a charged particle in an electromagnetic field. Such a theory thus had a reductionist structure, being based on interactions between the elementary components of the system: in a sense it was an electromagnetic version of mechanical theories. Lorentz's theory, which embodied the concept of ether, was quite successful: it predicted, from the electromagnetic nature of the interacting forces in matter, a contraction of bodies in the direction of their motion through the ether (*Lorentz contraction*), that exactly offset the "aether wind". This Lorentz contraction was not (as it is sometimes presented) an ad hoc hypothesis, but rigorously followed from electromagnetic theory: it is the same as predicted from Einstein's theory of special relativity, which retains electromagnetism while rejects Newtonian mechanics and Galileo transformations [3]. At the time, all experiments confirming the electron theory were also in agreement with special relativity, and vice versa. How could it happen, then, that the latter superseded Lorentz's electron theory? This question will be answered below.

Apart from the physical difficulties raised by the mechanistic philosophy, it is worth remarking that such an attitude suffered from an internal contradiction, and a crucial limitation of its own potentialities: in fact, once the recourse to models had been adopted as a powerful way to investigate and forecast new properties and phenomena, the limitation to mechanical models constituted an unjustified limitation of their full potential. We will analyse this aspect in more detail in the next sections.

Moreover, while physicists were making such an effort to complete this mechanical building, to try and reach the final explanation of natural phenomena, a series of completely new processes were being discovered. The discovery of X-rays, of radioactivity, of cathode rays and of the electron (which were only later recognized as the same thing), of the internal structure of atoms, and the determination of the complete spectrum of electromagnetic radiation created a need for new physical concepts and theoretical frames. This meant that difficulties and contradictions had to be overcome in a completely different context, and this was to introduce a further deep change in the very basis of science, signalling the first break with the mechanistic philosophy.

5. The Second Industrial Revolution and the anti-Mechanistic turn of the German Chemists

German chemists were the first to break with the mechanical foundation of science at the end of the nineteenth century. This happened with no dramatic discussions, and appeared to be a logical choice faced with problems that would have been too difficult, if not impossible, to solve using a mechanical approach.

In order to understand this evolution, one has to put it in the context of the economic and social changes taking place at the end of the nineteenth century. The end of the Civil War in the United States and the unification of the German Empire in 1871, followed by the economic crisis and the great depression of 1873–96, opened the door to the development of an industrialisation process so new and so rapid that it really was a revolution. The order of the most advanced industrial powers was completely upset in a short period of only a few decades: around 1850 Great Britain had an overwhelming lead, and France was in second place, but by the start of the twentieth century they had been overtaken by Germany and the US, when several of the leading productive sectors in Britain almost collapsed, faced with increasing German technical leadership and competition. Also, the material and technical bases of production radically changed during these decades: from coal, wood, some of the simplest chemical products, iron and some empirically produced steel, to electricity, industrially produced steel, an increasing quantity of complex chemical products and oil. The power of the newly unified German Empire (following the Austrian–Prussian war of 1866) immediately manifested itself by routing the imperial French army in the French–Prussian war (1870–71), when the recently founded Krupp ironworks provided the German army with 300 new guns.

Scientific development took quite different courses in different countries, as a consequence of variations in the respective socio-economic situations. German science acquired world leadership for half a century, in physics, chemistry and mathematics (until the "brain drain" under the Nazi regime). The US, on the other hand, thanks to its enormous natural resources, lagged behind in science until the first decades of the twentieth century, and developed a technicistic and pragmatic attitude, disregarding general theoretical frameworks and systematic investigation [31]. The eclipse of British science and technology in the second half of the nineteenth century was clearly perceived, denounced and analysed in Britain: several Parliamentary Commissions were appointed in these decades, to investigate the reasons for this worrying decline and to suggest solutions. A German chemist, A.W. Hoffmann, was appointed as director of the Royal College of Chemistry in London. Meanwhile, British physics remained tied to mechanical models. France also lost the leading scientific position it had enjoyed in the first half of the nineteenth century: there were some eminent French personalities (Poincaré for example), but in general French science lagged behind; in particular, for a long time French chemists did not accept the atomic theory, under the authority of M. Berthelot.

The German economy was characterised by the rapid growth of a modern industrial system, based on continually renewing processes, on intensive and programmed technological progress, and on scientific investigation, both fundamental and applied. The chemical industry (especially organic chemistry, and in particular the dye industry, which accounted for 85–90% of world output at the end of the nineteenth century) and the electric industry were the leading sectors. The recently established modern chemical firms (BASF, Hoechst, Geigy) grew very rapidly in these decades [22], built up huge laboratories and developed intensive scientific investigation. It was here that industry assumed its modern structure, based on programmes of team investigation, and gradually shifted direction towards the research laboratory. Modern research laboratories were shaped in these science-based industries. The percentage of workers with a university degree in the principal German chemical firms at the end of the century was comparable with today's figures (in 1900 the German chemical industry employed 3,500 chemists out of a total of 80,000 employees, 40% of whom worked in plants with more than 200 employees). Very close contacts and collaborations were established by these firms with leading university investigators. Major technical scientific programmes were undertaken, such as BASF's seventeen-year investigation, costing one million pounds, into the industrial synthesis of indigo, and the "fixation of nitrogen" (or ammonia synthesis) carried out in 1913 by the academic chemist Fritz Haber and BASF chemist C. Bosch. The latter process allowed Germany, while completely surrounded, to continue to resist for years during the First World War, since it could synthesise explosives and fertilisers (the only other source being guano from Chile, which was subject to the British control of the seas, and was moreover subject to exhaustion: this had been in fact the main motivation behind the Germans developing the new methods of synthesis). When the Allies inspected the Oppau plant for ammonia production in 1919, they discovered the great technical progress that had been made. In this connection, it is worth recalling Haber's direct involvement in war research, as the organiser, and the real father (violating international laws) of the German manufacture and use of chemical weapons (at Ypres in 1915): a role that he claimed in his Nobel Lecture (1918).

Much research was also taking place within the German electrical industry. In Berlin in 1884 the large and very advanced *Physicalische Technische Reichsanstalt* was founded, at which the fundamental measures on the spectrum of electromagnetic radiation were obtained in 1900.

One main factor supported all these developments: Germany took advantage of its very advanced educational system, which could satisfy the growing need for specialised and trained scientists and technicians. Besides the universities, which had radically modernised their laboratories and teaching methods, there was a system of polytechnics (*Technische Hochschulen*) unknown in other countries, that prepared highly qualified technicians with a university education in applied research. At both levels the above-mentioned collaboration with the main firms guaranteed close contact of the academic milieus with the concrete problems of production and technical innovation. The German schools of chemistry became the most advanced in the world, and almost all chemists who wanted a thorough training went to study in Germany.

Many crucial branches of physics arose from the dynamics of technical innovation in chemistry. The problem of radiant electromagnetic energy and of black-body radiation (Appendix, *10*) – which was to be at the basis of the quantum revolution at the beginning of the next century – derived directly from the spectroscopic method of chemical analysis introduced by Bunsen and Kirchhoff in 1860 (which led to the discovery of some chemical elements). The study of the thermodynamic properties of physical systems at high temperatures and pressures was encouraged by the discovery of complex chemical reactions (such as the synthesis of ammonia) that do not occur under normal conditions. The problem of catalysis stimulated the production of new alloys (such as the chromium–nickel alloys supplied by Krupp), and a better understanding of metallurgy. Research into the physics of very low temperatures was in turn pushed forward by problems in the liquefaction of gases and the fractional distillation of air, in order to obtain cheap oxygen and nitrogen for the chemical industry.

The main consequence of this situation in fundamental science was a change in the foundations of chemistry and thermodynamics. The search for increasingly refined compounds was becoming relentless: more complex molecules had to be synthesised through increasingly sophisticated operations of "molecular engineering" which involved more complex reactions under increasingly difficult conditions of chemical equilibrium, leading to unusual conditions of pressure and temperature, very far from the normal ones.

This situation inevitably affected not only the practical attitude of chemists, but also their way of shaping the scientific treatment of these processes. Chemists were confronting new problems of unprecedented difficulty, and were subject to increasing pressure and urgency. The mechanical approach, i.e. the construction of thermodynamic properties starting from the interactions of microscopic components of matter, revealed itself as inadequate to cope with both the complexity of these processes and the new rate of innovation and development of the chemical industry. In other words, mechanics had provided a natural reference frame as long as the technical basis of production and the organisation of labour had maintained a mechanical structure, but the development of the production cycle of the modern chemical (mainly organic) industry presented a complex, systemic structure, showing the limitations of the scientific elaboration that could be based on mechanics, and indicating the need for a new, more compact approach.

An example of this is the calculation of conditions for chemical equilibrium in complex reactions. The approach introduced by Guldberg and Waage could in principle be developed in terms of collisions between the molecules; however, the process was too cumbersome to achieve concrete results, especially since even the simplest reactions proceed through a sequence of partial decomposition and recombination stages, strongly dependent on thermodynamic conditions: this would lead to incredible complexity in the dynamics of molecular activation collisions.

In this situation, chemists were the first to abandon the mechanical approach, and turn instead towards thermodynamics, since it was more flexible. Thermodynamic laws are largely model-independent (recall how Sadi Carnot adopted a "hydraulic analogy" for heat, while Joule had referred to mechanical considerations: the analysis of this apparent contradiction led Kelvin and Clausius to the formulation of the second principle around 1850). They allow one not to be concerned with the specific course of the reactions, since a treatment in terms of state functions depends only on the initial and the final states. In the last decades of the nineteenth century the theory of the free energy state functions was developed (the term "thermodynamic potentials" coined for them was emblematic) in order to treat equilibrium conditions in heterogeneous systems, whose most typical applications were to chemical reactions. After partial contributions by Van't Hoff, Le Chatelier and others, the general theory was formulated in 1876 by Willard Gibbs [19], the same man who later formulated statistical mechanics.

The atomic and molecular constitution was obviously not rejected, and often there was the need to resort to microscopic considerations (for instance, to compute reaction speeds): this however was no longer done on mechanical grounds, but only after the general features of the processes had been established using a thermodynamic treatment. In some sense, thermodynamics took the place of mechanics as a guide and reference frame to work out the properties and laws of the processes: the way was open for conceiving a different, more general and flexible class of models and theories than the mechanical-reductionist ones. An explicit acknowledgement of this shift may be perceived, for instance, in Van't Hoff's approach to reaction kinetics, a dynamic property that cannot be deduced from thermodynamics:

> Within the limits of our knowledge of theoretical laws, reaction velocity may be approached from two different points of view. First of all, we find support from thermodynamics, since the laws governing the speed must be in agreement with the equilibrium laws which are established at the end. Secondly, one may, on the ground of simple kinetic concepts, foresee laws that have always had up to date a good experimental confirmation. We will develop successively: (a) the reaction speed and equilibrium; (b) the reaction kinetics [35].

6. The Take-off of the Twentieth-Century Revolution in Physics: Planck, Gibbs, Einstein and Nernst

This section analyses developments in physics at the beginning of the twentieth century. According to our analysis, the essence (or at least the pre-condition) of this revolution consisted of the superseding of the mechanical and reductionist methods. Martin J. Klein has discussed the role played by thermodynamics in Planck's and Einstein's method and concepts [24]; Navarro correctly adds that, as we will see, it was, more precisely, the statistical formulation of thermodynamics that was important to Einstein's work [30].

This revolution produced three physical theories: statistical mechanics, the special theory of relativity and quantum theory. Einstein played a crucial role in all three, both as a pioneer and as the author of some of the main ideas. We will not follow a strictly chronological order.

The new ideas derived from the acquisition of an anti-mechanistic and anti-reductionist attitude. The new principles, rather than being built up in a reductionist way, were extracted from phenomenological data by selecting some general property to be raised at the level of a fundamental principle. Often the same big problems that, as we have seen, were bothering physicists were turned upside down and simply eliminated by assuming them to be principles of general validity, confirmed by overall experimental evidence: agreement with traditional theoretical frameworks was thus thought to be unnecessary.

The most evident example of this are the two principles that in 1905 Einstein held at the basis of the special theory of relativity [16], i.e. that the velocity of light is independent from the velocity of the source, and that physical laws retain the same form for observers in relative uniform motion (Appendix, 9): in this way, the problems related to the hypothesis of the ether simply disappeared, this concept being inessential in the theory from the beginning. Thus special relativity simply gave up the ether. The anti-mechanistic stand is particularly evident in this theory, since Maxwell's electromagnetic theory is retained for its covariant character, while Newtonian mechanics is rejected and deeply modified. Special relativity, moreover, presents quite a different structure from Lorentz's electron theory: apart from the substantial equivalence of the results, the same problems are solved here in a much simpler way. After strong initial opposition, Einstein's theory was finally accepted by the scientific community [21], for its non-mechanistic character and its greater flexibility and simplicity, although Lorentz's theory concretely represents an alternative framework.

Einstein's scientific attitude changed the methodological criteria of a scientific explanation. One of his constant concerns and guiding criteria was with symmetry requirements of the physical laws and the physical aspect of nature. This was the starting point of both his 1905 papers on special relativity ("It is known that Maxwell's electrodynamics [...] when applied to moving bodies leads to asymmetries that do not appear to be inherent in the phenomena") and one on the light quantum [15] ("A formal difference of great importance exists between the conceptions that physicists support with respect to gases and the other material bodies, and Maxwell's theory concerning electromagnetic processes in the so called vacuum").

The basic methodological features in the formulation of the theory of special relativity had been anticipated some years before in the formulation of statistical mechanics, by Gibbs [20] in 1902, and by Einstein himself in a series of independent papers published in 1902–1904 [17]. In this case the basic role of thermodynamics was evident. The general laws of thermodynamics for heterogeneous systems had already been established, and had shown their usefulness in the treatment of complex systems, as compared with the reductionist approach, which remained essentially limited to rarefied gases. The time was thus ripe for a general theoretical advance, providing a deeper foundation for thermodynamics. Gibbs' and Einstein's substantially equivalent formulations of statistical mechanics overturned the mechanical approach to thermodynamics: the fundamental thermodynamic state functions and their properties (unifying the First and Second Principles) are retained as the basic relations, introduced in a compact form through the appropriate definition of a general probability function in an abstract space, having a number of dimensions of the order of the Avogadro's number. The expression of this probability is precisely what defines the characteristic thermodynamic function (see Appendix, *4*), which is therefore introduced and calculated in a purely formal way, instead of being "built up" from the interactions of the elementary components, or "reduced" to them.

With the formulation of statistical mechanics, the contradiction between thermodynamics (in its kinetic formulation) and mechanics simply disappears, and they become two independent and complementary theories. When a system composed of a number of smaller particles is considered, only one of two complementary descriptions can be adopted:

- *either* one determines its *microscopic*, or *mechanical state*, defined by the exact values of the positions and velocities of all the particles: such a state obeys the laws of mechanics, is *reversible*, and exhibits Poincaré's recurrence property; or
- one defines its *macroscopic*, *or thermodynamic state*, in terms of the probabilities that the particles fall into small but finite intervals (cells) of the coordinates: in such a case, the microscopic state is no longer determined, the laws of mechanics are therefore no longer applicable, the evolution of the macroscopic state is *irreversible* and obeys the laws of thermodynamics.

One may observe how this concept was anticipated by Boltzmann who, however, maintained an ambiguous position on it: in fact, he did not seem to appreciate the introduction of statistical mechanics (so he never acknowledged Planck's formula for blackbody radiation, although he had actively worked on it, deriving the so-called Stefan–Boltzmann law for total emission power).

As for Einstein, it is important to remark that, although his formulation of statistical mechanics is equivalent to that of Gibbs, he developed a much deeper appreciation of the probabilistic aspects of thermodynamics [29]. In fact, while Gibbs was satisfied with proof that fluctuations are negligible when systems are made up of a large number of particles, Einstein, on the other hand, was interested in singling out physical situations (characterised by a relatively small number of particles) in which fluctuations were no longer negligible, in order to trace them back to the atomic structure of the system. This led him to the treatment of Brownian motion (Appendix, 5) in his third 1905 memoir [14], and provided the basis for methods to determine Avogadro's number.

Einstein's contributions to quantum theory are discussed in Chapters 8 and 9, so we will end here with an appreciation of his originality. Planck is generally praised for the early introduction of the quantum hypothesis in his famous paper of 1900, but it seems difficult to maintain this thesis: while it must be acknowledged that he introduced the first non-mechanistic approach in the study of a physical problem. From the end of the nineteenth century, disagreeing with Boltzmann's probabilistic interpretation, Planck had developed an approach based on pure thermodynamics, and he had mastered the thermodynamic theory of the radiant field. In 1900, as soon as new experimental results on the full spectrum of cavity radiation had been presented, he derived his fundamental radiation formula in two successive papers [33]: in the first (by now completely forgotten) through pure parametrisation of a thermodynamic function of the radiation field (Appendix, *11*). Searching then for an explanation of his formula (not because of the failure of Rayleigh's approach, which he never quoted), he resorted in the second paper to Boltzmann formalism, used in a heterodox, non-reductionist way. He later called this "an act of despair" – assuming an exotic expression for probability (that a quarter of a century later became Einstein–Bose statistics [2], but in 1900 it did not make sense). Here, in order to reckon the different distributions of a continuum energy on discrete material oscillators, he used a discretisation procedure that was quite common at that time [10]: as for the "small parts" of energy – proportional to frequency for mere thermodynamic reasons, $\varepsilon_0 = h\nu$, the value of "Planck's constant" h being determined by a fit to the experimental curve – when the ratio $E_\nu / h\nu$ "is not an integer, one takes the nearest integer number". As he later wrote:

> "Since the creator of an hypothesis has *a priori* full freedom in its formulation, he has the faculty of choosing as he likes the concepts and propositions, provided that they do not contain logical contradictions" [32].

Einstein's 1905 hypothesis of the "quantum of light" was by no means a development of Planck's idea (which "seemed to me even opposite to mine", as he wrote the following year [13]), and should be acknowledged as the true introduction of the quantum as a fundamental physical entity.

The analogy we have drawn between the break with mechanism in chemistry and in physics is confirmed by Nernst's 1906 formulation of the third principle of thermodynamics: he was actually a chemist, but realising that the vanishing of specific heats at low temperatures, which follows from his principle, had been foreseen by Einstein in 1907 [12], he turned into a supporter of the new quantum theory, and went on to publish several important works with Einstein.

Acknowledgement

I am grateful to Silvio Bergia for information.

Appendix

In this Appendix we will summarise some aspects of the phenomena, scientific problems and theories we have discussed in this chapter: we will follow a logical instead of a historical order, grouping the arguments in different scientific fields.

Thermodynamics, Kinetic Theory and Statistical Mechanics

1. Kinetic theory.   The starting point of the kinetic model was the interpretation of the pressure exerted by gas as the average force impressed per unit surface and time by the elastic collisions of the molecules on the walls. The model interprets the absolute temperature T as proportional to the average kinetic energy of the molecules $\langle \varepsilon \rangle \equiv \frac{1}{2}m\langle v^2 \rangle$: for a monatomic gas $\langle \varepsilon \rangle = 3/2 \cdot k_B T$, where $k_B = R/N_A$ is the Boltzmann's constant ($R = 8.31$ J/K·mol is the ideal gas constant, and $N_A = 6 \cdot 10^{23}$ the Avogadro number, i.e. the number of molecules per mole). Another fundamental consequence is the *theorem of equipartition of energy*: the average energy $1/2 \cdot k_B T$ is associated with each *degree of freedom* of the molecules (the number of independent movements: three translational ones for monatomic molecules, two additional rotational movements for diatomic molecules: when they are considered respectively as a rigid sphere and a rigid dumbbell, see point 2). Note that the average molecular energy grows linearly with the (absolute) temperature.

2. Specific heats, quantisation and "frozen" degrees of freedom.   The application of the theorem of equipartition of energy gives the right results for the specific heats of monatomic and diatomic gases, provided that their molecules are treated respectively as rigid spheres or dumbbells. But atoms have an internal structure, and chemical bonds are far from rigid: this introduced further (internal) degrees of freedom, whose contribution on the basis of equipartition modify such results. This difficulty was resolved by quantum theory, according to which the energy states of a (bounded) physical system are quantised, and the separation between the quantised states depends on the specific system. Recalling that the average molecular energy $\langle \varepsilon \rangle$ grows linearly with temperature, the energy gap between the rotational states of a molecule is lower than $\langle \varepsilon \rangle$ at room temperatures, so that they may be excited by thermal motion. The electronic states of atoms and molecules and the vibrational states of a diatomic molecule are separated by energy gaps higher than $\langle \varepsilon \rangle$ at room temperatures, so that they cannot be excited, and appear to be "frozen": they may be excited at higher temperatures (sodium placed on a flame shows the yellow colour, or characteristic spectral line, due to excitation of its electron state; in the nucleus of a star, reaching millions of degrees, atoms are completely ionised, and the nuclear states are excited).

3. Boltzmann's probabilistic foundation of thermodynamics.   Boltzmann's reaction to criticism led him to the introduction of probabilistic concepts in kinetic theory: in 1877 he associated a probability W with the "thermodynamic" state of the gas, defined as the number of the corresponding, distinct "microscopic" states. He defined entropy as $S = k_B \cdot \ln W$, interpreted irreversibility as an evolution towards more probable states (entropy growth), and derived the thermodynamic state of equilibrium as the most probable one.

4. Thermodynamics and statistical thermodynamics (mechanics).   In order to appreciate the development in statistical thermodynamics (1902–04) with respect to the reductionist approach of kinetic theory, consider for example the *canonical ensemble*, which
describes a macroscopic system in thermal equilibrium with a thermostat, whose corresponding thermodynamic function is Helmholtz's free energy, $F = U - T \cdot S$, where
the state functions F, U, S depend on the state variables V, T, N. Let $E(q, p)$ be
the total energy of the system in terms of the set $\{q, p\}$ of canonical coordinates of the
N microscopic constituent particles (a number sN is given for each of the canonical
coordinates, s being the number of degrees of freedom of the particles: $s = 3$ for a
monatomic gas): $q_1, q_2, \ldots, q_{sN}$; $p_1, p_2, \ldots, p_{sN}$. The fundamental *probability function* in the $2sN$-dimensional abstract space (Γ-space) subtended by this set of coordinates
is *assumed* to have the expression

$$\rho(q, p) = \exp\left\{\frac{F(V, T, N) - E(q, p)}{k_B T}\right\}$$

$$\equiv \exp\left\{\frac{F(V, T, N)}{k_B T}\right\} \cdot \exp\left\{\frac{-E(q, p)}{k_B T}\right\},$$

in which the free energy F is directly inserted as a normalisation constant (the probability
being normalised by prescribing $\int \rho(q, p) \cdot d^{sN}q \, d^{sN}p = 1$) and is therefore *directly
linked* to (rather than "deduced from") this canonical probability through integration in
the abstract Γ-space

$$F(V, T, N) = \ln\left\{\int \exp\left[\frac{E(q, p)}{k_B T}\right] \cdot d^{sN}q \, d^{sN}p\right\}.$$

The expression of the abstract probability $\rho(q, p)$ is therefore such as to correspond
to free energy: notice that ρ is a function of the microscopic coordinates, while F depends on the macroscopic state variables (V, T, N), and is therefore "constant" in the
Γ-space. $\rho(q, p)$ is the generalisation of Boltzmann's probability W, and is most conveniently described in terms of an *ensemble* of identical systems in the same macroscopic
state, exhausting all the distinct microscopic states; such a treatment can be extended to
other macroscopic states (*microcanonical ensemble, gran-canonical ensemble*). In this
approach, the atomic constitution of the system is clearly assumed, but its thermodynamic properties are no longer built up from the dynamics of the interactions between
these microscopic constituents: mechanics plays no role in thermodynamic behaviour.
This direct correspondence of probability with the thermodynamic functions implies that
the former retains the same functional expression in quantum statistical thermodynamics.

5. Brownian motion.   While a *solution* contains molecules dispersed amid molecules, in
a *suspension* (*colloid*) the dispersed particles are aggregates of microscopic dimensions
but are composed of a considerable number of molecules (*Brownian particles*, from the
name of their discoverer Brown, 1836). Such particles show disordered zigzag motions.
In 1905 Einstein correctly described their behaviour as a random process, deriving from
the casual, asymmetrical collisions of the surrounding molecules on such particles (interpreted as deviations, i.e. fluctuations, from the statistically isotropic distribution).

Chemistry

6. Analogy between kinetic theory and the concept of chemical equilibrium. The basic tool in Boltzmann's gas theory is the function $f(r, v, t)$, specifying the number (or the probability) of molecules with position r and velocity v, and the basic hypothesis is that the rate of collisions between molecules with velocities v and u is simply proportional to the product of individual probabilities $f(r, v, t) \cdot f(r, u, t)$ (*molecular chaos* hypothesis: we have discussed the limitations and the problems raised by such hypothesis). Guldberg and Waage assumed that in a chemical reaction such as $A + B \leftrightarrow C + D$ the velocities of the direct and inverse reactions are respectively proportional to the products of concentrations (i.e. the numbers of molecules) $[A][B]$, and $[C][D]$ – the analogy with Boltzmann's assumption is evident – resulting in an equilibrium constant given by the ratio $[C][D]/[A][B]$. We will see that both formulations had to reveal the common limitation consisting in their substantially mechanical nature.

7. Kekulé's hexagonal model and organic chemistry. Organic chemistry deals with carbon compounds. The classification of such substances is based on the internal structure of their molecules. The basic distinction is between the *acyclic* (or *aliphatic*) *series* – such as methane or ethane, which have open-chain molecules – and the *cyclic* (or *aromatic*) *series* (besides two minor ones, the *alicyclic* and the *heterocyclic series*). Cyclic compounds are particularly stable, owing to the closed structure of their molecules, derived by the simplest one, i.e. the hexagonal structure of benzene, C_6H_6, which is a hexagon of six carbon atoms, each linked to an outer hydrogen atom.

Electromagnetism, Ether and Special Relativity

8. Maxwell's equations and electromagnetic waves. Maxwell's equations establish the fundamental laws of the electromagnetic field: they represent the first *unified theory* in physics, since they unify electrical and magnetic phenomena. It follows from them, for instance, that a variable electric (magnetic) field generates a variable magnetic (electric) field (Faraday's law, and its symmetrical). This is the principle of the generation of alternating electric currents and *electromagnetic waves*, consisting of an electric and a magnetic field oscillating in phase, reciprocally and continuously generating each other, perpendicular between them and to the direction of propagation, with velocity c.

9. "Ether wind", Lorentz's electron theory and Einstein's special relativity. Maxwell established his equations on the basis of a model, in which the electric and magnetic fields were identified with the states of a fluid (ether). However, the interpretation of Maxwell's theory in the context of mechanical philosophy generated several problems, the most generally known being so-called "ether wind": as a wind is experienced when moving through still air, one would expect the behaviour of the electromagnetic phenomena to be affected by a motion relative to ether. This, however, is not observed (Michelson–Morley experiments). In Lorentz's electron theory, the electromagnetic nature of the interactions inside matter results in a contraction in the direction of the uniform motion ("Lorentz contraction") which exactly counterbalances the "ether wind": in these investigations he obtained the so-called "Lorentz transformations", although he did not ascribe them a fundamental meaning.

On the contrary, Einstein's 1905 theory of special relativity *assumed* that the ether wind does not exist, and he based this belief on two principles: the principle of relativity, stating that no experiment performed inside a system can reveal a uniform motion; and the constancy of the velocity of light c, which is the maximum velocity attainable in nature. On this basis Einstein rejected Newton's concepts of absolute space and time, concluding that a uniform motion is described by Lorentz's instead of Galileo's transformations: contraction of distances (and dilatation of time intervals) in the direction of the motion follow, as well as the mass-energy equivalence $E = mc^2$. Maxwell's equations turn out to be correctly Lorentz covariant, while Newtonian dynamics has to be replaced by relativistic dynamics, which it approaches for small velocities compared to that of light c.

Radiant Energy and Planck's 1900 Papers

10. Radiant energy consists of electromagnetic waves, which also propagate in a vacuum. Every surface emits radiant energy, whose characteristic frequency spectrum in thermal equilibrium starts from zero and extends to higher frequencies as its absolute temperature T increases (the total energy emitted per unit surface and time being proportional to T^4; the Stefan–Boltzmann law): at room temperature it is limited to the infrared region, and only at thousands of degrees does it reach the visible band (we cannot see objects at night). The shape of this spectrum was determined experimentally in 1900, was heuristically explained by Planck, and was studied by Einstein, who introduced the "light quantum" (1905) and wave-particle dualism (1909).

11. Planck's 1900 first paper. Planck treated the radiation field on the basis of a thermodynamic function, the second derivative of the entropy s of an oscillator coupled with the field, with respect to its energy ε: $(\partial^2 S/\partial\varepsilon^2)^{-1}$. This expression may be shown to represent the mean square fluctuation of the oscillator energy. Planck had already identified Wien's commonly accepted expression for the spectrum, $E(v, T) = av^3 \cdot \exp(-b(v/T))$, with the simplest parametrisation, posing this function as simply proportional to the oscillator energy ε (integrate once, and identify the first derivative of entropy with the inverse of the absolute temperature T). When new experimental results disproved Wien's formula, in 1900, Planck simply tried the subsequent parametrisation, i.e. a linear combination of ε and ε^2, and got his formula: $E(v, T) = \frac{av^3}{e^{bv/T}-1}$, a and b being two constants that in his second paper he related to Planck's constant $h = 6 \cdot 10^{-34}$ J·s, and Boltzmann's constant k_B. It is interesting to note that this parametrisation can be interpreted *a posteriori* as an interpolation between Wien's and Rayleigh's formulas for the two extremes of the spectrum, the term in ε^2 giving just the second one: but Planck did not even mention Rayleigh's formula. Planck's parametrisation also implies wave–particle duality, as evidenced by Einstein in 1909.

12. Planck's 1900 second paper. The success of his formula led Planck to search for an explanation, resorting to Boltzmann's statistical approach (which he had previously not accepted), but he did not use Boltzmann's expression for the number of distinct partitions of N particles in k cells, $W = N! \cdot \prod_{i=1}^{k} \frac{1}{n_i!}$. Instead he assumed, without an explicit justification, the radically different expression: $W = \prod_{i=1}^{k} \frac{(n_i+z_i)!}{n_i!z_i!}$, z_i being the number of material oscillators in cell i, in which n_i quanta are distributed.

References

[1] A much more detailed analysis along these lines may be found in: A. Baracca, S. Ruffo and A. Russo, *Scienza e Industria 1848–1915*, Bari, Laterza, 1979.

[2] S. Bergia, C. Ferraro and V. Monzoni, "Planck's heritage and the Bose statistics", *Annales de la Fondation Louis de Broglie*, **10** (1985), 161; S. Bergia, "Who discovered the Bose-Einstein statistics?", in *Symmetries in Physics (1600–1980)*, edited by M.G. Doncel, A. Hermann, L. Michel, A. Pais, Servei de Publicacions UAB, Barcelona, 1987.

[3] For these aspects see for instance S. Bergia: "Einstein and the birth of special relativity", in: *Einstein, A Centenary Volume*, ed. by A.P. French, Heinemann, London, 1979; *Einstein: Quanti e Relatività, una Svolta nella Fisica Teorica*; series "I Grandi della Scienza", I, **6**, December 1988; R. McCormmach, "Einstein, Lorentz and the electron theory", *Hist. Stud. Physical Sciences*, **2** (1970), pp. 41–87.

[4] L. Boltzmann, "Bemerkungen über einige Probleme der mechanischen Wärmyheorie", *Wien Ber.*, **75** (1877), pp. 62–100; "Über die Beziehung zwischen dem Zweiten Hauptsatze der mechanischen Wärmtheorie und der Wahrscheinlichkeitsrechnung resp. den Sätzen über Wärmegleichgewicht", *Wien. Ber.*, **76** (1877), pp. 373–435.

[5] L. Boltzmann, *Vorlesungen über Gastheorie*, 2 vols., J.A. Barth, Leipzig, 1896–1898: foreword to the second volume (English translation with an introduction by Stephen G. Brush, *Lectures on Gas Theory*, University of California Press, 1964).

[6] L. Boltzmann, "Weitere Studien über das Wärmegleichgewicht unter Gasmolekülen", *Sitzungsberichte Akad. Wiss.*, Wien, II, **66** (1872), pp. 275–370.

[7] S. Cannizzaro, "Sunto di un corso di filosofia chimica, fatto nella R. Università di Genova", *Il Nuovo Cimento*, **VII** (1858), pp. 321–66.

[8] For a deep survey of Boltzmann's figure see C. Cercignani, *Ludwig Boltzmann, e la Meccanica Statistica*, La Goliardica Pavese, 1997 and *Ludwig Boltzmann, the Man who Turned Atoms*, Oxford University Press, 1998.

[9] F. Darwin, *More Letters of Charles Darwin*, London, 1903, Vol. 1, pp. 126, 139.

[10] Boltzmann had a "finitist" concept of mathematics: see R. Dugas, *La Théorie Physique au Sens de Boltzmann*, Griffon, Neuchâtel, 1959, pp. 25–29.

[11] "It was Mach who, in his *History of Mechanics*, shook this dogmatic faith [in mechanics "as the final basis of all physical thinking"]; this book exercised a profound influence upon me in this regard while I was a student" (A. Einstein, "Autobiographical notes", in P.A. Schilpp (ed.), *Albert Einstein: Philosopher–Scientist*, La Salle, Open Court, p. 21).

[12] A. Einstein, "Die Plancksche Theorie der Strahlung und die Theorie der spezifischen Wärme", *Annalen der Physik*, **22** (1907), pp. 180–90, 800 (Berichtigung).

[13] A. Einstein, "Theorie der Lichterzeugung und Lichtabsorption", *Ann. Phys.*, **20** (1906), pp. 199–206.

[14] A. Einstein, "Über die von molekularkinetischen Theorie der Wärme geforderte Bewegung von in ruhenden Flüssigkeiten suspendierten Teilchen", *Annalen der Physik*, **17** (1905), pp. 549–60. In a subsequent paper Einstein acknowledged the relationship of his theory with the Brownian motion: A. Einstein, "Zur Theorie der Brownschen Bewegung", *Annalen der Physik*, **19** (1906), pp. 371–81. See e.g. L. Navarro Verguillas, *Einstein Profeta y Hereje*, *cit.*, Chap. I (6–8).

[15] A. Einstein, "Über einen die Erzeugung und Verwandlung des Lichtes betreffenden hueristischen Gesichtspunkt", *Annalen der Physik*, **17** (1905), pp. 132–48: Einstein refers respectively to the atomic model, that implies a finite number of co-ordinates, and to the field description, in terms of continuous functions.

[16] A. Einstein, "Zur Elektrodynamik bewegter Körper", *Annalen der Physik*, **17** (1905), pp. 891–921. See S. Bergia, "Einstein and the birth of special relativity", in *Einstein, A Centenary Volume*, edited by A.P. French, Heinemann, 1979, pp. 65–90; T. Hirosige, "The ether problem, the mechanistic worldview, and the origins of the Theory of Relativity", *Hist. Stud. Physical Sciences*, **7** (1976), pp. 3–82.

[17] A. Einstein: "Kinetische Theorie des Wärmegleichgewichtes und des zweiten Hauptsatzes der Thermodynamik", *Annalen der Physik*, **9** (1902), pp. 417–33; "Eine Theorie der Grunlagen der Thermodynamik", *Annalen der Physik*, **11** (1903), pp. 170–87; "Zur Allgemeinen molekularen Theorie der Wärme", *Annalen der Physik*, **14** (1904), pp. 354–62.

[18] M.C. Gerhardt, *Précis de Chimie Organique*, Paris, 1844–45.

[19] J.W. Gibbs, "On the equilibrium of heterogeneous substances", *Trans. Connecticut Acad.*, **3**, 108, 343 (1875–1878); in *The Scientific Papers of J.W. Gibbs*, Longmans Green and Co., London, 1906, I, 55–349.

[20] W. Gibbs, *Elementary Principles in Statistical Mechanics, Developed with Special Reference to the Rational Foundations of Thermodynamics*, Yale University Press, 1902; reprinted by Dover, New York, 1960.

[21] See for instance: S. Goldberg, "In defence of ether: the British response to Einstein's special theory of relativity, 1905–1911", *Hist. Stud. Physical Sciences*, **2** (1970), pp. 89–125. As a matter of fact, opposition and criticisms against the special theory of relativity has never appeased, up to now; apart from the furious attacks the Einstein and his theory received by the nazi regime and scientific community.

[22] The annual compared rates of growth of the German chemical industry were (P.M. Hohenberg, *Chemicals in Western Europe 1850–1914*, Rand McNally & Co., Chicago, 1967):

1872–1913	Production:	chemical industry	6.2%
		global industrial	3.7%
1850–1913	Employment:	in chemistry	4.0%
		in industry	1.9%
1875–1913	Labour productivity:	in chemistry	2.3%
		in the whole industry	1.6%

[23] A. Kekulé, *Bulletin de la Société Chimique*, t. III (1865), 98; *Annalen der Chemie und Pharmacie*, **137** (1866), 129.

[24] M.J. Klein: "Thermodynamics and quanta in Planck's work", *Physics Today*, **19**, 27 (Nov. 1966); "Einstein's first paper on quanta", *The Natural Philosopher*, **2** (1963), pp. 59–86; "Einstein, specific heats and early quantum theory", *Science*, **148** (1965), pp. 173–80; "Thermodynamics in Einstein's thought", *Science*, **157** (1967), pp. 509–516.

[25] Our natural reference is to D.S. Landes, *Unbound Prometheus* (Cambridge University Press, 1969), whose analysis of technological changes coincides with our appreciation of the features and periods of scientific changes.

[26] H.A. Lorentz, *The Theory of Electrons* (Reprint) Dover, New York, 1952.

[27] J.C. Maxwell, *A Treatise on Electricity and Magnetism*, 1873 (reprinted by Dover, New York, 1954).

[28] J.C. Maxwell, "On Faraday's lines of force", *Transactions of the Cambridge Philosophical Society*, **10** (1856), 27, reprinted in *The Scientific Papers of J.C. Maxwell*, Paris, 1927, Vol. I, pp. 155–229.

[29] J. Mehra, "Einstein and the Foundation of Statistical Mechanics", *Physica*, **79A** (1975), pp. 447–477; A. Baracca and R. Rechtman, "Einstein's Statistical Mechanics", *Revista Mexicana de Fisica*, **31** (1905), pp. 695–722; L. Navarro Verguillas, *Einstein Profeta y Hereje*, Barcelona, Tusquets, 1990, Chap. 1.

[30] L. Navarro Verguillas, "On Einstein's statistical mechanical approach to early quantum theory", *Historia Scientiarum*, **43** (1991), pp. 39–58; *Arbor*, **CXLVIII, 581** (1994), 109.

[31] See for instance: D. Noble, *America by Design*, New York, Knopf, 1977; M. De Maria and R. W. Seidel, "The scientist and the inventor", *Testi e Contesti*, **4** (1980), pp. 5–32.

[32] M. Planck, *La conoscenza del mondo fisico*, Italian translation Torino, Boringhieri, 1964, p. 231.

[33] M. Planck, "Über eine Verbesserung der Wien'schen Spektralgleichung", *Verhandl. Deutsch. Physik Gesellschaft*, **2** (1900), 202–204; and "Zur Theorie des Gesetzes der Energieverteilung im Normalspektrum", *Verhandl. Deutsch. Physik Gesellschaft*, **2** (1900), pp. 237–245. The procedures of the two Planck's papers, and several aspects of the early quantum theories, are discussed in details in A. Baracca, *Manuale Critico di Meccanica Statistica*, CULC, Catania, 1976.

[34] See for instance A. Rossi, "The two paths of physics", *Scientia*, VII–VIII (1973), pp. 565–84 (first part), and IX-X-XI-XII (1974), pp. 1–26.

[35] Van't Hoff, *Leçons de Chimie-Physique*, Hermann, Paris, 1898, p. 175.

Physics Before and After Einstein
M. Mamone Capria (Ed.)
IOS Press, 2005

Chapter 4

The Origins and Concepts of Special Relativity

Seiya Abiko

Seirei Christopher College, 3453 Mikatahara-Town Hamamatsu-City, 433-8558, Japan

1. Einstein's 1905 Paper of Special Relativity

Albert Einstein published three epochal papers in 1905. The first, dated 17 March, is known as the theory of light quantum [21]. The second, dated May, is the theory of Brownian movement [25]. The last, dated June, is the special theory of relativity (referred to here as 'STR') [23]. It is this last paper that we will discuss in this chapter.

These three papers, therefore, were published within four months. What is more, he published on 30 April 1905, i.e. between the first and the second, his doctoral dissertation "A New Determination of Molecular Dimensions" [10]. The brevity of the period during which these four papers were published, suggests certain deep relationships among them. I will discuss this later, in Section 8, and begin here with a general overview of the STR paper.

As is well known, Einstein's paper on STR led to fundamental reformation in our conceptual framework of space and time. He wrote in 1905 to his close friend Conrad Habicht [13]:

> [The STR paper] is an electrodynamics of moving bodies, which employs a modification of the theory of space and time; the purely kinematic part of this paper will surely interest you.

As suggested above, this paper consists of two parts: "I. Kinematic Part" and "II. Electrodynamic Part". Besides them, he supplies at the start of the paper a short general introduction. In it, he first explains why he thinks it necessary to introduce the first postulate, i.e. the principle of special relativity (referred to as "the relativity postulate" below). Then he proceeds, without giving any reason, to the introduction of the second postulate, i.e. the principle of constancy of light velocity (referred to as "the light-velocity postulate" below).

He explains the relativity postulate as follows:

> Not only the phenomena of mechanics, but also those of electrodynamics have no properties that correspond to the concept of absolute rest. Rather, the same laws of electrodynamics and optics will be valid for all coordinate systems in which the equations of mechanics hold [i.e. the so-called "inertial systems"].

That is to say, the Galilean principle of relativity[1] known to be valid in classical mechanics should be extended to the electromagnetic and optical regions. The crucial point here is that, as is the case with the Galilean principle of relativity, the applicability of relativity postulate in STR is restricted to the inertial systems, which form a special class of coordinate systems having uniform mutual velocities.

Einstein supplies two reasons for introducing the relativity postulate. One is the asymmetry inherent in his contemporary Maxwell's electrodynamics, i.e. in the relative motion of a magnet and a conductor, an explanation of electric induction being quite different depending on which one of the two is in motion, while the observed phenomena depends only on the relative motion. The other is the failure of attempts to detect, in the electromagnetic and optical phenomena, an influence of the Earth's motion relative to the stationary ether at rest. The stationary ether had been stipulated to exist as a medium conveying electromagnetic or light waves by Hendrik A. Lorentz's theory of electrons, which was widely accepted then for the explanation of electromagnetic and optical phenomena.

On the other hand, Einstein does not supply any reason for the light-velocity postulate, which he states as follows:

> Another postulate, which is only seemingly incompatible with it [the relativity postulate], namely that light always propagates in empty space with a definite velocity c that is independent of the state of motion of the emitting body.

By stating "only seemingly incompatible with it", he seems to have meant the incompatibility of the light-velocity postulate with the classical additivity of velocities, which is a result of the Galilean transformation of coordinates.[2] As the latter transformation is based on the Galilean principle of relativity, the light-velocity postulate is in conflict with the relativity postulate in classical mechanics.

The "Kinematic Part" consists of five sections. Section 1 gives the "definition of simultaneity", where he first points out the close relationship between the judgments of time and of simultaneity as follows,

> [A]ll our judgments involving time are always judgments about *simultaneous events*. If, for example, I say that "the train arrives here at 7 o'clock," that means, more or less, "the pointing of the small hand of my watch to 7 and the arrival of the train are simultaneous events." [italics original]

He then proceeds to define simultaneity at different locations, A and B. Utilizing the light-velocity postulate, he stipulates that a "time" common to A and B can be established by the definition that the "time" required for light to travel from A to B ($= t_B - t_A$) is equal to the "time" it requires to travel from B to A ($= t'_A - t_B$). Thus, the two clocks placed at A and B are synchronous to each other, by definition, only if

$$t_B - t_A = t'_A - t_B.$$

---

[1] Galileo Galilei explained, in "The Second Day" of his book *Dialogue Concerning the Two Chief World Systems* published in 1632, the reason why the effects of the Earth's motion do not appear in the motion of bodies on the Earth. He suggested there the relativity principle named after him, which implies that the laws of motion are the same among the coordinate systems with uniform mutual velocities.

[2] To every choice of values x, y, z, t that determines the place and time of an event in the rest system K, there are corresponding values ξ, η, ζ, τ that fix this event to the system k (which has parallel spatial axes to those in K) moving with velocity v in the x-direction relative to the rest system. The Galilean transformation of coordinates states that the relation between these values are $\xi = x - vt, \eta = y, \varsigma = z, \tau = t$.

Times thus defined constitute "the time of the rest system," and permit the definition of "simultaneity at different locations" in this rest frame.

It is worth mentioning here that, with the rapid development of the railway network in Europe at that time, the synchronization of distant clocks had become urgent technological subjects in order to keep train services punctual. As a country famous for its clock-making industry, Switzerland was a centre for the unification of time. As a result, many applications for patents on this theme were arriving at the Patent Office at Bern, to which Einstein belonged [29].

In Section 2, he introduces, besides clocks, a measuring rod, and tries to utilize it to determine the length of a rigid rod in uniform translational motion (a) from the moving coordinate system and (b) from the rest system, as follows:

> (a) The observer moves together with the aforementioned measuring rod and the rigid rod to be measured, and measures the length of the rod by laying out the measuring rod in the same way as if the rod to be measured, the observer, and the measuring rod were all at rest.
> (b) Using clocks at rest and synchronous in the rest system as outlined in "Section 1," the observer determines at which points of the rest system the beginning and the end of the rod to be measured are located at some given time t. The distance between these two points, measured with the [measuring-]rod used before – but now at rest – is also a length that we can call the "length of the rod."

He states the invariance of the length of the rod in the case (a):

> According to the principle of relativity, the length determined by operation (a), which we call 'the length of the rod in the moving system,' must equal the length ℓ of the rod at rest.

On the other hand, he insists that the length of the rod changes in the case (b):

> [T]he length determined by operation (b), which we shall call 'the length of the moving rod in the rest system,' will be determined on the basis of our two principles, and we shall find that it differs from ℓ.

He then proceeds to examine more precisely the situation in (b) above, and to imagine that the two ends (A and B) of the moving rod (with velocity v) are equipped with clocks that are synchronous with the clocks of the rest system; hence, these two clocks are "synchronous in the rest system". A ray of light starting at time t_A on the clock at A adjusted to "the time of the rest system" is reflected from B at time t_B on the clock at B, which is similarly adjusted, and returns back to A at time t'_A. With r_{AB} designating the length of the moving rod found in operation (b), the ray of light travels, in the rest frame, between times t_A and t_B the distance $r_{AB}+v(t_B-t_A)$, and between t_B and t'_A the distance $r_{AB} - v(t'_A - t_B)$. Utilizing the light velocity c in the rest frame, he obtains

$$t_B - t_A = \frac{r_{AB}}{c - v}, \qquad t'_A - t_B = \frac{r_{AB}}{c + v}.$$

He then examines whether or not the observer moving with the rod finds that the synchronization criterion of "Section 1" applies to these clocks. From the viewpoint of this observer, this ray of light goes and returns for the same length ℓ of the rod, and, if the clocks were synchronous, would have required the same time-span each way. Hence, he states:

Observers co-moving with the rod would thus find that the two clocks do not run synchronously, while observers in the system at rest would declare them to be running synchronously.

The above statement also implies the so-called "relativity of simultaneity", i.e. the judgment of whether or not the two events occur simultaneously varies depending on the system of coordinates from which the two events are observed.

In Section 3, he derives the so-called "Lorentz transformation". For every choice of values x, y, z, t that determine the place and time of an event in the rest system K, there are corresponding values ξ, η, ζ, τ that fix this event to the system k (which has parallel spatial axes to those in K) moving with velocity v in the x-direction relative to the rest system. The problem to be solved is to find the system of equations (i.e. "transformation") connecting these two sets of quantities. He points out at the start, "It is clear that these equations must be linear because of the properties of homogeneity that we attribute to space and time." Utilizing this criterion, he derives the transformation by evaluating, from the rest system, the synchronization condition of the clocks in the moving system k, i.e. $\tau_1 - \tau_0 = \tau_2 - \tau_1$; and the light-velocity postulate applied for the propagation of rays of light emitted at $t = \tau = 0$, e.g. $x/t = \xi/\tau = c$, $y/t = \eta/\tau = c$ and $z/t = \zeta/\tau = c$. Thence, he arrives at the result,

$$\xi = \beta(x - vt), \qquad \eta = y, \qquad \zeta = z,$$

$$\tau = \beta\left(t - \frac{v}{c^2}x\right), \quad \text{where } \beta = \frac{1}{\sqrt{1 - (v/c)^2}}.$$

Although he does not name these equations, we call them the "Lorentz transformation", because about a year earlier Lorentz wrote down very similar transformation equations and stated that Maxwell equations remain unchanged under this transformation [35]. But, as will be explicated in Section 5 of this chapter, the equations Lorentz wrote contained some errors, and did not keep their form totally unchanged. Furthermore, unlike Einstein, Lorentz only stipulated them *a priori* without supplying any derivation.

Einstein then derives, in Section 4, several results from the above Lorentz transformation equations. First is the so-called "Lorentz contraction". He shows that, seen from the rest system, every rigid body at rest relative to the moving system (with velocity v) appears to be contracted in the direction of motion in the ratio $1 : \sqrt{1 - (v/c)^2}$. Next, he shows that a clock placed at the origin of the moving system k, i.e. at $x = vt$ in the rest system K, delays each second by $1 - \sqrt{1 - (v/c)^2}$ sec, or by about $(v/c)^2/2$ sec, relative to those clocks synchronous in the rest system.

Concerning this latter result, he points out:

> If there are two synchronous clocks in A, and one of them is moved along a closed curve with constant velocity v until it has returned to A, which takes, say t sec, then this clock will lag, on its arrival at A, $t(v/c)^2$ 2 sec behind the clock that has not been moved.

This is the origin of the famous "twin paradox", the problem in which is that, seen from the viewpoint of the moved clock, it is the clock at rest at A that appears to be moved and, thus, that should appear to be delayed. This paradox was later resolved by the remark that, in order to move a clock along a closed curve to its starting point, an acceleration process is inescapable, and this constitutes the cause of the delay [14]. To avoid such confusion, Einstein gives just after the quoted part above an experimentally verifiable fact:

From this we conclude that a balance-wheel clock that is located at Earth's equator must be very slightly slower than an absolutely identical clock that is located at one of the Earth's poles.

The last section of part 1 is entitled "Section 5. The Addition Theorem for Velocities." He evaluates there, from the rest system, the velocity U of a point moving with a constant velocity w in the ξ-direction on the moving system k (with velocity v in the x-direction with respect to the rest system). Utilizing the Lorentz transformation, he obtains

$$U = \frac{v + w}{1 + vw/c^2},$$

which substitutes the classical additivity of velocities (i.e. $U = v + w$), and eliminates the incompatibility of the two postulates stated before. He shows, from the above, that the light velocity c is the greatest possible velocity.

As for the "Electrodynamic Part", we restrict ourselves here to a brief survey. In Section 6, he derives the transformation law for electric and magnetic forces, by applying the relativity postulate and the Lorentz transformation to Maxwell's electromagnetic equations. One result from this is that the electromotive force arising upon motion in a magnetic field, i.e. the so-called "Lorentz force", is nothing but the electric force appearing in the moving system. In Section 7, he derives, by applying the results of Section 6 to the equations representing electromagnetic waves, the so-called Doppler's principle for light waves and the law of aberration of a ray of light, i.e. the transformation law for the frequency and direction of electromagnetic waves.

In Section 8, he considers the transformation from rest to a moving system, of the energy in a ray of light enclosed within a closed surface, and comments:

> It is noteworthy that the energy and the frequency of a light complex vary with the observer's state of motion according to the same law.

Although he avoids mentioning it, this result is in exact agreement with the energy value $h\nu$ of the light quantum, which he showed in the paper he wrote three months earlier.

Utilizing the equations thus obtained, he further considers the reflection of light by a perfect mirror. He calculates the difference in energy between the incident and the reflected light-rays, and identifies this difference as the work done by the radiation pressure on the mirror. Thus, he finally arrives at an expression for the pressure of radiation, to which we will return in Section 8 below.

2. The Origins of the Relativity Postulate

As stated in the preceding section, Einstein supplies in the introduction to his STR paper two reasons for introducing the relativity postulate. One is the asymmetry inherent in his contemporary Maxwell's electrodynamics; the other is the failure of attempts to detect motion of the Earth relative to the stationary ether.

As to this latter reason, Einstein explained it in his Kyoto Address delivered in 1922 during his visit to Japan as follows [1]:

> It is certain, however, that the idea [of the principle of relativity] was contained in the problems concerning the optics of moving bodies. Light propagates through the sea of the ether. The

Earth also moves in this same ether. If seen from the Earth, ether flows against it. Nevertheless, I could not find the facts verifying this flow of ether in all physics literature.

Therefore, I wanted somehow to verify this flow of ether against the Earth, namely, the movement of the Earth. When I posed this problem to my mind at that time, I never doubted the existence of the ether and the movement of the Earth. Thus, I wanted, by appropriately reflecting light from one source by mirrors, to send one light along the motion of the Earth and the other in the opposite direction. Anticipating that there should be some difference in the energy of these beams, I wanted to verify this by the difference of heat caused by them in terms of two thermocouples. This idea was just of the same sort as that of the Michelson's experiment, but I did not know this experiment very well then.

While I had these ideas in mind as a student, I came to know the strange result of Michelson's experiment. Then, I came to realize intuitively that, if we admit this as a fact, it must be our mistake to think of the movement of the Earth against the ether. That was to say the first route that led me to what we now call the principle of special relativity. Since then I have come to believe that, though the Earth moves around the Sun, we cannot perceive this movement by way of optical experiments.

The above quotation may clarify the details of the second reason.

Michelson's experiment, mentioned above, means the light-interference experiment carried out by Albert A. Michelson and Edward W. Morley in 1887. They performed a very precise measurement in order to detect the Earth's motion against the ether. However, even though they changed the directions of the two beams of light with respect to the Earth's motion, no change in interference pattern was observed. Einstein seems to have come to know of this experiment due to Wihelm Wien's paper of 1898 [51], about which Einstein wrote to his fiancée on 28 September 1899 (when he was a student) [55]. Wien indicated in this paper ten attempts that all had negative results to detect the Earth's motion against the ether, where the Michelson–Morley experiment was included as the last of ten attempts.

Nevertheless, later in 1952, Einstein denied the direct influence of this experiment on his construction of STR. He said in a reply to a question when being interviewed by Robert Shankland [46]:

> I am not sure when I first heard of the Michelson experiment. I was not conscious that it had influenced me directly during the seven years that relativity had been my life. I guess I just took it for granted that it was true.

However, Shankland continues:

> Einstein said that in the years 1905–1909 he thought a great deal about Michelson's result, in his discussions with Lorentz and others in his thinking about general relativity. He then realized (so he told me) that he had also been conscious of Michelson's result before 1905 partly through his reading of the papers of Lorentz and more because he had simply assumed this result of Michelson to be true.

Therefore, the "ten attempts with negative results" listed in Wien's paper seem enough for Einstein in 1905. The reason he mentioned only Michelson's experiment in his Kyoto Address was that he regarded it the most important and famous of the ten at the time of the Address.

The relativity postulate had another root precedent to the two stated in the STR paper. P.A. Schilpp, the editor of *Albert Einstein: Philosopher–Scientist*, persuaded Einstein earnestly to contribute to his book an article, which explains how Einstein arrived at his various theories. As a result, Einstein's "Autobiographical Notes" (abbreviated below as

"Notes") appeared. He described in it a paradox, which he hit upon at the age of 16, in 1895, and which contained the germ of the relativity postulate.

He imagined himself pursuing a ray of light with light velocity c and seeing a spatially periodical electromagnetic field at rest. He stated [57]:

> From the very beginning it appeared to me intuitively clear that, judged from the standpoint of such an observer, everything would have to happen according to the same law as for an observer, who, relative to the Earth, was at rest. For how should the first observer know, i.e., be able to determine, that he is in a state of fast uniform motion?

According to my interpretation, the above statement is a consequence of Einstein's conjecture that, owing to the Galilean principle of relativity, there would be no way of determining whether one is in a state of fast uniform motion or not [2].

The above view is consistent with the description in Einstein's small essay on the state of the ether in a magnetic field, which he wrote to one of his uncles in the same year of 1895 [4]. In it, Einstein did not adopt Lorentz's view of the stationary ether, which he encountered much later in 1901 [54]. In fact, he discussed in this essay, "The motion of the ether produced by an electric current" and "the deformation produced by the motion of the ether" [22]. Therefore, as far as Einstein viewed the ether as a movable (i.e. draggable) mechanical entity, he had no reason to doubt the validity of the Galilean principle of relativity in this case, which was then known to be valid for mechanical phenomena.

I have pointed out one more possible origin of the relativity postulate, consistent with the above, buried in his doctoral dissertation mentioned in Section 1 [3]. There he solved the viscous hydrodynamic equation in the coordinate system of a suspended solute molecule. At this point he utilized the Galilean transformation of the hydrodynamic equation from the rest frame of the solvent to the moving coordinate of the solute. Although he published his dissertation in 1905, it is certain that he was familiar with this method from his student years of 1896–1900 [61]. Similarly, he treated in his STR paper the coordinate transformation of the electromagnetic equations from the rest to the moving frame. Moreover, in 1910 he utilized the latter result to solve the electromagnetic equation in the moving coordinate of a small mirror suspended in a cavity filled with black-body radiation [28].

Therefore, the first step leading to the relativity postulate seems to have been the application of the Galilean transformation in the continuous medium.

3. Revision of the Concept of Time

Einstein's Kyoto Address, mentioned in Section 2, tells almost the same story of the construction of STR as that given by Max Wertheimer [49] based on his conversation with Einstein in 1916. In both these accounts, Einstein began with the conviction that (a) Maxwell's equations are valid and that (b) Maxwell's equations – and all other laws of nature – must have the same form in all inertial systems, i.e. the relativity postulate.

Maxwell's equations lead to the deduction of the equation for electromagnetic waves and, thus, also light velocity. Therefore, the application of the relativity postulate to Maxwell's equations brings forth the constancy of light velocity, i.e. light velocity should be constant irrespective of the motion of the system of coordinates observing it. The latter proposition, however, seemed inconsistent with the classical additivity of velocities, which required that light velocity in vacuo c should depend on the velocity of the observer. Einstein tried to keep Maxwell's equations valid for all inertial systems while allowing c to vary, but in vain.

It was around this point that he became aware of the Michelson–Morley experiment [56], which implied the conclusion to which Einstein's thinking had already led him: that c is constant for all observers. Gradually, he focused on the question of the meaning of the measurement of a moving body and, finally, on the meaning of the judgment of simultaneity involved in such experiments. Thus, Einstein arrived at his operational definition of distant simultaneity, described in Section 1, in terms of simultaneity in the same place using the presumed constancy of light velocity.

Einstein's seemingly innocuous requirement that simultaneity be operationally defined led to the rejection of the concept of an absolute time valid in all coordinate systems. On this matter, his Kyoto Address testifies to the important role played by his friend Michele Besso [1, p. 14]:

> This invariance of light velocity conflicted with the law of the additivity of velocity well known in mechanics. Why on earth did these two contradict each other? I felt I had come up against a serious difficulty. Expecting to modify the Lorentz's way of thought somehow, I spent almost one year in useless thoughts. Then, I could not but think that this mystery would be too hard for me to solve.
>
>   Nevertheless, a friend [Michele Besso] of mine in Bern relieved me by chance. It was a beautiful day. I visited him and began to talk to him like this.
>
>   "I have a problem that I cannot solve for the life of me. Today, I've brought with me the battle to you."
>
>   I discussed various things with him. Thereby, I felt inspired and was able to reach enlightenment. The next day, I revisited him and said to him,
>
>   "Thanks a lot. I have completely interpreted my problem now."
>
>   My interpretation was really about the concept of time. Namely, time could not be defined absolutely, but is in an inseparable relationship with the signal velocity. Thus, the previous extraordinary difficulty had been solved completely for the first time. Within five weeks of this realization, the principle of special relativity as we know it was established.

Besso is the person who had introduced Einstein to Ernst Mach's *The Science of Mechanics* (*Die Mechanik in Ihrer Entwicklung Historisch-Kritisch Dagestellt*) [37], and to whom Einstein acknowledged his debt for "many a valuable suggestion" in his 1905 STR paper.

4. Poincaré's Analysis of Space and Time

If the accomplishment of Einstein's STR were no more than what has been presented so far, the distance between Einstein's and other contemporary reflections and methods would not have been as great as is often claimed. In fact, it is well known that Einstein read around 1903, together with his friends in the "Olympia Academy" ("Akademie Olympia"), Poincaré's book *Science and Hypothesis* (*La Science et l' Hypothèse*), which had been published in 1902 [47]. As explained below, this book and Poincaré's paper of 1898 "Measure of Time" ("Mesure du Temps") cited therein, include such topics as the revision of the concept of time, the relativity postulate, the constancy of light velocity, and the unobservability of the Earth's motion against the ether.

In Chapter 6, "The Classical Mechanics", of that book, the following description is found concerning the concept of space and time [45]:

1. There is no absolute space, and we only conceive of relative motion; and yet in most cases mechanical facts are enunciated as if there were an absolute space to which they can be referred.
2. There is no absolute time. When we say that two periods are equal, the statement has not meaning, and can only acquire a meaning by a convention.
3. Not only have we no direct intuition of the equality of two periods, but we have not even direct intuition of the simultaneity of two events occurring in two different places. I have explained this in an article entitled "Measure of Time."
4. Finally, is not our Euclidean geometry in itself only a kind of convention of language? Mechanical facts might be enunciated with reference to a non-Euclidean space which would be less convenient but quite as legitimate as our ordinary space; the enunciation would be more complicated, but still would be possible.

As concerns the relativity postulate, Chapter 7, "Relative and Absolute Motion", refers to it as "the principle of relative motion" [43]:

The movement of any system whatever ought to obey the same laws, whether it is referred to fixed axes or to the movable axes which are implied in uniform motion in a straight line. This is the principle of relative motion; it is imposed upon us for two reasons: the commonest experiment confirms it; the consideration of the contrary hypothesis is singularly repugnant to the mind.

The constancy of light velocity is referred to in his 1898 paper "Measure of Time" [42]:

When an astronomer tells me that some stellar phenomenon, which his telescope reveals to him at this moment, happened nevertheless fifty years ago, I seek his meaning, and to that end I shall ask him how he knows it, that is, how he has measured the velocity of light.

He has begun by *supposing* that light has a constant velocity, and in particular that its velocity is the same in all directions. That is a postulate without which no measurement of this velocity could be attempted. [italics original]

With regard to the effect of the motion of the Earth against the ether, Chapter 10, "The Theories of Modern Physics", states [41]:

Suppose we discover that optical and electrical phenomena are influenced by the motion of the Earth. It would follow that those phenomena might reveal to us not only the relative motion of material bodies, but also what would seem to be their absolute motion. Again, it would be necessary to have an ether in order that these so-called absolute movements should not be their displacements with respect to empty space, but with respect to something concrete. Will this ever be accomplished? I do not think so, and I shall explain why.

Thus, just like Einstein, Poincaré also believed that one could not discover the motion of the Earth against the ether. What is more, Einstein read the above much earlier than 1905.

5. The Lorentz–Einstein Problem

What then did Lorentz think about it? His 1895 book, written in German, which Einstein had certainly read, as testified by his letter of December 1901 to his fiancée, and by his Kyoto Address, contains Lorentz's explanation of the Michelson–Morley experiment. According to it, although the effects proportional to v/c (v: velocity of the Earth against the ether) compensate for each other, those proportional to its square should be observable. However, this was not detected experimentally.

In order to explain this, Lorentz could only assume that the whole of the experimental apparatus was contracted in the direction of the Earth's movement (i.e. Lorentz contraction). He gave the reason why such a contraction should occur through his assumption on the inter-molecular forces, as follows [36]:

> Surprising as this hypothesis may appear at first sight, yet we shall have to admit that it is by no means far-fetched, as soon as we assume that molecular forces are also transmitted through the ether, like the electric and magnetic forces of which we are able at the present time to make this assertion definitely. [. . .] Now, since the form and dimensions of a solid body are ultimately conditioned by the intensity of molecular actions, there cannot fail to be a change of dimensions as well.

Poincaré criticized Lorentz's explanation as being "ad hoc" in Chapter 10 of his book subsequent to that quoted above [44]:

> I must explain why I do not believe, in spite of Lorentz, that more exact observations will ever make evident anything else but the relative displacements of material bodies. [. . .] He showed that the terms of the first order should cancel each other, but not the terms of the second order. Then more exact experiments were made, which were also negative; neither could this be the result of chance. An explanation was necessary, and was forthcoming; they always are; hypotheses are what we lack the least. [. . .] No; the same explanation must be found for the two cases, and everything tends to show that this explanation would serve equally well for the terms of the higher order, and that the mutual destruction of these terms will be rigorous and absolute.

The above is what Einstein knew about Lorentz–Poincaré's theory.

However, that was not the whole story. In response to Poincaré's criticism, in 1904 Lorentz constructed a more general theory entitled "Electromagnetic phenomena in a system moving with any velocity less than that of light", which was published in Holland and Einstein could not obtain. As stated in Section 1, it was in this paper that the Lorentz transformation was first introduced.

He starts with the assumption that Maxwell's equations only hold on the stationary ether. Then he introduces variables that are obtained by the Lorentz transformation from the true coordinates and time on the stationary ether. And he showed that, if one regards variables thus obtained as the coordinates and time on the moving bodies, Maxwell's equations on the moving bodies also retain the same form. He named the variable obtained by the Lorentz transformation from true time as "local time".

Moreover, Poincaré referred to this "local time" in his address delivered before the International Congress of Arts and Science in St Louis in 1904, saying that if we adjust our watches utilizing light signals, the watch on the moving body will show the "local time" [40]. After this address, Poincaré constructed, in his homonymous articles of 1905 and 1906, his own theory of electrons by introducing the relativity postulate ("le postulat de relativité") into Lorentz's theory [39].

At first sight, Lorentz–Poincaré's theory, as described above, seems to have informed Einstein's STR in advance. In fact, in 1953, the distinguished mathematician Sir Edmund T. Whittaker published a chapter "The Relativity Theory of Poincaré and Lorentz" in his

The History of Ether and Electricity, Vol. 2: Modern Theories 1900–1926 [50]. Einstein's contribution is only commented on in a section of the chapter.

Even before its publication, Max Born, a physicist friend both of Whittaker and of Einstein, tried to dissuade Whittaker from describing Einstein's work in that way, but in vain [60]. Born soon found an occasion (the fifteenth anniversary of the discovery of STR, held in Bern in 1955) to criticize Whittaker's account in public [5]:

1. Lorentz himself regarded Einstein as the discoverer of the principle of relativity, and was reluctant to abandon the ideas of absolute space and time to the end of his life.
2. The exciting feature of Einstein's STR is his audacity to challenge Newton's established philosophy, the traditional concepts of space and time.

The question of whether Lorentz–Poincaré's theory was STR or not is sometimes called the Lorentz–Einstein problem. In 1960, Japanese historian Tetu Hirosige developed Born's criticism through an analysis of Poincaré's works. He dealt with Poincaré's Sorbonne lecture of 1899, his address to the International Congress of Physics of 1900 at Paris, his book *Science and Hypothesis* of 1902, and his St Louis address of 1904. Hirosige stressed that [31]:

1. In contrast to Einstein, Poincaré lacked the concept of electromagnetic field as an independent and dynamical physical entity, as is shown by his indication of the crisis of the action–reaction principle in the interaction between matters and electromagnetic waves.
2. In spite of his indication that space and time need not be absolute, Poincaré did not attempt to reconsider the meaning of space and time in the theory of physics and to dispense with the concept of the ether.

At the same time, Gerald Holton criticized Whittaker from an historical point of view [32]:

1. Poincaré's 1904 address, which Whittaker cited, did not enunciate the new principle of relativity, but summarized the difficulties which contemporary physics opposed to six classical laws or principles, including the Galilean–Newtonian principle of relativity.
2. Lorentz's paper of 1903, which Whittaker cited as containing most of the basic results of Einstein's 1905 paper, was in fact published in 1904 in Holland, which Einstein could not obtain. Moreover, he did not need to know of it, because Einstein *derived* the transformation equations that Lorentz assumed *a priori*.
3. Lorentz's 1904 paper was not on STR, as we understand it since Einstein. There, Lorentz used nonrelativistic addition law for velocities ($v = V + u$), and, contrary to what Whittaker wrote, Lorentz's transformation equations of 1904 is valid only to small values of v/c, due to Lorentz's miscalculation.

Later on, Holton's position was further developed by his former student Stanley Goldberg [30].

6. Introduction of the Light-Velocity Postulate

As we have seen, Poincaré revised Lorentz's theory of electrons based on the Maxwell equations to include the relativity postulate. As noted in Section 3, a combination of these two (Maxwell's equations and the relativity postulate) allowed the constancy of light velocity to be deduced. Therefore, they felt no need to put forth independently the light-velocity postulate.

The situation was much different for Einstein, who had expressed his doubts about the existence of the ether as early as 1899 in a letter to his fiancée, as follows [53]:

> I'm convinced more and more that the electrodynamics of moving bodies as it is presented today doesn't correspond to reality, and that it will be possible to present it in a simpler way. The introduction of the term "ether" into theories of electricity has led to the concept of a medium whose motion we can describe, without, I believe, being able to ascribe physical meaning to it. I think that electrical forces can be directly defined only for empty space [...]

The above quotation shows that Einstein already doubted then the existence of the ether and the validity of Maxwell's electrodynamics based on it. Therefore, we should regard Einstein's STR as a theory constructed upon these doubts from the start.

In fact, Einstein's letter of 1955, which Max Born quoted in his Bern lecture mentioned in Section 5, states [58]:

> The new feature of [STR] was the realization of the fact that the bearing of the Lorentz-transformations transcended their connection with Maxwell's equations and was concerned with the nature of space and time in general. A further new result was that the "Lorentz invariance" is a general condition for any physical theory. This was for me of particular importance because I had already previously found that Maxwell's theory did not account for the microstructure of radiation and could therefore have no general validity [...]

Born commented on the above, "The last sentence of this letter is of particular importance. For it shows that Einstein's paper of 1905 on relativity and on the light quantum were not disconnected."

The above account is also confirmed by Einstein's comment on Max Laue's book [34] on STR. Einstein wrote to Laue on 17 January 1952 [33]:

> When one looks over your collection of proofs of [STR], one becomes of the opinion that Maxwell's theory is unquestionable. But in 1905 I already knew for certain that Maxwell's theory leads to false fluctuations of radiation pressure and hence to an incorrect Brownian motion in a Planck cavity.

What is more, in his later paper of 1916 on the quantum theory of radiation [26], he calculated the fluctuation of radiation anew. In this recalculation, he utilized the transformation equations corresponding to the optical Doppler effect and to the stellar aberration, which he had deduced in Section 7 of his STR paper of 1905 from the Maxwell equations. Concerning these relations, he states [9]:

> One could object that [these] equations [...] are based upon Maxwell's theory of electromagnetic field, a theory that is incompatible with quantum theory. But, this objection touches the form more than the essence of the matter. Because, in whichever way the theory of electromagnetic processes may develop, it will certainly retain Doppler's principle and the law of aberration [...] According to the theory of relativity, the transformation law applies, for example, also to the energy density of a mass that moves with the (quasi) speed of light.

The above quotations testify that STR was constructed, or at least thought of, as a theory applicable beyond the realm of the applicability of Maxwell's theory.

In fact, as distinct from Lorentz's 1904 paper [35], which starts with Maxwell's and Lorentz-Force's equations, the Kinematical Part of Einstein's STR paper contains neither. Therefore, in spite of its title "On the Electrodynamics of Moving Bodies" ("Zur Electrodynamik bewegter Körper"), his STR paper does not premise Maxwell's electrodynamics.

However, in order to derive the Lorentz transformation equations, Einstein required the constancy of light velocity, which had already led him to the revision of the concept of time. Thus, in order to transcend Maxwell's electrodynamics, he had no choice but to elevate the constancy of light velocity deduced from the latter to the status of the light-velocity postulate [3, p. 21]. Thus, the essential difference between the theory of Lorentz–Poincaré and that of Einstein lies in the fact whether or not the light-velocity postulate is put forth independently of the relativity postulate. In other words, Lorentz–Poincaré's theory lacks the "Kinematical Part" essential for STR.

Einstein's STR was later utilized in the explanation of the Compton effect (i.e. scattering of a photon by an electron) by Arthur Compton [6], and in the introduction of the matter-wave by Louis de Broglie both in 1923 [7], while Lorentz–Poincaré's theory was not. Moreover, Einstein's STR gave rise to Paul Dirac's relativistic quantum theory of electrons in 1928 [8], whereas Lorentz's and Poincaré's electron theories were useless. In short, while Lorentz–Poincaré's theory remained the classical theory, Einstein's STR survived the quantum revolution.

7. Einstein's Autobiographical Error

Why did so many excellent researchers of the history of STR overlook the obvious fact that STR was constructed as a theory transcending Maxwell's electrodynamics? The reason might lie in a crucial error contained in "Notes" in the first and the second editions of *Albert Einstein: Philosopher–Scientist*, published in 1949 and 1951 respectively. The problem occurred in the following lines [19]:

> Reflections of this type made it clear to me as long ago as shortly after 1900, i.e., shortly after Planck's trailblazing work, that neither mechanics nor *thermodynamics* (except in limiting cases) claim exact validity. By and by I despaired of the possibility of discovering the true laws by means of constructive efforts based on known facts. [italics added]

Einstein intended to write "electrodynamics", not "thermodynamics." The above paragraph hides the fact that the correct version was published first in 1955, the year of Einstein's death, as the German edition in Stuttgart [12]. This edition identifies itself as "The only authorized transcription of the volume published in 1949 [*Einzig autorisierte Übertragung des 1949 erscheinenen Bandes*]." Einstein himself approved the correction. This error misled students of the history of STR to believe that Einstein regarded electrodynamics as holding good instead of thermodynamics.

I inquired of The Albert Einstein Archives at Jerusalem how the error and the correction took place. The answer from an archivist (Barbara Wolff) was as follows [2, p. 204]:

> When "Autobiographisches" was published in 1949, someone found several errors in the printed version (we do not know who) and Helen Dukas [Einstein's secretary] marked the corrections in Einstein's copy of the book. We not only recognize her handwriting, but also have a letter in which she explains that she corrected the errors in Einstein's copy of the book. [. . .] In addition, she typed a "list of errata" (undated, supposedly just after the 1949 edition). [. . .] One copy of the list was given to Peter Bergmann [Einstein's assistant] and we were convinced that the first additions to the list were his.

Dukas' errata list is reproduced in Figure 1; the relevant passages in the manuscript and the correction on the printed version are shown in Figures 2a and 2b.

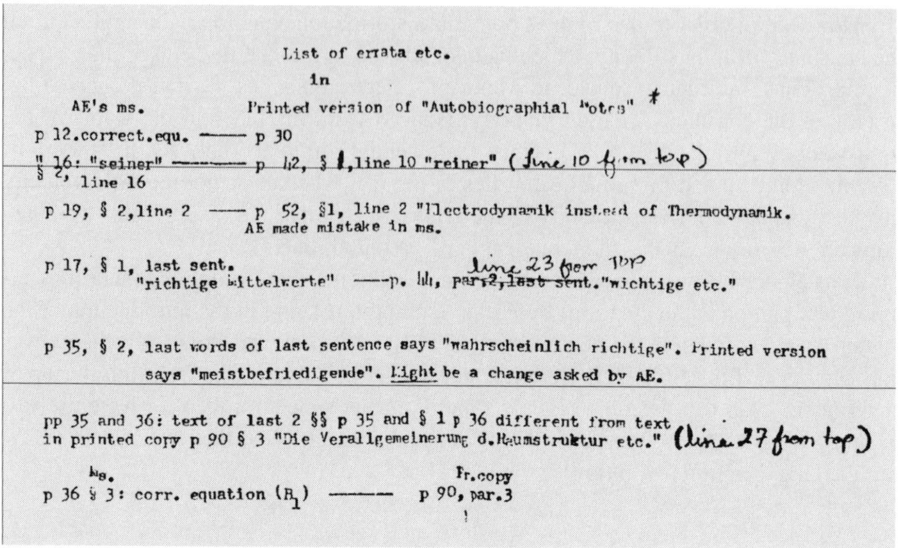
List of errata etc.
in
AE's ms. Printed version of "Autobiographial Notes" *

p 12.correct.equ. —— p 30

" 16: "seiner" —— p 42, § 1,line 10 "reiner" (line 10 from top)
§ 2, line 16

p 19, § 2,line 2 —— p 52, §1, line 2 "Electrodynamik instead of Thermodynamik.
AE made mistake in ms.

p 17, § 1, last sent.
"richtige Mittelwerte" ——p. 44, par. 2, ~~last sent~~. "wichtige etc." line 23 from top

p 35, § 2, last words of last sentence says "wahrscheinlich richtige". Printed version says "meistbefriedigende". Might be a change asked by AE.

pp 35 and 36: text of last 2 §§ p 35 and § 1 p 36 different from text in printed copy p 90 § 3 "Die Verallgemeinerung d.Raumstruktur etc." (line 27 from top)

Ms. Pr.copy
p 36 § 3: corr. equation (B_1) —— p 90, par.3

Figure 1. "List of errata" for Einstein's "Notes". The relevant mistake is on p. 19 of Einstein's manuscript and on p. 52 of the printed version of the first (1949) and the second (1951) editions, which was corrected in the German (1955) and the third (1969) editions. Two other important corrections are also stated: 'seiner' (his) on p. 16 of the manuscript was misread as 'reiner' (pure) on p. 42 of the printed version, which was also corrected. 'richtge' (exact) on p. 17 of the manuscript was misread as 'wichtige' (important) on p. 44 of the printed version, which was not corrected. Permission granted by the Albert Einstein Archives, The Jewish National & University Library, The Hebrew University of Jerusalem, Israel.

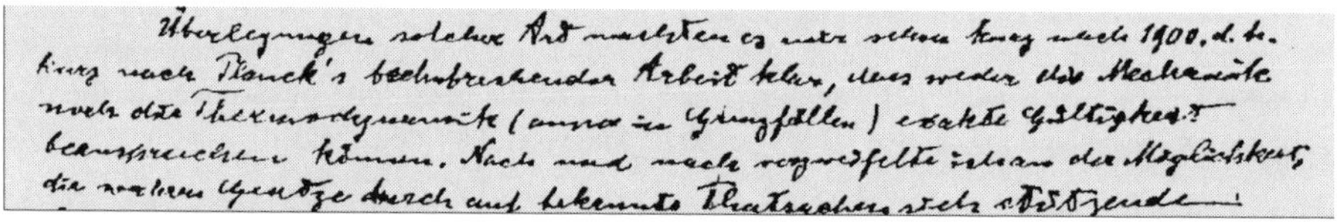

(a)

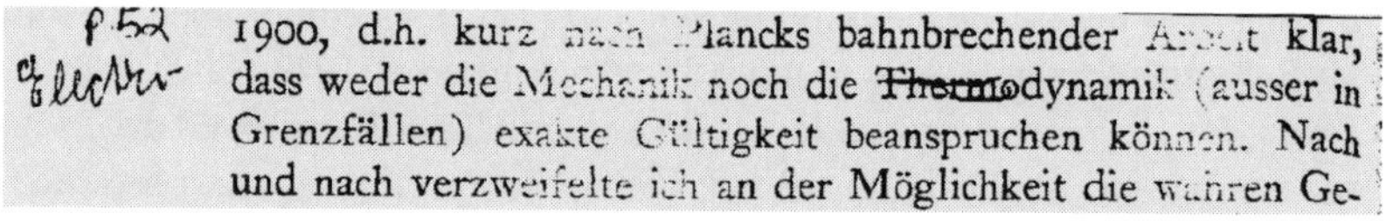
p 52 1900, d.h. kurz nach Plancks bahnbrechender Arbeit klar,
of Electro dass weder die Mechanik noch die ~~Thermo~~dynamik (ausser in
Grenzfällen) exakte Gültigkeit beanspruchen können. Nach
und nach verzweifelte ich an der Möglichkeit die wahren Ge-

(b)

Figure 2. (a) Einstein's manuscript, (b) the correction on the printed version. The third line of (a) reads as 'Thermodynamik'. In (b), 'Thermo' is corrected to 'Electro', where the handwriting is Dukas's. Permission granted by the Albert Einstein Archives, The Jewish National & University Library, The Hebrew University of Jerusalem, Israel.

The necessity of the correction is evident from three other passages in the "Notes" [18]:

> A theory is the more impressive the greater the simplicity of its premises is, the more different kinds of things it relates, and the more extended is its area of applicability. Therefore the deep impression which classical *thermodynamics* made upon me. It is the only physical theory of universal content concerning which I am convinced that, within the framework of

the applicability of its basic concepts, *it will never be overthrown*. [italics added]

  This form of reasoning [in Planck's derivation of his radiation formula] does not make obvious the fact that it contradicts the mechanical and *electrodynamical* basis, upon which the derivation otherwise depends. Actually, however, the derivation presupposes implicitly that energy can be absorbed and emitted by the individual resonator only in "quanta" of $h\nu$, [...] in contradiction to the laws of mechanics and *electrodynamics*. [italics added]

  The longer and the more despairingly I tried, the more I came to the conviction that only the discovery of a universal formal principle could lead us to assured results. The example I saw before me was *thermodynamics*. [italics added]

The correction was essential in order to make the text consistent.

One of those stymied by Einstein's mistake was Arthur I. Miller. He wrote in his book of 1982 [38],

  Of this paper on Brownian motion, Einstein wrote (1907d) that its results convinced him of *the insufficiency of mechanics and of thermodynamics* to account for properties of all systems of matter in motion. Moreover, Einstein continued, this state of affairs should be enough to convince everyone of the necessity for making fundamental changes in the basis of theoretical physics. [italics added]

The italicized part recalls the wording of the mistake in "Notes". Indeed, the statement Miller quotes agrees with the error in "Notes", but not with Einstein's paper of 1907 [24]. In that paper, Einstein meant, it is only in the absence of "the molecular-kinetic theory of heat", i.e. his statistical thermodynamics to be discussed in the next section, that the "fundamental changes" would have been required.

Because of this misinterpretation, Miller obscures the nature of STR [38]:

  Nevertheless, as long as one was not studying the instantaneous state of a system, in regions of space small enough so that fluctuation phenomena must be taken into account, then "equations of mechanics and thermodynamics can be employed," and with this restriction in mind one could also make use of Maxwell's equations. This Einstein did in the third paper that he published in vol. 17 of the *Annalen*, the relativity paper.

The above passage shows that Miller regards STR as a theory strictly restricted within the realm of classical physics applicable only to the macroscopic domain. If one takes this position, however, one can explain neither the necessity of setting up the light-velocity postulate as an independent postulate nor the difference between the theories of Lorentz–Poincaré and Einstein.

8. The Relationship among the Four Papers of 1905

As emphasized by Born, Einstein's papers on STR and on the light quantum are closely related. Japanese physicist–philosopher Mitsuo Taketani, former student and colleague of Hideki Yukawa (Japanese first Nobel Prize laureate physicist), wrote in a book on the history of quantum mechanics published in 1948 [48]:

  Einstein rejected the ether due not only to the Michelson–Morley experiment, but also to the standpoint of the light-quantum theory. [...] Einstein pointed out in the former paper that the classical electromagnetism contradicts the light quantum theory. In other words, light was not considered as vibrations of the ether in that paper. Therefore, Einstein did not reject the ether at first in his paper of relativity theory, but in his paper of light quantum. Having rejected positively the medium ether, it became necessary for him to set up a kinematics that does not rely upon the ether [hence the STR paper].

The above quotation reflects Taketani's three-stage theory for the development of scientific theories. According to it, Planck's radiation formula corresponds to the phenomenological stage, the light quantum theory to the substantial stage, and STR to the essential stage. This explains the relationship between the first and the last of the four papers written in 1905.

As stated in Section 1, the second of the four papers is his dissertation, "New Determination of Molecular Dimensions." In this paper, Einstein estimates the viscosity coefficient of a liquid solution by solving the viscous hydrodynamic equation in the coordinate system of the solute molecule. As stated in Section 2, this estimation has to do with one of the sources of the relativity postulate. He also estimates the diffusion coefficient of the solute molecules utilizing the expression for the osmotic pressure of a dilute solution. Combining these two expressions of viscosity and diffusion coefficient together with their experimental data available on tables, he could calculate the diameter of molecules and Avogadro's number. Although Einstein did not publish his dissertation until 1905, it is certain that he started to work on it earlier than 1903 [52]. In my view, this paper constitutes the common origin of his three epochal works of 1905. I will give my reason below.

Before 1905, he published three papers on statistical physics (referred to below as "the statistical trio"). The first, published in 1902, was his attempt to close the gap in Boltzmann's kinetic theory of heat [16]. By this, Einstein meant the lack of derivations of the law of thermal equilibrium and the second law of thermodynamics from the equations of mechanics and the probability calculus. On the other hand, the second and the third ones, published in 1903 and 1904, do not rely upon mechanics [11]. The system there treated is a generalized thermodynamical system expressed by state-variables (*Zustandsvariabeln*), and, therefore, as I pointed out elsewhere [3, p. 11], the theory presented in these papers should more properly be called 'statistical thermodynamics' rather than 'statistical mechanics'. Thus, leaving the realm of mechanics, Einstein could safely apply his theory to black-body radiation in the third of the statistical trio [59].

In this last part of the statistical trio, he showed that the obtained expression of energy fluctuation for material systems also applies to black-body radiation (i.e. thermal radiation at equilibrium with the emitting body kept at constant temperature). Therefore, it is conceivable that he examined the consequence of replacing osmotic pressure in his dissertation with radiation pressure. This may have been one of the routes by which he arrived at the theory of light quantum. In fact, the light-quantum theory paper argues the similarity of thermodynamic and probabilistic behaviours among monochromatic radiation, the ideal gas, and the dilute solution. This constitutes the relation between the first and the second of the four papers.

The general applicability of his statistical thermodynamics allows the application of his treatment of the diffusion of solute molecules in his dissertation to the diffusion of microscopically visible small particles. Therefore, if we replace the solute molecules in the dissertation with the microscopically visible small particles, the latter particles are expected to execute rapidly varying random motion just like molecules do. It might be in this way that he arrived at the theory of Brownian movement, which treats the diffusion process of small particles and the osmotic pressure exerted by them.

Einstein was stimulated by the success of the replacement of the equations for material systems with those of radiation in the last of the statistical trio and in the light quantum theory. Therefore, as stated in Section 2, he tried to consider the result of replacing the liquid solution in his dissertation with black-body radiation, and the hydrodynamic equation with electromagnetic field equations. This seems to have been one of the routes by which he arrived at STR. Thus, his dissertation seems the common source of Einstein's three epochal works of 1905.

A more definite relationship among the three epochal papers is exhibited in his "Notes". In it, Einstein describes a thought experiment,[3] which applies his theory of Brownian movement to a small mirror suspended in a cavity filled with black-body radiation [20]. As will be explained in Chapter 8, Einstein and quantum theory, of this book, in order to investigate more closely the structure of radiation, a precise expression of the radiation pressure and, therefore, the Maxwell's equations in the moving coordinate of the suspended mirror became necessary. Therefore, the urgent purpose of the construction of STR at that time seems the investigation into "the structure of radiation" and more generally into "the electromagnetic foundation of physics" [2, pp. 209–210].

9. Einstein's Route to General Relativity

In September 1905, as a sequel to the STR paper, Einstein published a paper entitled "Does the Inertia of a Body Depend upon Its Energy Content?" [15]. He derived in it the famous law of mass-energy equivalence,

$$E = mc^2.$$

His procedure of derivation was as follows. First, he identified the energy difference of a body measured from the rest and from the moving (with the velocity v) frames of reference as the kinetic energy $\frac{1}{2}mv^2$ of that body. Second, he considered the case that, measured from the rest frame, light with energy L is emitted from that body. Third, he calculated the energy of the emitted light measured from the moving frame, utilizing the transformation equation of light energy obtained in Section 8 of the STR paper. The resultant energy difference of the light energy measured from the two frames means the reduction in kinetic energy of that body after the emission of light. This reduction corresponds to that of its inertial mass by L/c^2 after the loss of its energy L.

In 1907, Johannes Stark, the editor of the *Jahrbuch der Radioaktivität und Elektronik*, asked Einstein to contribute a review on the recent development of STR to his journal. In writing this review, Einstein took the first step of his investigations into the gravitational field and into the general theory of relativity. On this process, Einstein said in the Kyoto Address:

> It was just then [preparing the "review"] that I realized that, despite the fact that all the rest of the laws of nature fitted into the discussion by the special theory of relativity, solely the law of universal gravitation did not. I felt deeply that I wanted somehow to find the reason why. But, I could not fulfill this purpose easily. Above all, what is most unsatisfactory to me was that, while the relationship between inertia and energy was given excellently by the special theory of relativity, that between this and weight, namely, between energy and the gravitational field,

---

[3]A thought-experiment is an imaginative experiment considered in order to make inferences theoretically.

was left quite uncertain. I imagined that its explanation could not be accomplished in terms of the special theory of relativity.

As he admitted in the above, Einstein does not seem to have thought seriously about the problem of gravitation until that time. One of the reasons might be that, in the chemico-thermal tradition (to be explained in Chapter 8) to which Einstein belonged, gravitation was not a common topic compared with more practical problems such as thermal radiation, physical chemistry and electromagnetic theory. Therefore, 26-year-old patent officer Einstein concentrated on the latter problems in 1905. Moreover, in that tradition, neglecting the problem of gravitation was not regarded as a fault of the paper, and thus, scholars of this tradition accepted and welcomed Einstein's STR [2, pp. 210–214].

On the other hand, in the case of Poincaré, who belonged to the other more idealistic tradition (i.e. the particle-dynamical tradition), the negligence of gravitation was a serious fault. What is more, when he wrote his paper of 1905, Poincaré was already a 51-year-old established scholar, and thus not allowed to commit such a fault. Thus, in Poincaré's homonymous articles of 1905 and 1906, he was at pains to explain in mathematical detail how one might account for gravitation in his new dynamics.

In the introduction to his 1907 paper, Einstein wrote as follows [27]:

> The most important result of the fourth part is that concerning the inertial mass of energy. This result suggests the question whether energy also possesses *heavy* (gravitational) mass. A further question suggesting itself is whether the principle of relativity is limited to *nonaccelerated* moving systems. In order not to leave these questions totally undiscussed, I added to the present paper a fifth part that contains a novel consideration, based on the principle of relativity, on acceleration and gravitation. [italics original]

And his review ends with the remark [27, p. 311]:

> Thus the proposition derived in Section 11 that to an amount of energy E there corresponds a mass of magnitude E/c^2, holds not only for the *inertial* but also for the *gravitational* mass, if the assumption introduced in Section 17 is correct. [italics original]

The "assumption" stated above is the so-called principle of equivalence, i.e. the equivalence of the laws of nature between the accelerated system and the system at rest in certain homogeneous gravitational fields.

From the above quotations we can see, as I noted elsewhere [1, p. 10], that "the heavy mass of energy" was a more urgent problem for Einstein at that time than the limitation of relativity postulate to nonaccelerated moving systems. He explained the reason for this in his Gibson lecture, given at Glasgow in 1933, which also gives the reason why he did not attempt to treat the problem of gravitation within STR [17]:

> In the theory I advanced, the acceleration of a falling body was not independent of its horizontal velocity or the internal energy of a system. [. . .] This did not fit in with the old experimental fact that all bodies have the same acceleration in a gravitational field. This law, which may also be formulated as the law of the equality of inertial and gravitational mass, was now brought home to me in its all significance. [. . .] I now abandoned as inadequate the attempt to treat the problem of gravitation [. . .] within the framework of special theory of relativity.

Thus, he was led to the 'principle of equivalence' in the way he remembered in the Kyoto Address:

> I was sitting in a chair in the patent office at Berne. Suddenly at that time, an idea dawned on me.

"If a man falls freely, he should not feel his weight himself."

I felt startled at once. This simple thought left me with a deep impression indeed. It was this deep impression that drove me to the theory of gravitation. I went on thinking and thinking.

When a man falls, he has the acceleration. The judgments he makes must be those in the system of reference with acceleration.

Thus, I determined to extend the principle of relativity, so as to be applicable not only to the system of reference moving with uniform velocity, but also to that moving with acceleration. By doing so, I expected that the problem of gravitation could be solved at the same time. I expected so because we can interpret the reason why a person in a free fall does not feel his weight, as that there is, other than the gravitational field caused by the Earth, another gravitational field compensating it. In other words, in a system of reference moving with acceleration, it is required that there should appear a new gravitational field.

Yet, I could not solve the problem completely at once. It was after another eight years that I found out the true relationships.

As we have seen, there was an undoubted continuity of his research on both the special and the general theories of relativity.

Acknowledgements

I am grateful to the Albert Einstein Archives, Jewish National & University Library, Jerusalem, for furnishing me with materials and information.

References

The following abbreviations are used: *AEPS,* P. A. Schilpp ed., *Albert Einstein, Philosopher–Scientist,* 1st ed. (Evanston, 1949), 2nd ed. (New York, 1951), 3rd ed. (La Salle, 1969): *AJP, American Journal of Physics*; *Brownian*, R. Fürth ed., *Investigations on the Theory of Brownian Movement* (London 1926, reprint New York 1956); *CPEE, The Collected Papers of Albert Einstein*, English Translation, vol. 1– (Princeton, 1987–); *HSPS, Historical Studies in the Physical and Biological Sciences; LCP, H. A. Lorentz, Collected Papers* (9 vols., The Hague, 1934–1939); *Letters*, J. Renn & R. Schulmann eds., *Albert Einstein–Mileva Marić, Love Letters* (Princeton, 1992); *Miraculous*, J. Stachel ed., *Einstein's Miraculous Year*, Princeton, 1998; *Origins*, G. Holton, *Thematic Origins of Scientific Thought* (Cambridge MA; 1973); *PR, The Principle of Relativity* (London, 1923, reprint New York, 1952).

[1]   S. Abiko: "Einstein's Kyoto Address: "How I Created the Theory of Relativity,"" *HSPS*, 31 (2000), 1–35 on 13.
[2]   S. Abiko: "On Einstein's Distrust of the Electromagnetic Theory," *HSPS*, 33 (2003), 193–215 on 201.
[3]   S. Abiko: "On the Chemico-Thermal Origins of Special Relativity," *HSPS*, 22 (1991), 1–24 on 22.
[4]   S. Abiko: "Reply to Darrigol," *HSPS*, 35 (2005), to be published.
[5]   M. Born: "Physics and Relativity," A. Mercier & M. Kervaire eds., *Fünfzig Jahre Relativitätstheorie* (Basel, 1956), pp. 244–260 on p. 247 & p. 250; *Physics in My Generation*, 2nd ed. (New York NY: Springer, 1969), pp. 100–115 on p. 103 & p. 105.
[6]   A.H. Compton: "A Quantum Theory of the Scattering of X-rays by Light Elements," *Phys. Rev.* 21 (1923), 483–502.
[7]   L. de Broglie: "Waves and Quanta," *Nature* 112 (1923), 540.
[8]   P. A. M. Dirac: "The Quantum Theory of the Electron," *Proc. Roy. Soc.* 117 (1928), 610–624; 118 (1928), 351–361.
[9]   A. Einstein: *op. cit.* ref. [**??**], on pp. 228–229.
[10] A. Einstein: "A New Determination of Molecular Dimensions," *Brownian*, pp. 36–62; *CPEE*, 2, Doc. 15, pp. 104–122; *Miraculous*, pp. 45–69.

[11] A. Einstein: "A Theory of the Foundations of Thermodynamics," *CPEE*, 2, Doc. 4, pp. 48–67; "On the General Molecular Theory of Heat," *CPEE*, 2, Doc. 5, pp. 68–77.

[12] A. Einstein: "Autobiographisches," P.A. Schilpp ed., *Albert Einstein als Philosoph und Naturforscher* (Stuttgart: Kohlhammer, 1955), pp. 1–36 on p. 19. I am indebted to Professor Masakatsu Yamazaki of the Tokyo Institute of Technology for having notified me the existence of this German edition.

[13] A. Einstein: *CPEE*, 5, Doc. 27, pp. 19–20.

[14] A. Einstein: "Dialog about Objections to the Theory of Relativity," *CPEE*, 7, Doc. 13, pp. 66–75; see also S. Abiko: "On the Origin of Asymmetric Aging in the Twin Problem: Comments on M. Harada's and M. Sachs' Views," *Physics Essays*, 15 (2002), 172–175.

[15] A. Einstein: "Does the Inertia of a Body Depend upon its Energy?" *PR*, pp. 67–71; *CPEE*, 2, Doc. 24, pp. 172–174; *Miraculous*, pp. 161–164.

[16] A. Einstein: "Kinetic Theory of Thermal Equilibrium and of the Second Law of Thermodynamics," *CPEE*, 2, Doc. 3, pp. 30–47.

[17] A. Einstein: "Notes on the Origin of the General Theory of Relativity," Carl Seelig ed., *Ideas and Opinions* (New York, 1954 & 1982), pp. 285–290 on p. 287.

[18] A. Einstein: "Notes," *AEPS*, pp. 32–33; pp. 44–45; pp. 52–53.

[19] A. Einstein: "Notes," *AEPS*, pp. 50–53.

[20] A. Einstein: "Notes," *AEPS*, 1, 50–51.

[21] A. Einstein: "On a Heuristic Point of View Concerning the Production and Transformation of Light," *CPEE*, 2, Doc. 14, pp. 86–103; *Miraculous*, pp. 177–198.

[22] A. Einstein: "On an Investigation of the State of the Ether in a Magnetic Field," *CPEE*, 1, Doc. 5, pp. 4–6 on p. 5.

[23] A. Einstein: "On the Electrodynamics of Moving Bodies," *PR*, pp. 35–65; *CPEE*, 2, Doc. 23, pp. 140–171; *Miraculous*, pp. 123–160. The English translations of this paper quoted below are from *Miraculous*.

[24] A. Einstein: "On the Inertia of Energy Required by the Relativity Principle," *CPEE*, 2, Doc. 45, pp. 238–251 on p. 239.

[25] A. Einstein: "On the Movement of Small Particles Suspended in Stationary Liquids Required by Molecular-Kinetic Theory of Heat," *Brownian*, pp. 1–18; *CPEE*, 2, Doc. 16, pp. 123–134; *Miraculous*, pp. 85–98.

[26] A. Einstein: "On the Quantum Theory of Radiation," *CPEE*, 6, Doc. 38, pp. 220–233; see also S. Abiko: "Einstein's Theories of the Fluctuation and the Thermal Radiation: the First Quantum Theory through Statistical Thermodynamics," *Histria Scientiarum*, 10 (2000), 130–147, on 141–143.

[27] A. Einstein: "On the Relativity Principle and the Conclusion Drawn from It," *CPEE*, 2, Doc. 47, pp. 252–311 on pp. 254–255. It should be noted here, that, in *CPEE*, 2, the last sentence quoted is mistranslated as "In order not to leave this question." The German words "Um diese Fragen" in the original should be translated into plural, i.e. "these questions." This mistake seems to be caused by the prevailing attitude, which disregard the importance of the first question.

[28] A. Einstein & L. Hopf: "Statistical Investigation of a Resonator's Motion in a Radiation Field," *CPEE*, 3, Doc. 8, pp. 220–230.

[29] P. Galison: *Einstein's Clocks, Poincaré's Maps*, New York: Norton, 2003.

[30] S. Goldberg: "Henri Poincaré and Einstein's Theory of Relativity," *AJP*, 35 (1967), 934–944.

[31] T. Hirosige: "Factors in the formation of the Special Theory of Relativity," *Kagakusi Kenkyu*, 55 (1960), 14–19; S. Nisio ed., *The Formation of Relativity Theory* (Tokyo, 1980), pp. 76–87 on p. 79, in Japanese.

[32] G. Holton: "On the Origins of the Special Theory of Relativity," *AJP*, 28 (1960), 627–636 on 633–636; *Origins*, 165–183 on 175–179.

[33] Unpublished letter quoted in G. Holton: "Influences on Einstein's Early Work," *The American Scholar*, 37 (1967–1968), 59–79; *Origins*, 197–217 on 201–202.

[34] M. Laue: *Das Relativitätstheorie* (Vieweg, 1911).

[35] H.A. Lorentz: "Electromagnetic Phenomena in a System Moving with any Velocity Less than That of Light" (1904), *LCP*, 5, pp. 172–197; *PR*, pp. 9–34.

[36] H. A. Lorentz: "Michelson's Interference Experiment," *PR*, pp. 5–6.

[37] E. Mach: *Die Mechanik in Ihrer Entwicklung Historisch-Kritisch Dagestellt*, 1883; the Engl. trans. *The Science of Mechanics*, La Salle: Open Court, 1893. In "Notes," Einstein wrote, "It was Ernst Mach who, in his *History of Mechanics*, shook this dogmatic faith [in mechanics as the final basis of all physics]; this book exercised a profound influence upon me in this regard while I was a student."

[38] A.I. Miller: *Albert Einstein's Special Theory of Relativity* (Reading, 1981; reprint New York, 1998), p. 128.

[39] H. Poincaré: "Sur la dynamique de l'electron," *Comptes Rendus de l'Academie des Science*, 140 (1905), 1504–1508 & *Rendiconti del Circolo Mathematico di Palermo*, 21 (1906), 129–176.

[40] H. Poincaré: "The Principles of Mathematical Physics," *The Monist*, 15 (1905), 1–24 on 11.

[41] H. Poincaré: *Value*, p. 129.

[42] H. Poincaré: *Value*, p. 220.

[43] H. Poincaré: *Value*, p. 87.

[44] H. Poincaré: *Value*, pp. 129–130.

[45] H. Poincaré: *Value*, pp. 73–74.

[46] R. Shankland: "Conversations with Albert Einstein," *AJP*, 31 (1963), 47–57 on 55.

[47] M. Solovine: "Introduction," A. Einstein: *Letters to Solovine*, New York: Citadel, 1993, pp. 5–13 on p. 9.

[48] M. Taketani: *The Formation and the Logic of the Quantum Mechanics*, vol. 1 (Tokyo, 1948; reprint Tokyo, 1972), p. 62, in Japanese. My translation from Japanese.

[49] M. Wertheimer: *Productive Thinking*, enlarged ed. (New York, 1959), pp. 213–233. He was one of Einstein's colleagues at the University of Berlin and is renowned as the father of Gestalt psychology.

[50] E. Whittaker: *A History of Aether and Electricity, the Modern Theories 1900–1926* (London, 1953), pp. 27–77.

[51] W. Wien: "Ueber die Fragen, welche die translatorische Bewegung des Lichtäters betreffen," *Annalen der Physik und Chemie*, 65 (1898), no. 3, Beilage: i–xvii on xv–xvi.

[52] Letter from Einstein to Michele Besso, 17 March 1903, *CPEE*, 5, Doc. 7, pp. 11–12 on p. 11.

[53] *Letters*, pp. 10–11 (10 Aug. 1899).

[54] *Letters*, p. 72 (28 Dec. 1901), where he expresses his intention to read Lorentz's 1895 book written in German.

[55] *Letters*, p.15 (28 Sept. 1899).

[56] *Letters*, p. 15 (28 Sept. 1899), which cited Wien's 1898 paper referring to the Michelson–Morley experiment; *Letters*, p. 72 (28 Dec. 1901), which referred to Lorentz's 1895 book explaining the result of this experiment.

[57] "Notes," *AEPS*, 1, pp. 52–53. I have omitted, in the last sentence of the quoted part in Schilpp's translation, the word "otherwise" which has no German counterpart in Einstein's original text and seems a mistranslation.

[58] Quoted in Born (1956), *op. cit.* note 26 on pp. 248–249; Born (1969) on p. 104. This letter was first published on *Technische Rundschau* N. 20, Jg. 47, Bern, 6 Mai 1955.

[59] See Section 3 of Chap. 8 of this book.

[60] *The Born–Einstein Letters* (London, 1971), pp. 197–198.

[61] This method of calculation is on G. Kirchhoff (W. Wien ed.): *Vorlesungen über Mechanik* (Leiptzig, 1987), which Einstein cited in his dissertation. He also states, while he was a student at ETH, "The balance of time I used in the main in order to study at home the works of Kirchhoff, Helmholtz, Hertz, etc.," on "Notes," *AEPS*, 1, pp. 14–15.

Physics Before and After Einstein
M. Mamone Capria (Ed.)
IOS Press, 2005

Chapter 5

General Relativity: Gravitation as Geometry and the Machian Programme

Marco Mamone Capria

> To put it in another way, if only a relative meaning can be attached to the concept of velocity, ought we nevertheless to persevere in treating acceleration as an absolute concept?
>
> *Albert Einstein, 1934*

The path that led to the birth of general relativity is one of the most fascinating sequences of events in the whole history of theoretical physics. It was a tortuous path, to be sure, not without a fair amount of wandering and misunderstanding. The outcome was a theory with a curious destiny. On one hand it elicited the admiration of generations of theoretical physicists and provided the ground for several observational programs in astronomy and astrophysics up to the present day. On the other hand, the theory was judged inadequate by its very creator, who worked to the end of his life to devise a radical improvement of it, which eventually he did not achieve to his own satisfaction.

The very name of the theory contains a degree of ambiguity: 'general relativity' can be read as a shortening of either 'general theory of relativity – meaning a generalization of the previous *theory* of relativity – or 'theory of general relativity – meaning something more specific, i.e. a generalization of the *principle* of relativity. The latter is what Einstein had in mind. In fact his primary aim was to generalize the principle of relativity which he had introduced in his 1905 theory, the validity of which was restricted to inertial reference frames. Whether he succeeded or not in this endeavor and – more importantly – whether he *could* possibly succeed is just one of the main issues of a foundational debate on the theory which began when the theory was still far from completion, and which has not yet come to an end. This is not surprising, given the depth of the questions which found a provisional accommodation in the Einsteinian synthesis: the nature of space, time, motion, matter, energy.

Was there a need to 'generalize' the original theory of relativity? According to some of the most eminent physicists of the time, there was not. Max Planck himself, whose approval had been crucial in earning recognition for the special theory, warned Einstein against wasting time in further speculations on that thorny topic. A different question is: why did Einstein insist on searching for a theory which had to supersede special relativity?

Historians and textbook authors have often emphasized, rather too hastily, that special relativity was 'incompatible with gravitation', and that this is the very reason Einstein was looking for a more general theory. This view is not correct, however, the clearest evidence to the contrary being that in his major paper on relativity, in 1906, Henri

Poincaré had already offered not just one, but several alternatives to translate Newton's attraction law into a Lorentz-invariant expression, that is, into a special-relativistic law. He did not think that his proposals lacked physical reasonableness, but he did not feel competent either, as a theoretician, to decide between them; for this he soberly referred to future astronomical observations (the great conventionalist was in fact more inclined to submit to experiment than historiographical folklore portrays him). Poincaré was not alone in this attempt: in the following decades many other gravitational theories, even after the rise of general relativity, were advanced.[1]

What one can surmise, in the face of this preliminary evidence, is that theories of gravitation which were compatible with special relativity might have been in some sense *not enough*. As is so often the case in the history of science, the only safe way to know why a certain theory has been created is to pursue as far as possible the real path followed by its creators; and in this case, much more than for special relativity, the theory was the creation of one man (though by no means unaided).[2]

1. Absolute Space, Relative Motion and Relativity

The 1905 article introducing the special theory of relativity is remarkable for some of the things it contains, but it is rarely considered remarkable, as it should be, for one thing it does *not* contain: gravitation. In fact it should be *prima facie* surprising that a young physicist upturning the very foundations of physical science simply does not pay any attention to gravitation – there is no announcement of work in progress, not even an incidental remark. Even more surprising is that he can neglect such a crucial topic and still get away with it, that is, without his paper being rejected by such a prestigious journal as *Annalen der Physik*. The basic explanation for this curious circumstance lies in the fact that at the end of the nineteenth century mechanics and gravitation had lost their primacy, and a new view of the physical world, based on electromagnetism, had taken their place. It was widely thought that gravitation would have been explained in terms of electromagnetism rather than the reverse. So the received wisdom of the epoch considered it acceptable to redefine space and time in order to best accommodate electrodynamics. Nevertheless, sooner or later, a viable new theory of gravitation just *had* to be produced.

The reasons for Einstein's pursuit were mainly theoretical in nature, although also based on empirical indications. He was deeply dissatisfied with the absoluteness of the *space–time* structure in special relativity, which did not seem to him, and was not in fact, a big philosophical improvement with respect to classical physics with its absolute *space* and *time*. In his 1921 lectures at the Institute for Advanced Study at Princeton he was quite explicit on this:

> Just as it was consistent from the Newtonian standpoint to make both the statements, *tempus est absolutum, spatium est absolutum*, so from the standpoint of the special theory of relativity we must say, *continuum spatii et temporis est absolutum*. In this latter statement *absolutum* means not only 'physically real', but also 'independent in its physical properties, having a physical effect, but not itself influenced by physical conditions' [38, p. 55].

---

[1]Chapter 7 in [76] bears the title "Incompatibility of gravity and special relativity" and describes in which sense one might say that gravitational theories based on Minkowski space–time are not satisfactory; see [106] for a survey of the many alternatives which have been put forward up to 1965.

[2]Whenever convenient, in this chapter and in the following I use the translations of Einstein's writings contained in [49] and [50].

It has been disputed many times to what extent the views of philosopher–scientist Ernst Mach were important in the building of the special theory.[3] What is beyond dispute is Mach's influence in the birth of the general theory. Mach's criticism of the concepts of absolute space and absolute motion in his *Science of Mechanics* [72], a book first published in 1886 which Einstein read carefully and discussed with his two friends of the "Olympia Academy", was the starting point of Einstein's line of thought.

Particularly decisive was Mach's reply to the "bucket argument", put forward by Newton in order to demonstrate that absolute acceleration (that is, acceleration with respect to the absolute space), as opposed to absolute uniform motion, can be physically recognized. Newton had remarked that the surface of the water contained in a bucket becomes curved inwards if and only if the water rotates with respect to an inertial system – it is not enough if there is just a relative rotation between the water and the bucket. Mach replied that what Newton considered 'absolute' rotation was in fact rotation with respect to matter, though admittedly not the matter of the bucket itself, but the *huge* amount of astronomical matter which exerted a centrifugal force on the water. Mach's powerful answer implied that the whole material universe and the relative motions of its different parts, rather than 'absolute' space and motions, are responsible for the forces which in classical mechanics are called 'inertial' (or, even more dismissively, 'apparent').

It is clear that a theory in which motion is truly relative should not single out *a priori* a class of privileged systems, such as the inertial systems of Newtonian physics or special relativity. The inertial systems are all relatively at rest or moving with constant relative velocity, and any system having a constant velocity with respect to any of them is itself inertial. Now suppose – as is in fact the case – that the laws of physics according to special relativity (for example) can be expressed by certain formulas *which are the same within that class*; this would provide a sufficient condition to decide whether a given system S is accelerated with respect to one (and therefore to all) of the inertial systems. To do that, one would have to check experimentally whether or not certain phenomena, as described in S, satisfy those formulas. If they don't, then S is accelerating with respect to the inertial frames, and so it certainly *moves* with respect to them.

This concept can be aptly illustrated by Galileo's famous ship experiment, vividly described in his *Dialogue Concerning the Two Chief World Systems – Ptolemaic and Copernican*, of 1632. If we are below deck, can we decide whether the ship is moving with respect to the shore without having to look (or otherwise communicating with) outside? Can we settle this question by just performing experiments inside? The answer, as Galileo (and Giordano Bruno before him) claimed, is negative if the ship is moving with constant velocity – which of course normally requires that the sea or the river is calm. On the other hand, during a tempest (or if for whatever reason the ship is gaining or losing speed), passengers below deck can usually establish quite easily whether the ship is moving: they may point, for instance, to an hanging rope or a lamp which 'no one has touched' and which has begun to oscillate, and this is sufficient proof that the ship is moving. There is no need to 'look outside' to be certain of this fact.

Thus, if we wish to rule out such a possibility, we need to formulate laws which maintain the same form not only in inertial systems, but also in accelerated ones. But is this a reasonable programme? As we have just shown, by testing the validity of certain

---

[3]One of the reasons for this perplexity is that Mach did not accept special relativity, of which he wrote in 1913 that he did not want to be considered a "forerunner" and which in his opinion was "growing more and more dogmatical".

laws in a system we *can* infer whether we are moving or not with respect to the inertial systems. The only possible answer to this objection from a Machian point of view is to concede that one can indeed have laws which are valid only for selected systems, and that therefore can be used to detect one's state of motion with respect to them, but that this fact does not rule out the possibility of finding other, more fundamental laws which are just the same in all systems.

Einstein endorsed this viewpoint, and embodied it in his *principle of general covariance* (which he also called "Principle of Relativity" in 1918 [33]), and that in the outline of general relativity he published in 1916 is stated as follows:

> *The general laws of nature are to be expressed by equations which hold good for all systems of coordinates, that is, are covariant with respect to any substitutions whatever (generally covariant)* [31, p. 117].

We shall discuss in due course what this formulation has to do with a 'principle of relativity'. However, before arriving at the confidence expressed in this statement, Einstein had to travel a long way. He started studying elementary accelerated reference frames: a freely falling chest and a rotating disk. In fact the simplicity of these examples was to a large extent deceptive.

2. The Principle of Equivalence

The concept of an absolute space–time continuum is unavoidable insofar as one wishes to maintain the ordinary law of inertia. But there are two reasons, according to Einstein [38], to abandon this law, at least as usually construed.

The first reason, for which Einstein refers to Mach's criticism of classical mechanics, is that such an absolute entity would enjoy the property, which runs "contrary to the mode of thinking in science", of acting without being acted upon by anything. One might reformulate this objection by saying that in classical physics the action and reaction principle does not hold in noninertial systems (cf. quotation in §1).

The other reason is the numerical equality, experimentally established with "very high accuracy" by Roland von Eötvös, between *inertial mass* and *gravitational mass*, which still remain conceptually separate in classical physics.

The latter point had arisen early in Einstein's thought; it was first introduced in a survey of special relativity published in 1907 [22] and then presented in a paper published in 1911, which contains the basic physical insights of general relativity [23].

The equality of the two kinds of mass (after suitable choice of the units) follows from Newton's force law and Galileo's law of free fall which, put in a general form, states that all bodies in a homogeneous gravitational field have the same constant acceleration. Using a torsion balance Eötvös had checked in 1896 a simple consequence of this equality, namely, that in a reference frame attached to the surface of a rotating homogenous sphere (modelling the Earth) the total (i.e. gravitational plus inertial) force acting on a body must have one and the same direction, independent of the mass and constitution of the body. In other terms, any difference between the two kinds of mass would result in different directions for the 'plumb lines' determined by using differently constituted bodies. Eötvös's experiments gave a negative outcome: no difference could be detected (cf. Chap. 7). The rôle of his experiments with respect to general relativity was compared by Einstein to that played by Michelson's experiments for special relativity [24, Sect. 1].

Assuming that the two kinds of mass are equal means that, for gravitational phenomena, a uniformly accelerated reference system, far from all bodies, cannot be distinguished from one at rest in a homogeneous gravitational field. This was illustrated by the ideal experiment of "a spacious chest resembling a room with an observer inside who is equipped with apparatus" [40, p. 68]: such an observer cannot distinguish, by making mechanical experiments inside this container, whether it is moving with uniform acceleration or it is at rest but in a constant gravitational field (like the gravity field near the Earth's surface). If this is true, then a freely falling chest in a constant gravitational field *will behave as an inertial reference frame*, at least as far as mechanical phenomena are concerned. This idea is what Einstein called, in an unpublished manuscript of 1920, "the happiest thought of my life". From the Kyoto address we learn that 'Einstein's apple' fell down "all of a sudden" when he was "sitting in a chair in the patent office at Bern" and the following remark occurred to him: "If a person falls freely, he should not feel his weight himself" [1, p. 15].

Einstein proposed that the impossibility of distinguishing experimentally between these two types of coordinate systems should be raised to a fundamental principle holding *for all physical phenomena*, not just the mechanical ones.

In a theory where this *principle of equivalence* (as Einstein called it) holds, the inertial and gravitational masses are not just *empirically* equal: they are *identical* – the very same physical property as measured by different procedures. We recognize in this move a close methodological resemblance to the principle of special relativity. What experience suggested in 1905 was that the laws of electromagnetism do not differ appreciably across different inertial frames: and Einstein (and Poincaré before him) assumed *that these laws do not differ at all*. The importance of the asserted identity of the two masses was stressed by Einstein in unambiguous terms:

> The possibility of explaining the numerical equality of inertia and gravitation by the unity of their nature gives to the general theory of relativity, according to my conviction, such a superiority over the conceptions of classical mechanics, that all the difficulties encountered must be considered as small in comparison with this progress [38, p. 58].

The 'explanation' mentioned here is of the same kind as the 'explanation' of the Michelson–Morley and other optical ether-drift experiments by the constancy postulate plus the principle of relativity. One might say as well that if these postulates are valid, then *there is nothing to explain* in those experimental results, except for their deviating from a perfectly 'null' outcome.

3. Two Early Empirical Consequences

Already in his 1907 survey [22] Einstein derived two of the three effects that were to provide the main testing ground for general relativity during the next fifty years (cf. Chap. 7): the gravitational redshift and the deflection of light rays.

If system S is uniformly accelerated, its natural time coordinate τ is linked to a suitable inertial time t by the following approximate formula:

$$t = \tau\left(1 + \frac{gh}{c^2}\right),$$

where h is the height in the direction of the acceleration. Now the principle of equivalence enables us to infer that in an inertial system embedded in a uniform gravitational field the same formula should be applied, the term gh being interpreted as the gravitational potential ϕ:

$$t = \tau \left(1 + \frac{\Phi}{c^2} \right).$$

This result implies that a clock at rest in a place with a higher potential goes *faster* than a clock at the origin (which is the zero level in this case).

If in the last statement we traslate 'clock' into 'every physical process', we can apply it to the "generators of the spectral lines". Thus, assuming that the same argument can also be applied, to some extent, to inhomogeneous gravitational fields, Einstein concluded that the radiation coming from material on the surface of the Sun and other heavenly bodies must arrive to us redshifted (i.e. with a lower frequency) with respect to radiation produced by similar material on the Earth.

From the equivalence principle Einstein derived in 1907 a second consequence which was at variance with one of the postulates of the special theory. In a gravitational field the speed of light is *not* constant, but it depends on the place according to the law:

$$c' = c_0 \left(1 + \frac{\Phi}{c^2} \right),$$

where c_0 is the speed of light at the origin. Formally, this is tantamount to interpreting the gravitational field as a medium with varying refraction index, which implies that the paths of the light-rays are in general not rectilinear.

In a paper of 1911 ("On the influence of gravitation on the propagation of light" [23]) Einstein came back to these results, and by using Huyghens's principle derived, for light passing close to an heavenly body, the following approximate formula for the total deflection angle:

$$\alpha \approx \frac{2GM}{c^2 \Delta}, \tag{1}$$

where Δ is the distance between the ray and the centre of the body. In the case of the Sun he obtained the following estimate: $\alpha = 4 \cdot 10^{-6} = 0.83$ seconds of arc. He ended this paper by inviting astronomers to check "apart from any theory [...] whether it is possible with the equipment at present available to detect an influence of gravitational fields on the propagation of light". In fact Einstein had to correct this prediction a few years later, when the theory reached its final form.

4. Special Relativity as a Geometric Theory

Before moving on, it is important to mention a circumstance that turned out decisive in directing Einstein's efforts towards a geometric theory of gravitation. A geometric formulation of special relativity had been developed a couple of years after the *Annalen* article by Hermann Minkowski [75], a former teacher of Einstein at the Zürich Polytechnic. In fact Minkowski followed in the steps of Henri Poincaré, who had anticipated the basic idea in his paper of 1906 [84], but he studiously omitted to mention the great French scientist.

Euclidean geometry can be studied, formally speaking, with any number of dimensions: what makes dimensions 2 and 3 so special is that they allow for a simple physical interpretation of the axioms and theorems, and therefore for a natural visualization. The axiomatic presentation of Euclidean geometry as can be found in classical textbooks is somewhat cumbersome. In fact it is equivalent to a more logically transparent (if less intuitive) presentation putting at centre-stage the axioms of the set of real numbers $\mathbb{R}$ and the assumption of the existence of an inner product[4] on the space of translations of $\mathbb{R}^n$ (the space of n-tuples of real numbers). The main difference between Euclidean and Minkowskian 4-dimensional geometry is that in the former the inner product of two vectors $u = (u^1, u^2, u^3, u^4)$ and $v = (v^1, v^2, v^3, v^4)$ has the appearance:

$$\langle u, v \rangle = u^1 v^1 + u^2 v^2 + u^3 v^3 + u^4 v^4,$$

which when $u = v$ gives the well-known Pythagorical form:

$$\langle u, u \rangle = \left(u^1\right)^2 + \left(u^2\right)^2 + \left(u^3\right)^2 + \left(u^4\right)^2,$$

while in Minkowskian geometry the inner product is

$$g(u, v) = -u^1 v^1 - u^2 v^2 - u^3 v^3 + c^2 u^4 v^4, \tag{2}$$

where c is the speed of light in empty space (or, formally, any positive constant). An inner product with this *signature* $(-, -, -, +)$ is called *Lorentzian*.[5] The Minkowski inner product can also be written as

$$g(u, v) = \sum_{\mu, \nu} \eta_{\mu\nu} u^\mu v^\nu, \tag{3}$$

where:

$$(\eta_{\mu\nu}) = \begin{pmatrix} -1 & 0 & 0 & 0 \\ 0 & -1 & 0 & 0 \\ 0 & 0 & -1 & 0 \\ 0 & 0 & 0 & c^2 \end{pmatrix}. \tag{4}$$

An immediate consequence of (2) is that $g(u, u)$ can be positive, negative or zero – and when it is zero this does *not* imply that u is the zero vector, contrary to Euclidean intuition. In fact one can classify all nonzero vectors according to whether $g(u, u) > 0$ (these vectors are called *timelike*), $g(u, u) < 0$ (they are called *spacelike*), or $g(u, u) = 0$ (they are called *ligthlike* or *null*).[6] The timelike vectors fill two solid cones, symmetric with respect to the zero vector, and the lightlike vectors together with the zero vector

---

[4] An inner product is a non-degenerate symmetric bilinear form on a real vector space. Details on the concepts of this section can be found in textbooks on linear algebra, e.g. [68].

[5] Several authors (e.g. Pauli [82]) use the signature $(+, +, +, -)$, also called Lorentzian. Here the opposite convention is adopted throughout to conform to Einstein's usage. Of course in the passage from one to the other convention many definitions have to be slightly modified.

[6] The zero vector is decreed to be spacelike.

form the boundary of the set of all timelike vectors. The set of all lightlike vectors is called the *lightcone*.

A curve whose 4-dimensional velocity vector (or *4-velocity*) is always either timelike or lightlike is called a *worldline*. It represents any ordinary physical process, and the condition on the 4-velocity means that its ordinary 3-dimensional velocity, *according to every inertial system*, does not exceed the speed of light. Suppose the timelike worldline Γ from p to q is parametrized as $x^\mu = x^\mu(s)$, with $\mu = 1, 2, 3, 4$ and s in $[a_1, a_2]$; then its Lorentzian 'length' is

$$T = \int_{a_1}^{a_2} \sqrt{\sum_{\mu,\nu} \eta_{\mu\nu} \frac{dx^\mu}{ds} \frac{dx^\nu}{ds}} \, ds, \tag{5}$$

and is called the *proper time* between p and q along Γ; the physical interpretation of this quantity is that T represents the time measured by a clock whose worldline is Γ. As expected, the timelike line segments are critical points of the proper-time functional, symbolically:

$$\delta\left(\int_p^q d\tau\right) = 0,$$

but they do not minimize proper time: they *maximize* it. This is one important point where Lorentzian geometry is sharply at variance with Euclidean geometry.

The fact that all physical processes known to us are endowed with a specific, objective time-direction is represented geometrically by selecting one of the timelike cones – the *future* cone – together with the corresponding lightlike vectors, and prescribing that the correct time order along every worldline is that defined by any parametrization with 4-velocity vector always in the future cone. This means that a *time orientation* is fixed.

Thus special relativity can be reformulated as the theory according to which the space of all events – physical space–time – is represented by a *Minkowski space*, i.e. an affine space of dimension 4, with an inner product on its associated vector space, and a time orientation; a space–time orientation is usually also added to the definition.

5. The Rotating Disk and the Physical Meaning of Coordinates

As to the effects of rotation, Einstein discussed in several papers, starting in 1912, the case of an uniformly rotating circular platform of radius R [22]. He had begun to think about it much earlier, as documented by a letter to Sommerfeld of 1909 [97].

What is the length L of the rim of this disk according to the rotating observer, as compared with the length L' measured by an inertial observer? Clearly $L' = 2\pi R$, since it is assumed that the platform is circular according to the latter observer. On the other hand, if the speed of a point on the rim with respect to the centre is v, the length of a rotating element of arc ds should suffer a Lorentz contraction according to the inertial observer, who will measure it as:

$$ds' = \sqrt{1 - \frac{v^2}{c^2}} \, ds.$$

Thus, by integrating all over the rim, one obtains

$$2\pi R = L' = \sqrt{1 - \frac{v^2}{c^2}} L,$$

that is, the length L of the rim of the disk according to the rotating observer is *bigger* than $2\pi R$. Since the radius of the disk suffers no contraction (because it is orthogonal to the velocity), its length must be R even according to the rotating observer; it follows that for this observer the ratio of the circumference to the radius is more than 2π: in other words, a basic theorem of Euclidean geometry fails.

Similarly, because of time dilatation, a clock on the rim of the disk will appear to run slower to an observer at the centre, and therefore one cannot assume that all clocks on the platform have the same rate, independent of position (as was assumed on special relativity).

This argument was taken by Einstein to imply that the coordinates in a noninertial frame cannot be interpreted as directly associated with measurement operations:

> In the general theory of relativity, space and time cannot be defined in such a way that differences of the spatial coordinates can be directly measured by the unit measuring-rod, or differences in the time coordinate by a standard clock [31, p. 117].

In his Princeton lectures Einstein used the example of the disk together with the principle of equivalence to infer that, as rotation can produce a non-Euclidean physical space, so also "the gravitational field influences and even determines the metrical laws of the space–time continuum" and that "in the presence of a gravitational field the geometry [of ideal rigid bodies] is not Euclidean" [38, p. 61]. In the 'context of discovery' this is perhaps the crucial link – though paradoxically based on a conceptual confusion, as we shall see (§14) – from the initial speculation about freedom of coordinate choice to geometrization of gravitation.

The rotating disk argument has been much discussed; in fact, perhaps surprisingly, the physical status of rotating coordinate systems in general relativity is still not a matter of universal agreement among scholars.[7] However, some firm points can be established.

First of all, if all coordinate systems are allowed, then no guarantee exists that the coordinates of an arbitrary system possess, individually, *any* physical meaning at all. For instance, in special relativity it is also very easy to define coordinate systems where no coordinate can be interpreted as either 'timelike' or 'spacelike'. Notice that, even in terms of such unphysical coordinate systems, one can define expressions corresponding to measurable physical quantities: in a sense this is the whole point of introducing tensors and the absolute calculus (§7). The upshot is that in general relativity all measurable physical quantities should be expressed in tensor form (or by using a generalization of tensors, called *geometric objects*, cf. [91]). For the most general quantities this has been done (with important exceptions, cf. §13), but use of coordinates can hardly be avoided in concrete examples, and controversy over the physical meaning of different coordinate systems has raged among relativists from the beginning.

Second, in special relativity all the inertial observers agree upon the Euclidean character of space geometry (though they definitely disagree on the numerical values of lengths and angles!), but the introduction of non-inertial, even physically reasonable, co-

---

[7]See the book review [70].

ordinate systems inevitably leads to a variety of non-Euclidean space geometries – *even in special relativity*. Given our first point, this means not that one is forced to admit non-Euclidean space geometries, but that space geometry (as contrasted to *space–time* geometry) should not be given a big importance in general relativity.

Finally, if space–time is *not* of the Minkowskian type, then *for no possible choice of the coordinate systems* can one have *all* lengths and times (along coordinate lines) to be measured as the differences of the coordinates of points. Here the situation can be compared properly with the one holding on an ordinary sphere, where the longitude-latitude coordinate system (for instance) allows one to measure distances between same-longitude points as the difference of their latitudes, but not to measure distances between same-latitude points as the difference of their longitudes (except at the equator). In fact on the sphere there exist *no* coordinate systems which make it possible to interpret in this direct metrical sense the differences of *both* coordinates. This is connected with the issue of the curvature of a surface, as we shall see in the next section.

The fact that arbitrary coordinate systems cannot be endowed with a direct physical meaning was the main stumbling block in Einstein's path to general relativity, which delayed him until 1912, according to his own recollection: "I was much bothered by this piece of knowledge, for it took me a long time to see what coordinates at all meant in physics. I did not find a way out of this dilemma until 1912 [...]" [41, p. 316].

6. Gauss' Theory of Surfaces

Einstein had been taught the differential geometry of surfaces in Zürich [81, p. 212], and now that he had to deal with both non-Euclidean geometries and general coordinate systems, this was the place he was to look for tools and inspiration.

The German mathematician Carl Friedrich Gauss (1777–1855) had established a basic distinction between those properties of a surface which depend on the way it lies in the Euclidean 3-space, and those which pertain to its *intrinsic geometry*. The fundamental intrinsic property is the length of a curve lying on the surface.

If $x^a = x^a(u)$ (with $a = 1, 2, 3$ and $u = (u^1, u^2)$ varying in an open subset of $\mathbb{R}^2$) is a parametrization of a piece of a surface,[8] and $u^1 = u^1(s), u^2 = u^2(s)$ with s varying in the interval $[a_1, a_2]$ are the parametric equations of this curve in the surface coordinates u^1, u^2, then the length of the curve is given by the integral:

$$L = \int_{a_1}^{a_2} \sqrt{\sum_{i,j=1}^{2} g_{ij}(u(s)) \frac{du^i}{ds} \frac{du^j}{ds}} ds,$$

where

$$g_{ij} = \sum_{a=1}^{3} \frac{\partial x^a}{\partial u^i} \frac{\partial x^a}{\partial u^j}.$$

---

[8]Notice that it is not always the case that a single parametrization suffices to cover the whole surface. For instance the ordinary sphere requires at least two.

Now the g_{ij} can be computed from data which are all available to a dweller on the surface, so can L. In the traditional infinitesimal language of calculus one can say that the basic metric information about a surface is given by the distance of two 'infinitesimally close' points, whose square is:

$$ds^2 = \sum_{i,j=1}^{2} g_{ij}(u)\, du^i\, du^j. \tag{6}$$

This defines a quadratic form (or, equivalently, a scalar product) on each tangent space, and is called a *metric* on S.

By definition, all intrinsic properties of a surface are those which, in any given coordinate system, can be expressed in terms of the g_{ij}. Two surfaces that can be put into a one-to-one correspondence which preserves the length of curves are called *isometric*, and such a correspondence is called an *isometry*. Clearly, two isometric surfaces, different as they may be in other ways, have the same intrinsic geometry.

For instance, if we cut a 'square' out of a cylinder, this is in intrinsic terms indistinguishable from a square (with the same side) cut out of a plane, even though, as is obvious, the 'square' extracted from the cylinder is 'bent', and the other is not. The fact is that the bending of the cylindric 'square' has to do with its extrinsic, not with the intrinsic, geometry: indeed, the two 'squares' are isometric surfaces. In ordinary terms, this can be reformulated as follows: it is possible to have a perfect 'geographic' map of the cylindric square. On the other hand such a map is impossible for any region, no matter how small, of a *sphere*. So when we see that the geographic maps of a sufficiently big portion of the Earth's surface give curiously disproportionate features to some regions (e.g. Greenland), we are shown a phenomenon which can be proven to occur, to varying degrees, even for the smallest portions of a spherical surface. We shall see in a moment how this can be proven.

A natural question that can be asked for two points p, q on a connected surface is whether some of the curves linking them *minimize* the length. The problem was first given a solution in 1728 by the great Swiss mathematician Leonhard Euler [51]. Such curves, if any exist, will be (some of the) critical points of the length functional; in the formalism of the calculus of variations the condition defining the length-critical curves is symbolically indicated by:

$$\delta\left(\int_p^q ds\right) = 0,$$

which can be shown to be equivalent, in an arbitrary system of coordinates, to the second-order system of 2 ordinary differential equations:

$$\frac{d^2 u^i}{ds^2} + \sum_{j,k=1}^{2} \Gamma^i_{jk} \frac{du^j}{ds} \frac{du^k}{ds} = 0 \quad (i = 1, 2), \tag{7}$$

where the functions Γ^i_{jk} are called *Christoffel symbols* and turn out to have the following expression:

$$\Gamma^i_{jk} = \frac{1}{2} \sum_{r=1}^{2} g^{ir}\left(\frac{\partial g_{rk}}{\partial u^j} + \frac{\partial g_{rj}}{\partial u^k} - \frac{\partial g_{jk}}{\partial u^r}\right), \tag{8}$$

where (g^{ir}) is the inverse matrix of (g_{ir}). Notice that $\Gamma^i_{jk} = \Gamma^i_{kj}$, that is, this 3-indices family of functions is symmetric with respect to the lower indices.

Curves satisfying the system of Eqs (7) are called *geodesics* of the surface. Well-known examples are the straightlines (and their segments) in the plane, and the arcs of the big circles on a sphere. As is clear from (8), the geodesics depend on the g_{ij} only, hence they belong to the intrinsic geometry of the surface.

Gauss introduced a concept which was to prove crucial in subsequent developments. It is a number which is defined at each point of a surface, called the *total* (or *Gaussian*) *curvature*, or simply the *curvature* of the surface at that point. Though its original definition did not suggest it, Gauss proved what he himself called "a most excellent theorem", which can be stated as follows: if there is an isometry between two surfaces, then corresponding points have equal curvature. In other words, the total curvature is an intrinsic geometric quantity.

Comparison of curvature functions provides one easy way of proving our previous statement on the non-existence of perfect maps of spherical regions, no matter how small. In fact it can be shown that a sphere of radius R has constant curvature equal to $1/R^2$, while the Euclidean plane has constant curvature equal to zero (it is *flat*); therefore no isometry can exist between any region of the plane and any region of the sphere, as claimed.

The fact that the curvature of the Euclidean plane vanishes everywhere suggests the natural question whether the converse implication holds: are flat surfaces isometric to the plane? The answer is positive, with an obvious restriction: if on the surface Σ the curvature vanishes, then for each point of Σ there exists *in a neighbourhood of that point* an Euclidean coordinate system (that is, such that $g_{ij} = \delta_{ij}$).[9] In other terms, the curvature is *the* local obstruction to the existence of Euclidean coordinates or, equivalently, the obstacle to the existence of a coordinate system with everywhere vanishing Christoffel symbols. As we shall see, it is this formal characterization of curvature that lends itself most directly to generalization.

7. The "Absolute Calculus"

In order to have a space–time analogue of Gaussian surfaces the concepts we have described in Sect. 6 had to be generalized to dimensions higher than two. This required, first of all, to develop the concept of an n-dimensional space, or *manifold*, which was to surfaces what n-dimensional Euclidean space is to the plane. Half a century before Einstein started to think about gravitation, in 1854, this task was accomplished by the German mathematician Bernhard Riemann (1826–1866) in a famous lecture entitled "On the hypotheses which lie at the basis of geometry" and published in 1868, posthumously.[10] This lecture is by common agreement one of the high points in the history of mathematical thought, and it provided the foundation for the development of differential geom-

---

[9]As usual, δ_{ij} is 1 if $i = j$, 0 if $i \neq j$.

[10]Though Riemann has been rightly qualified as "the man who more than any other influenced the course of modern mathematics" [99, p. 157], he was also deeply interested and knowledgeable in physics and philosophy, and this circumstance had an influence on his mathematical work. The history of relativity is the work of a number of scholars for whom the formal distinctions between different disciplines, which are so dear to bureaucrats, had little or no relevance.

etry over the next hundred years, including the tensor calculus (or "absolute calculus", where 'absolute' means 'coordinate-independent') to which the Italian geometers gave a decisive contribution (cf. [85]).

Thus, contrary to Newton's case, the mathematics required for general relativity was not Einstein's creature. It was ready-made when he needed it, and he had to ask assistance from his friend, the mathematician Marcel Grossmann, to reach a reasonable mastery of it. Indeed, he declared his theory to be "a real triumph of the general differential calculus as founded by Gauss, Riemann, Christoffel, Ricci, and Levi-Civita" [27], and the theory itself was to have an influence on the development of differential geometry [105]. Let us review here the main concepts and notation, for future reference.[11]

Essentially, an n-dimensional manifold is a set which can be covered with compatible coordinate systems with values in $\mathbb{R}^n$. Once the concept of a n-dimensional manifold is clarified, it is easy to generalize all the notions we have defined in the case of surfaces to spaces of higher dimension: it is sufficient to substitute everywhere n to 2. For instance, the basic concept of the metric is simply generalized by defining:

$$ds^2 = \sum_{\mu,\nu=1}^{n} g_{\mu\nu}(x)\, dx^\mu\, dx^\nu. \tag{9}$$

From now on we shall adopt the convention (introduced by Einstein) that when an index occurs twice in some expression (normally once in an upper, the other in a lower position), the sum over it is automaticaly understood: for Latin indices $(a, b, \ldots, i, j, \ldots)$ from 1 to 3; for Greek indices $(\alpha, \beta, \ldots, \mu, \nu, \ldots)$ from 1 to 4. Thus (9) can be re-written as:

$$ds^2 = g_{\mu\nu}(x)\, dx^\mu\, dx^\nu.$$

In modern terms, this is a symmetric rank 2 covariant tensor, such that $(g_{\mu\nu})$ is a positive definite matrix at each point in the domain of the coordinate system. This is also called a *Riemannian metric*, and a manifold with a Riemannian metric is a *Riemannian manifold*.

The equations of geodesics can be written in a similar way:

$$\frac{d^2 u^\alpha}{ds^2} + \Gamma^\alpha_{\mu\nu} \frac{du^\mu}{ds} \frac{du^\nu}{ds} = 0, \tag{10}$$

where the $\Gamma^\alpha_{\mu\nu}$ are the Christoffel symbols, which can be proven to be expressed exactly as in the case of surfaces (cf. (8)):

$$\Gamma^\alpha_{\mu\nu} = \frac{1}{2} g^{\alpha\beta} \left(\frac{\partial g_{\beta\nu}}{\partial x^\mu} + \frac{\partial g_{\beta\mu}}{\partial x^\nu} - \frac{\partial g_{\mu\nu}}{\partial x^\beta} \right). \tag{11}$$

The concept of a tensor generalizes that of a vector and is fundamental in the differential geometry of manifolds. Tensors (or multilinear functions) enjoy the important property that their multi-indices expression (or *set of components*) in every coordinate

---

[11] Most textbooks on general relativity present a more or less detailed account of differential geometry and tensor calculus (e.g. [38,60]). Two useful textbooks are [14] and (more comprehensive) [18]; a wide-ranging account of pseudo-Riemannian geometry is contained in [79]. Weyl's masterpiece [104] is still well worth studying.

system is dictated by their very nature, in the sense that knowing the set of components of a tensor in a given coordinate system allows one to derive its set of components in any other coordinate system. Tensors can be multiplied, contracted and, on a Riemannian manifold, differentiated.

Generalizing Gaussian curvature to n-dimensional Riemannian manifolds is a little more tricky, but not exceedingly so. As we know, the curvature of a surface can be considered as that function which must vanish if a local coordinate system exists with vanishing Christoffel symbols. Thus we can ask under which condition a local coordinate system $(x^1, \ldots, x^n)$ on an n-manifold exists such that the Christoffel symbols vanish identically. A computation shows that a necessary condition, which in fact also turns out to be a sufficient one, is the vanishing of the following entity:

$$R^{\alpha}_{\mu\nu\beta} = \frac{\partial \Gamma^{\alpha}_{\mu\beta}}{\partial x^{\nu}} - \frac{\partial \Gamma^{\alpha}_{\mu\nu}}{\partial x^{\beta}} + \Gamma^{\alpha}_{\nu\tau}\Gamma^{\tau}_{\mu\beta} - \Gamma^{\alpha}_{\beta\tau}\Gamma^{\tau}_{\mu\nu}. \tag{12}$$

This object, with 4 indices, is another, very important example of a tensor: the *Riemann–Christoffel* (or *curvature*) *tensor*. It expresses the local deviation of the manifold from Euclidean n-space; when it vanishes everywhere the manifold is called *flat* (in agreement with the usage introduced above in the case of surfaces). Riemann first defined it exactly the way we have just shown, in a paper he submitted in 1861 for a competition at the French Academy of Sciences (the paper did not win, and was eventually published, posthumously, in 1876).

The *Ricci tensor*[12] is defined as the contracted tensor

$$R_{\mu\nu} = R^{\alpha}_{\mu\nu\alpha}.$$

By using the formula:

$$\Gamma^{\alpha}_{\alpha\mu} = \frac{\partial}{\partial x^{\mu}} \log \sqrt{|g|}, \quad \text{where } g = -\det(g_{\mu\nu}), \tag{13}$$

the Ricci tensor can be expressed in the following useful way:

$$R_{\mu\nu} = -\frac{\partial \Gamma^{\alpha}_{\mu\nu}}{\partial x^{\alpha}} + \Gamma^{\alpha}_{\nu\beta}\Gamma^{\beta}_{\mu\alpha} + \frac{\partial^2 (\log \sqrt{|g|})}{\partial x^{\mu} \partial x^{\nu}} - \Gamma^{\alpha}_{\mu\nu}\frac{\partial (\log \sqrt{|g|})}{\partial x^{\alpha}}, \tag{14}$$

which clearly displays the Ricci tensor as a *symmetric* tensor (i.e. $R_{\mu\nu} = R_{\nu\mu}$). Formula (14) was one of the most frequently used in earlier computations in general relativity. It also explains why Einstein consistently assumed that the coordinate changes satisfy the condition $g = -1$; in such a case, indeed, we obtain a rather simple-looking formula (which, nonetheless, is the sum of 20 terms!):

$$R_{\mu\nu} = -\frac{\partial \Gamma^{\alpha}_{\mu\nu}}{\partial x^{\alpha}} + \Gamma^{\alpha}_{\nu\beta}\Gamma^{\beta}_{\mu\alpha}. \tag{15}$$

---

[12]Einstein used the following notation: $B^{\alpha}_{\mu\nu\beta}$ instead of $R^{\alpha}_{\mu\nu\beta}$, $G_{\mu\nu}$ instead of $R_{\mu\nu}$, and G instead of S for the scalar curvature; moreover he used the symbols $\Gamma^{\sigma}_{\mu\nu}$ to indicate the *negative* of what is today normally so denoted. Other authors today define the curvature tensor as the *negative* of that here defined. Here and in the following chapter we shall stick to Einstein's conventions (not to his symbols) except for the Christoffel coefficients. When comparing different accounts of relativity, readers are warned to check carefully both the notation and the definitions of the main mathematical entities; they might find it useful to consult the "Table of Sign Conventions" in [76], where thirty-odd works by different authors are classified.

When the Ricci tensor vanishes, the manifold is called *Ricci-flat*. It is clear that a flat manifold is also Ricci-flat, but the converse fails (this technical fact will turn out to be of the utmost importance in the debate on the extent to which general relativity can be considered to satisfy Mach's principle).

A further invariant quantity can be obtained by contracting the Ricci tensor itself; it is called *scalar curvature* and is defined as

$$S = g^{\mu\nu} R_{\mu\nu}. \tag{16}$$

Clearly in a Ricci-flat manifold, we have also $S = 0$.

On a Riemannian manifold it is possible to define several differential operators acting on tensors. These are all built from the coordinate independent version of the partial derivative, which is called a *covariant derivative*, and on a vector field V acts in the following way:

$$\nabla_\mu V^\nu = \frac{\partial V^\nu}{\partial x^\mu} + \Gamma^\nu_{\mu\lambda} V^\lambda. \tag{17}$$

One tensor operator which is very important in the formalism of general relativity is the *divergence*. In ordinary Euclidean 3-space and in orthonormal coordinates, the divergence of a vector is defined in the familiar way:

$$\text{div }\vec{v} = \frac{\partial v^1}{\partial x^1} + \frac{\partial v^2}{\partial x^2} + \frac{\partial v^3}{\partial x^3} = \frac{\partial v^a}{\partial x^a}.$$

If general coordinates are used, the concept of covariant derivative allows us to discover immediately the correct divergence expression, for any dimension of the space, and for any Riemannian metric ds^2: div $V = \nabla_\mu V^\mu$, which can be rewritten in a more transparent fashion by exploiting (13) again:

$$\text{div } V = \frac{1}{|g|^{1/2}} \frac{\partial(|g|^{1/2} V^\mu)}{\partial x^\mu}. \tag{18}$$

One can define similarly the divergences of a tensor of any order.

8. Spaces of Constant Curvature and the Relativity of Position

The curvature tensor is related to the Gaussian curvature as follows: if we select two independent tangent vectors A, B at a point p in M, we can consider the set of all points of M which are joined to p by a geodesic issuing from p with velocity vector lying in the subspace Π generated by A, B. In the neighbourhood of p this is a surface $\Sigma(\Pi)$, and we can compute its Gaussian curvature $K(\Pi)$ in p. This is given in an arbitrary coordinate system by:

$$K(\Pi) = -\frac{g_{\lambda\mu} R^\lambda_{\nu\rho\sigma} A^\mu B^\nu A^\rho B^\sigma}{(g_{\mu\rho} g_{\nu\sigma} - g_{\mu\sigma} g_{\nu\rho}) A^\mu B^\nu A^\rho B^\sigma}.$$

If $K(\Pi)$ is independent of the tangent 2-plane Π, then $K(\Pi)$ is called the (sectional) curvature at p and is denoted simply by $K(p)$. It is an important theorem that if this independence holds for all points p (and $n > 2$, of course), then $K(p)$ is also independent of p, that is, M is a manifold *with constant curvature*.

In such a manifold, the curvature K and the scalar curvature S are proportional according to the formula:

$$S = -n(n-1)K,$$

where n is the manifold's dimension. (For instance for $n = 3$ we have $S = -6K$, and for $n = 4$, $S = -12K$.) Note that, vice versa, from the fact that S is constant (for instance, $S = 0$) it does *not* follow that K also is.

Let us illustrate these concepts with some examples which are of crucial importance in cosmology: *the simplest Riemannian manifolds of dimension 3 and constant curvature.*

The metric of the 3-sphere of radius R_0, that is of the subspace of $\mathbb{R}^4$ defined by the equation:

$$\left(\xi^1\right)^2 + \left(\xi^2\right)^2 + \left(\xi^3\right)^2 + \left(\xi^4\right)^2 = R_0^2, \tag{19}$$

can be expressed in spherical coordinates as:

$$d\sigma^2 = R_0^2\left(d\theta^2 + \sin^2\theta\left(d\phi^2 + \sin^2\phi\, d\chi^2\right)\right),$$

and a standard computation shows that $K = 1/R_0^2$. If we introduce the new coordinate $r = R_0 \sin\theta$, we can rewrite the metric as

$$d\sigma^2 = \frac{dr^2}{1 - Kr^2} + r^2\left(d\phi^2 + \sin^2\phi\, d\chi^2\right). \tag{20}$$

Now this formula happens to give a metric of constant curvature for *any* value of K, including negative numbers and zero, so it turns out to be quite useful in computations (cf. Chap. 6).

Interestingly, the issue of physical geometry is relevant to the issue of absolute vs. relative space, and thus to the relativity of motion. In fact the argument that it is inherently impossible to know one's own position in absolute space has been used to deprive this notion of any empirical content, and as a consequence of any scientific relevance. In his booklet on the foundations of mechanics, *Matter and Motion*, Maxwell wrote, anticipating in 1876 the kind of analysis made familiar fifty years later by the logical positivist school:

> All our knowledge, both of time and place, is essentially relative. When a man has acquired the habit of putting words together, without troubling himself to form the thoughts which ought to correspond to them, it is easy for him to frame an antithesis between his relative ignorance and a so-called absolute knowledge, and to point out our ignorance of the absolute position of a point as an instance of the limitation of our faculties. Any one, however, who will try to imagine the state of a mind conscious of knowing the absolute position of a point will ever after be content with our relative knowledge [74, p. 12].

However, this argument, including its final touch of mockery, relies heavily on an assumption on the nature of physical geometry of which Maxwell was apparently unaware. The assumption is that physical space is *homogeneous*, that is, that for any two points p, q there is an isometry of the whole space sending p to q; in particular the sectional curvature must be constant. Thus any two points in Euclidean 3-space or on the 3-sphere are, indeed, geometrically indistinguishable, but the same would not be true for two points, if such exist, of physical space where curvature assumes different values.[13]

It is perfectly conceivable that physical space has no isometries other than the identity map. In this case it would be justified to speak of space as an absolute entity. This shows that the homogeneity of space is in fact a basic principle of relativity.

A final fact which deserves to be emphasized is that two spaces which are both of constant curvature K can be proven to be isometric in the small, but not necessarily in the large, where *topological properties* enter. For instance, a plane and a cylinder are locally isometric (§6), but not (globally) isometric. The main instance where this difference becomes important is in the study of the large-scale structure of the universe, where topological notions must be given special attention.

9. The Generalized Space–Time

When Minkowski introduced his geometric approach he neatly set the stage for further generalization by Einstein, since he expressed the inner product as the pseudo-distance between two 'infinitesimally close' points:[14]

$$ds^2 = -dx^2 - dy^2 - dz^2 + c^2 dt^2 = \eta_{\mu\nu}\, dx^\mu\, dx^\nu,$$

where we have put $x^1 = x$, $x^2 = y$, $x^3 = 3$, $x^4 = t$. This expression generalizes very naturally to a Riemannian metric in the form:

$$ds^2 = g_{\mu\nu}\, dx^\mu\, dx^\nu.$$

What was unusual in the mathematics of general relativity, even to mathematicians, was the *signature* of the metric of the relativistic space–time, which was Lorentzian rather than Euclidean (§4). However, the definitions we have briefly listed for Minkowski space–time transfer with no great difficulty to the case of Lorentzian manifolds.

In a general-relativistic space–time there are several (local) time functions, and therefore several (local) spatial sections. As we have emphasized (§5), even in Minkowski space–time these spaces are not all Euclidean, and need not share the same geometry. So in general relativity, just as in Riemann's geometry, a plurality of possible space geometries has to be admitted from the start; we shall see in Chapter 6 some striking and relevant examples.

Although Einstein is often identified as the man who introduced the non-Euclidean geometries in physics, this is far from true. Other scientists before him investigated the possibility that the geometry of the physical space could be non-Euclidean, and thought that it was partly an empirical matter to establish which was the true physical geometry [37]. In fact both claims had been discussed in the nineteenth century, notably by Lobachevsky, Riemann and Clifford; Gauss himself, who had been involved in geodetic surveys, was not alien to this line of thought. It has been estimated [77, p. 73] that at least 80 papers were published in the fifty years before 1915 on the physics valid in a world where the true geometry is non-Euclidean!

---

[13]This refutation of Maxwell's argument was presented by W.K. Clifford in [15, pp. 193–204].

[14]One should not take the square in 'ds^2' literally, as if it meant that Lorentzian distance may sometimes be an imaginary number: it only means that ds^2 is a quadratic form.

The British mathematician William Kingdom Clifford, who translated Riemann's famous lecture, made some bold conjectures, and in his popular *The Common Sense of the Exact Sciences*, posthumously published in 1885, stated:

> We may conceive our space to have everywhere a nearly uniform curvature, but that slight variations of the curvature may occur from point to point, and themselves vary with the time. These variations of the curvature with the time may produce effects which we not unnaturally attribute to physical causes independent of the geometry of our space. We might even go so far as to assign to this variation of the curvature of space 'what really happens in that phenomenon which we term the motion of matter' [15, pp. 202–3].

In 1900 Karl Schwarzschild, an astronomer and applied mathematician of the school of Göttingen, discussed the possibility of identifying the geometry of physical space by measurements of stellar parallaxes, and also fixed numerical bounds on the curvature, based on the then available data ([93]; cf. [87]).

On the other hand, two years later in his *Science and Hypothesis* [83], Poincaré famously argued that, though different sets of geometric axioms could be assumed in physics, none could be held to be the 'true' one: geometry was to a large extent a matter of reasonably choosing one's conventions. This sobering approach, however, was accompanied by a risky prediction, namely, that for the foreseeable future physicists would have stuck to Euclidean geometry, not for any confidence in the 'certainty' of this geometry, but because of its superior simplicity. However, the actual development of physics took a different direction – a good cautionary tale for historical prophecy.

10. What is Space–Time, Really?

In 1913 Einstein coauthored with his old friend Marcel Grossmann an essay, *Outline of a Generalized Theory of Relativity and of a Theory of Gravitation* [42], which was divided into one "Physical Part", by Einstein, and a "Mathematical Part", by Grossmann. In this work, we find several anticipations of things to come and basic ingredients which were to stay in the subsequent versions, all embedded in a fog of misunderstandings and also mathematical errors.

One decisive point which was to remain was the form of the field equations. In Sect. 5 Einstein suggested that his intended adaptation of Poisson's equation:

$$\Delta\phi = 4\pi G\rho, \tag{21}$$

which is the differential form of the Newtonian attraction law,[15] would likely have the form:

$$\kappa T^{\mu\nu} = G^{\mu\nu},$$

where $G^{\mu\nu}$ was to be a rank 2 contravariant tensor derived from $g_{\mu\nu}$ and containing up to the second-order derivatives of the $g_{\mu\nu}$. The tensor also had to be symmetric, since $T^{\mu\nu}$ was the "stress-energy tensor of the material flow"; the form of $T^{\mu\nu}$ was derived in the case of "continuously distributed incoherent masses" – that is, a *dust* – as:

$$T^{\mu\nu} = \rho\frac{dx^\mu}{ds}\frac{dx^\nu}{ds}, \tag{22}$$

---

[15]More about this point will be said in Chap. 6, §§1 and 4.

and it was shown that, under the assumption of infinitesimal conservation of the energy-momentum, it satisfied a law which Einstein postulated for the stress-energy tensor of *any* matter distribution:[16]

$$(\text{div } T)^{\mu} = \nabla_{\lambda} T^{\lambda\mu} = 0. \tag{23}$$

A major setback was that at this stage Einstein had come to be convinced that the general covariance principle (§1) had to be given up, and that one should be satisfied with a theory in which the admissible coordinate changes are *linear*. He explicitly wrote: "[. . .] we have no basis whatsoever for assuming a general covariance of the gravitational equations". This opinion was subsequently strengthened, when he stumbled upon the following conceptual difficulty, first presented in Sect. 12 of the article, which bears the unambiguous title "Proof of a Necessary Restriction in the Choice of Coordinates".

The argument, known as 'the hole problem' and which has spanned over the last two decades a vast historical and epistemological literature ([78], cf. [71]), can be formulated as follows. Suppose that space–time is identified with a certain region S of R^4, and that we change coordinates only inside a bounded domain D, leaving the standard coordinates unchanged outside. In modern terms this is equivalent to defining a diffeomorphism Φ of S, which sends D into itself and is equal to the identity everywhere except in D. Now consider two distributions of matter on S, as represented by the stress-energy tensors T and T', with T vanishing in D and $T' = \Phi_*(T)$, i.e. T' is the result of transforming T by means of Φ. Suppose $g = g_{\mu\nu}\, dx^{\mu}\, dx^{\nu}$ is the space–time metric, and let $g' = \Phi_*(g)$. Then it is clear that T', *as a function of the standard coordinates*, must be identical with T, while g' is certainly not the same as g. The unavoidable consequence seems to be that the stress-energy tensor does not, and cannot, determine by itself the space–time metric. Or one may say, with Einstein, that there are *two* different solutions for the $g_{\mu\nu}$ in the *same* coordinate system, which coincide on the boundary of D: "In other words, *the course of events in this domain cannot be determined uniquely by general-covariant differential equations*" ([25]; italics in the original).

Einstein discussed this problem with several colleagues, and received crucial help from them (cf. for instance [64]). At first, as we have seen, he decided to restrict co-ordinate changes only to linear transformations. Within this special class of coordinate changes it is clear that a diffeomorphism such as Φ cannot exist.

Eventually, however, he came back to general covariance and solved his difficulty by admitting that the triples (S, T, g) and (S, T', g') must be considered as representing *one and the same space–time*. In other words, space–time as a mathematical model of the physical space–time is uniquely determined only up to isometry class: so it is not a single space, but an equivalence class of spaces.[17] On the other hand, the physical space–time, as the set of all events (or physical coincidences), can obviously be given an intrinsic meaning; as Einstein wrote to Ehrenfest at the end of his journey to general relativity (December 26, 1915):

> The physically real in the world of events (in contrast to that which is dependent upon the choice of a reference system) consists in *spatiotemporal coincidences* [footnote: "and in nothing else!"] (cit. from [64, p. 37]).

---

[16] What Einstein actually wrote would be, in our notation: $|g|^{1/2}(\text{div } T)_{\mu} = 0$.

[17] This point is emphasized in modern textbooks, such as [60, p. 56] and [90, p. 27].

It was in the autumn of 1915 Einstein that came to the conclusion that the field equations he had written down in his paper with Grossmann [24] were incorrect, and that the restriction to linear coordinate changes was misconceived. The final work took place from October to November – "one of the most exciting and strenuous times of my life, but also one the most successful ones", as he wrote to Sommerfeld [81, p. 250]. On 25 November he presented to the Royal Prussian Academy the paper containing the revised field equations with matter. During the month of November he submitted in all 4 papers to the Academy (on the 4th, 11th, 18th, 25th), all quite short (two dozens pages in all), and with a number of changes and second thoughts from one to the other. In a letter of January 1916 to Lorentz, Einstein honestly described them as follows: "The series of my papers about gravitation is a chain of false steps [Irrwegen] which nevertheless by and by led to the goal" [81, p. 271]. It is clear that Einstein was under pressure not to be outdone in the search for the field equations by other scientists he knew were working on the same problem, and particularly by David Hilbert, one of the greatest mathematicians of his age.[18]

11. The Newtonian Limit

We shall not dwell on the several "false steps" documented in the first two papers [27,28] of this breathless series, and shall concentrate on the two final papers.

The paper of 18 November [29] deals with two applications of the new field equations to the study of the planetary motion, and corrects by a factor of 2 formula (1), that is, the prediction on the deflection of light rays made four years before in [23]. Notice that at this stage he has arrived at an incomplete version of the final field equations:

$$R_{\mu\nu} = \kappa T_{\mu\nu}. \tag{24}$$

Einstein uses (24) to show that

> this most fundamental theory of relativity [...] explains qualitatively and quantitatively the secular rotation of the orbit of Mercury (in the sense of the orbital motion itself) which was discovered by Leverrier and which amounts to 45 sec of arc per century.

This he calls "an important confirmation" of the theory. He adds:

> Furthermore, I show that the theory has as a consequence a curvature of light rays due to gravitational fields twice as strong as was indicated in my earlier investigation.

Einstein links this correction of (1) directly to the hypothesis that the trace of the matter tensor must vanish, an assumption he is making at this time and which implies, given (24), *that the scalar curvature of the metric must be zero*. This hypothesis was discarded just a few days later, in the fourth paper.

The method used to solve the planetary problem is an approximation scheme, which takes the metric to be of the form:

$$g_{\mu\nu} = \eta_{\mu\nu} + h_{\mu\nu},$$

where $\eta_{\mu\nu}$ is the Minkowski metric (with units such that $c = 1$), and $h_{\mu\nu}$ is "small" with respect to 1.

---

[18]On the 'race' Einstein–Hilbert see the update [17].

Moreover it is assumed that 1) the metric $g_{\mu\nu}$ is time-independent and time-orthogonal (the time coordinate being $t = x^4$); 2) it reduces to the Minkowski metric at spatial infinity; and 3) the spatial part of the metric is spherically symmetric. This gives the following approximate form of the metric:

$$ds^2 = -\left(\delta_{ab} + \alpha\frac{x^a x^b}{r^3}\right)dx^a\,dx^b + \left(1 - \frac{\alpha}{r}\right)dt^2. \tag{25}$$

The field Eqs (24) are used *under the hypothesis of empty space*, so that they can be written simply as:

$$R_{\mu\nu} = 0. \tag{26}$$

The main point of the mathematical development is to arrive at an expression for the potential which contains a term in r^{-3}, so that the equations of motion for a planet can be written, in second approximation, as:

$$\frac{d^2x^a}{ds^2} = \frac{\partial\Phi}{\partial x^a}, \quad \text{with } \Phi = -\frac{\alpha}{2r}\left(1 + \frac{B^2}{r^2}\right).$$

It is in fact the cubic term which is responsible for the perihelion advance effect, as had already been shown twenty years before by one Paul Gerber, who had also given the very same final formula obtained by Einstein for the advance:

$$\epsilon = \frac{24\pi^3 a^2}{T^2 c^2(1 - e^2)}, \tag{27}$$

where e is the excentricity, a is the semi-major axis and T is the period of the unperturbed orbit. This circumstance, and even more the sense in which general relativity "explains quantitatively and qualitatively the secular rotation of the orbit of Mercury" deserves a digression.

12. Einstein and Mercury's Perihelion

It is certainly remarkable that a German high-school teacher (because such was Gerber (1854–?)), had proposed in a renowned physics magazine in 1898 a different potential for the gravitational interaction which gave, in approximation, precisely the "r^{-3}" correction (with the right coefficient!), and provided a striking connection between the gravitational interaction and the speed of light. The result did not go unnoticed at the time, notwithstanding the obscurity of the author, as is proved by the fact that Ernst Mach mentioned it in his *Mechanik*, for the first time in 1901, in the fourth edition. Interestingly, Mach stressed that "only Paul Gerber" had succeeded in connecting the propagation of gravitation with the speed of light.[19]

Doubts as to the correctness of the relativistic derivations were already being voiced by some eminent contemporaries, such as the mathematicians C. Burali-Forti,

---

[19]"Only Paul Gerber [and here reference is made to [55]], studying the motion of Mercury's perihelion, which is 41 [sic] seconds of arc per century, did find that the speed of propagation of gravitation is the same as the speed of light. This speaks in favour of the aether as the medium of gravity" (Mach 1901, p. 199).

S. Zaremba, C.L. Poor, between 1922 and 1923. This was not a side issue, since the lack of rigour in the derivation was the main argument used by Einstein to discredit Gerber's priority when he was challenged on this score:

> The other way [to explain Mercury's perihelion advance without the theory of relativity] is to quote a paper by Gerber, who gave the correct formula for the movement of the Mercury perihelion before I did. Yet experts not only agree that Gerber's derivation is faulty from beginning to end, but also that the formula cannot be obtained as a consequence of the assumptions from which Gerber started out. The paper of Herr Gerber is, therefore, completely worthless, a miscarried and irreparable theoretical attempt. I state that the general theory of relativity provided the first real explanation of the movement of the perihelion of Mercury. I had not mentioned the Gerber paper because I was not aware of it when I wrote my paper of the movement of the Mercury perihelion; but even if I had known about it, there would have been no reason to mention it [36].

So for Einstein the fact that the right formula (27) had been derived from Gerber's theory could not count in favour of that theory as long as the derivation was logically objectionable: indeed it was not enough to give that theory even a modicum of respectability (cf. "completely worthless").[20]

The question of the problematic aspects of the relativistic derivation of Mercury's advance surfaced even at the level of the Nobel committee in 1921–1922: one of the experts, A. Gullstrand, from Uppsala, "a scientist of very high distinction", had objected that "other, long-known deviation from the pure two-body Newtonian law should be re-evaluated with general relativistic methods before there could be even an attempt to identify the residual effect to be explained" [81, p. 509].[21] John L. Synge, who was the author of one of the classic reference books on relativity [101], wrote half a century after Einstein's first formulation of general relativity:

> [...] when one examines some proofs in the Neo-cartesian spirit, too often they seem to dissolve completely away, leaving one in a state of wonder as to whether the author really thought he had proved something. Or is the reader stupid? It is hard to say. *In any case I am still waiting for a rational treatment of the dynamics of the solar system according to Einstein's theory* [[100, p. 14] italics added].

Things are made even more confusing by the circumstance that the total observed secular advance is $5599''.74 \pm 0.41$, most of which is accounted for in terms of inertial effects (i.e. the Earth not being an inertial frame).[22] Moreover the general relativistic derivation requires that the Sun is strictly a sphere, and a deviation from the spherical form is enough to destroy the ideal Keplerian ellipse and to produce a perihelion advance even in Newtonian physics.

Einstein's winning opinion that his derivation, relying so heavily on the Newtonian explanation of all but the residual anomalous advance of about $43''$ – an argument, not surprisingly, termed "intellectually repellent" by Synge – gave general relativity

---

[20]In referring to the "experts", Einstein is alluding to Seeliger, who in 1906 had advanced the accepted Newtonian explanation of Mercury's anomalous advance. When in 1917 Gerber's 1902 paper was republished, this time in *Annalen der Physik*, Seeliger claimed that in his calculations Gerber had made an elementary mistake. But the mistake was Seeliger's, according to [88, p. 139].

[21]Pais – who, incidentally, does not mention Gerber's work – comments that this objection is "not very weighty"; but he does not justify in any way this dismissive remark, which seems to me not just cavalier but incorrect.

[22]As explained in [101, p. 296, note 4]; see also [76, p. 1113, Box 40.3].

such an exalted status, while Gerber's anticipation, which had been appreciated by Mach, did not even deserve to be cited, is surely an interesting topic for sociologists of science.

13. The 'Final' Field Equations

The final November 1915 [30] paper begins with a pithy account of the roundabout development of the field equations in the first two papers:

> Historically they evolved in the following sequence. First, I found equations that contain the Newtonian theory as an approximation and are also covariant under arbitrary substitutions of determinant 1. Then I found that these equations are equivalent to generally-covariant ones if the scalar [i.e. the trace] of the energy tensor of "matter"vanishes. The coordinate system could then be specialized by the simple rule that $\sqrt{-g}$ must equal 1, which leads to an immense simplification of the equations of the theory. It has to be mentioned, however, that this requires the introduction of the hypothesis that the scalar of the energy tensor of matter vanishes.

Now Einstein says he has found how to avoid "this hypothesis about the energy tensor of matter", and the method is "merely by inserting it into the field equations in a slightly different manner". The field equations for the vacuum, which were used in the previous paper, are untouched by this change, which leads to the final version:

$$R_{\mu\nu} = -\kappa \left(T_{\mu\nu} - \frac{1}{2} T g_{\mu\nu} \right), \tag{28}$$

which in today's textbooks is more frequently found in the (equivalent) form:

$$R_{\mu\nu} - \frac{1}{2} S g_{\mu\nu} = -\kappa T_{\mu\nu}. \tag{29}$$

The left side is known today as the *Einstein tensor*. (We shall come back in Chapter 6, §11, to the rationale of this expression.)

The Newtonian limit is elaborated in general in [31, Sect. 21]. The assumptions are the following: $g_{\mu\nu} = \eta_{\mu\nu} + h_{\mu\nu}$ with $h_{\mu\nu}$ "small compared with 1" (in ordinary units this means 'small compared with the speed of light c'), and quantities of second and higher order are to be neglected; at spatial infinity the $h_{\mu\nu}$ must go to zero; the 3-velocity of the particles is "small"; and the time derivatives of the $g_{\mu\nu}$ must be negligible. In the case of matter modelled as a dust, we obtain the Poisson's equation in the form:

$$\Delta\phi = \frac{\kappa c^4}{2} \rho \tag{30}$$

whence it follows, by comparison with the ordinary Poisson's Eq. (21), that $\kappa c^4/2 = 4\pi G$, that is,

$$\kappa = \frac{8\pi G}{c^4} = \frac{1.87 \times 10^{-27} \text{ cm/g}}{c^2}, \tag{31}$$

where G is the gravitational constant. As is clear, the 'coupling constant' κ between 'geometry' and 'matter' is very small.

Even in the last 1915 version, the stress-energy tensor of matter is *postulated* to have zero divergence. This is evidence enough that Einstein ignores the so-called *(second) Bianchi identities*, which imply that the left-hand side of (28), and thus also the energy tensor, has zero divergence for *any* metric. It is interesting to mention that the Bianchi identities had been discovered in 1880, but until the end of 1915 Einstein was not aware of them, and, surprisingly, neither was Hilbert.[23]

By using what to Einstein is an independent assumption, i.e. div $T = 0$, a "conservation theorem of matter and gravitational field" is obtained, in the form:

$$\frac{\partial}{\partial x^\mu}\left(T_\nu^\mu + t_\nu^\mu\right) = 0. \tag{32}$$

It must be pointed out, however, that t_ν^μ, which Einstein calls "the 'energy tensor' of the gravitational field" in [30], is *not* a tensor, as he acknowledges in [31]; today it is called the *pseudotensor* of the gravitational field. In particular, it is always possible to choose a coordinate system such that, at a given point, it vanishes. The fact is that in general relativity *one cannot derive genuine integral conservation laws*, notwithstanding all the emphasis placed on the Lagrangian formalism by Einstein and his contemporaries. There is not even a coordinate-independent notion of 'inertial mass' of a system. At most, one can interpret the vanishing of the stress-energy tensor as a *local* conservation law, which can be applied in regions where the space–time curvature happens to be negligible. Given the importance of the (global!) laws of conservation in the historical development of physics, one might have expected a more widespread dissatisfaction amongst physicists as regards these features of general relativity, but eventually most of them accepted even this departure from classical physics.[24]

The conclusion of the paper is that, though

> the postulate of relativity in its most general formulation (which makes space–time coordinates into physically meaningless parameters) leads with compelling necessity to a very specific theory of gravitation that also explains the movement of the perihelion of Mercury [. . .]

nonetheless that postulate

> cannot reveal to us anything new and different about the essence of the various processes in nature that what the special theory of relativity taught us already.

Except for gravitation itself, apparently, if "compelling necessity" is to be taken seriously. In a sense this statement represents an anticipated, if puzzling, answer to Kretschmann's criticism of the principle of general covariance. In the next section we shall discuss at some length the subtle and important issues involved.

The paper in which Einstein formulated for the first time in a systematic fashion the foundations of his general relativity only appeared in 1916 in *Annalen der Physik*, and with its sixty-odd pages was the size of a booklet (in fact it was also published as such,

---

[23][81, p. 275; 89]. In the best tradition of mathematical misattributions, the Bianchi identities were not discovered by Bianchi. It seems that the first to find them was Ricci, who communicated his result to E. Padova, who published it in 1889; however, the *contracted* Bianchi identities, which are sufficient for the proof that the Einstein tensor always vanishes, had been published by Aurel Voss *nine* years earlier.

[24]Of course objections were already raised at the time, among others by Lorentz, Levi-Civita and Schrödinger. A recent criticism of general relativity from the viewpoint of its failure of admitting genuine laws of conservation is contained in [69]; see also the discussion in [9].

with added "Einleitung" and "Inhalt"). It consists of 22 sections grouped into five parts; the longest one is the second (more than one-third), which provides a (very essential) introduction to Riemannian geometry and tensor calculus. This part was certainly not easy going for whoever was not already acquainted with the basic mathematical concepts. It was precisely the novelty of the mathematics – not in itself, but as to its use in a physical theory – which explains the awe which many of Einstein's colleagues felt at their first encounter with the equations of the theory. Though many a mathematician could boast to know better and handle differential geometry more skilfully than Einstein,[25] the same could not be said of the average physicist. It is this circumstance that earned Einstein, widely though incorrectly, the fame of being a mathematician more than a physicist.

On the other hand, apart from Grossmann's crucial collaboration, Einstein had been supported and helped by the mathematicians of Göttingen, including Felix Klein, Hilbert and Hermann Weyl. He stayed there between June and July 1915, and also gave a series of six lectures on the still developing general relativity; these lectures were very favourably received. It is interesting that the eminent algebraist Emmy Noether wrote in 1915 that she was in a team *performing difficult computations for Einstein*, and that "none of us understands what they are good for" [81, p. 276]. He corresponded on technical difficulties also with other mathematicians and physicists, including Lorentz [65], Tullio Levi-Civita [12] and Paul Hertz [64]. So much for the enduring myth of the genius who works his way in solitude.

14. Was the Copernican Controversy a Pseudo-Problem?

One widespread interpretation of general relativity, and the one that may be most responsible for its fame among non-specialists, was that it gave the final verdict on the Copernican controversy. The verdict – it is reported – is that the controversy was rooted in a conceptual confusion: there is no fact of the matter as to whether the Earth or the Sun is at rest, since all depends on one's (arbitrary) choice of coordinate system. This earned Einstein the honour of being compared to Copernicus, and many popular books appeared which traced the development of physics "from Copernicus to Einstein" (cf. [86]).

This interpretation was to some extent endorsed by Einstein himself, in the book he coauthored in 1938 with Leopold Infeld [43, p. 212]:

> Can we formulate physical laws so that they are valid for all CS [= coordinate systems], not only those moving uniformly, but also those moving quite arbitrarily, relative to each other? The struggle, so violent in the early days of science, between the views of Ptolemy and Copernicus would then be quite meaningless. Either CS would be used with equal justification. The two sentences, 'the sun is at rest and the earth moves', or 'the sun moves and the earth is at rest', would simply mean two different conventions concerning two different CS. Could we build a real relativistic physics valid in all CS; a physics in which there would be no place for absolute, but only for relative motion? This is indeed possible!

Similarly Max Born wrote: "Thus from Einstein's point of view Ptolemy and Copernicus are equally right. What point of view is chosen is a matter of expediency" [8, p. 345].

---

[25] Cf. Felix Klein's comments quoted at pp. 33, 36 of [64].

Though one is entitled to be a little surprised by the rash approach to the history of science that these passages reveal, one should not in general under-rate the importance of such oversimplifications in getting public and professional recognition for a certain scientific theory. Historical deformation is not the smallest factor in the attraction a new scientific theory exerts on lay persons and professionals alike.[26] However, in this case the trouble is not just with the way general relativity was advertised: it had to do with Einstein's own perception of the contribution he had given by adopting "general covariance". On several occasions he stated that in his theory all coordinate systems are on an equal footing, and that *therefore* he had succeeded in generalizing the special principle of relativity to all accelerated systems.

In order to clarify this point, it is useful to remember that Newtonian mechanics has a very effective means for dealing with non-inertial systems. For instance, if we have an inertial system $(\vec{r}, t)$ and a uniformly rotating system with same origin $(\vec{r}\,', t)$, the coordinate change is given by $\vec{r}\,' = A\vec{r}$, where A is a function of time with values in the group of the special orthogonal matrices. It can be shown, by computation, that if the second principle of dynamics $\vec{F} = m\vec{a}$ is satisfied for a given particle of mass m, then $\vec{F}' = m\vec{a}'$ also holds, with

$$\vec{F}' = A\vec{F} + \vec{F}_1 + \vec{F}_2, \tag{33}$$

where $\vec{F}_1$ is the Coriolis force and $\vec{F}_2$ is the centrifugal force, both depending on the angular velocity of the rotating system with respect to the inertial one. So one might say that the second principle of dynamics is satisfied also in rotating systems, at least if one is willing to consider (33) as a reasonable transformation law for a force. Why did this undisputed fact not stop the search for a theory without absolute space and time? The answer, of course, is that the *form* of the forces in the two systems is *not* preserved: for instance, if $\vec{F}$ depends just on the distance of the particle from the origin, then it is clear that the same property does not, and cannot, hold for $\vec{F}'$. In other words, the *notational similarity* of the original and the primed force laws cannot be taken as indicating *law invariance* in the spirit of the principle of (Galilean or special) relativity.

So there is no doubt that the mere acceptance of 'all coordinate systems' in the formulation of the physical laws cannot be read as a generalization of the principle of relativity – otherwise classical mechanics should also be considered 'generally relativistic', at least in the sense of giving equal mathematical treatment to inertial and non-inertial systems.

In 1917 mathematician Erich Kretschmann [66] made a similar point, arguing that "general covariance" in Einstein's sense was not a substantive physical principle, since it could be adopted in the formulation of virtually *any* physical theory. Einstein replied by saying that he concurred, adding only that the principle "carries considerable heuristic weight", and explaining what he meant as follows [33]:

> Among two theoretical systems, both compatible with experience, one will have to prefer the one that is simpler and more transparent from the point of view of absolute differential calculus. One just should bring the mechanics of Newtonian gravitation in the form of absolute-covariant equations (four-dimensional) and one will certainly become convinced that princi-

---

[26]It is interesting that one of the first times Einstein was likened to Copernicus was by no less than Max Planck, in his (successful) recommendation of Einstein in 1910 for a professorship in Prague [81, p. 192].

ple (a) [i.e. general covariance] excludes this theory, not on theoretical grounds, but on practical ones!

It is clear that Einstein meant this as a rhetorical challenge. In fact just a few years later the first generally covariant four-dimensional formulations of Newtonian gravitation were published [11,53]. If 'simplicity' and 'transparency' are to be judged, it is at least doubtful if the covariant 4-dimensional version of Newtonian physics should be considered as more obscure or difficult to handle than general relativity.[27]

Coming back to the main physical issue which was at stake in the Copernican controversy, does general relativity make any difference to the distinction between locally inertial and non-inertial (e.g. rotating) systems? Not much. As two Italian physicists pointed out in 1929 [58], in general relativity a locally rotating system, for instance, is as easily distinguishable from a locally inertial system as it is in classical mechanics. In other words, inertial effects exist also in general relativity![28] If this was not the case, general relativity would simply contradict ordinary empirical evidence.[29]

However, it can be argued that as to absolute motion, general relativity does introduce a novelty in the spirit of Mach's criticism: the locally inertial systems are *determined* by the metric (they are the freely falling systems), and the metric, in turn, is *linked* to the energy–matter distribution through the field equations, so the inertial effects turn out to be *linked* to that distribution, rather than to absolute motion. However, general relativity is not a completely 'Machian' theory, for the simple but inescapable mathematical reason that in the preceding statement 'linked' cannot be changed to 'determined'. This is proven, for instance, by the existence of nontrivial (i.e. different from Minkowski space–time) solutions of the vacuum field Eqs (26) (see next section). The debate on Mach's arguments against the basic Newtonian notions is still going on, and papers and books adopting a 'Machian' standpoint, sometimes seriously departing from general relativity and going back to classical notions, appear from time to time [7,6,56].

Our conclusion is that, though repeatedly and authoritatively endorsed, the claim that general relativity finally 'solved' the Copernican controversy by showing that Copernicus and Ptolemy were 'both right', is based on several mistakes, concerning both the gist of their disagreement and the scope of 'general covariance'. However such a claim was probably decisive in winning for Einstein's theory a cultural and philosophical place second to no other physical theory of his time.[30]

15. The First Exact Solution

The metric (25) used by Einstein to give his explanation of the anomalous advance of Mercury's perihelion was not an exact solution of his equations. The problem of finding the exterior metric corresponding to the field produced by a spherically symmetric body

---

[27]For a more recent outline, see [21].

[28]The mathematical tool by which this can be most easily shown was introduced by the 21-year-old Enrico Fermi in 1922 [52].

[29]Incidentally, it was this basic reason, rather than "a great and unfortunate talent for creating difficulties for himself" [81, p. 232] that made Max Abraham [4,5], as well as many other physicists sceptical about the very programme of generalizing the principle of relativity.

[30]This privilege was later stolen by quantum mechanics, for similarly unwarranted claims concerning its radical philosophical implications. For more on the topic discussed in this section [57] is worth reading.

was solved within weeks from the publication of [29] by Schwarzschild (cf. §9), who was then serving as a volunteer on the Russian front, and knew Einstein's third paper, not the fourth. The metric he found ([94], communicated to the Prussian Academy by Einstein himself on January 13), was a solution of the *vacuum* field Eqs (26), and this was a happy circumstance, since the full field equations of 18 November (24) do not coincide with the full final Eqs (28).

This is the metric:

$$ds^2 = -\frac{dr^2}{1 - \alpha/R} - R^2\left(d\theta^2 + \sin^2\theta d\phi^2\right) + \left(1 - \frac{\alpha}{R}\right)c^2\, dt^2, \tag{34}$$

where (r, θ, ϕ) are standard spherical coordinates, R is an auxiliary variable, defined as $R = (r^3 + \alpha^3)^{1/3}$, and α is a positive expression containing the mass of the body ($\alpha = 2Gm/c^2$).

Slightly more than a month later, and a few months before dying aged 43 of an illness, Schwarzschild submitted a second paper ([95], also communicated by Einstein, on 24 February), where he described the exact solution for the interior of a homogeneous sphere of incompressible fluid with (constant) density ρ_0. The metric was:

$$ds^2 = -\frac{3}{\kappa\rho_0}\left(d\chi^2 + \sin^2\chi d\theta^2 + \sin^2\chi \sin^2\theta d\phi^2\right)$$

$$+ \left(\frac{3\cos\chi_a - \cos\chi}{2}\right)^{-2}c^2\, dt^2, \tag{35}$$

where the subscript a indicates that the quantity is computed at the surface of the sphere. The conclusion of the second article is:

> For an observer measuring from outside [...] a sphere of a given gravitational mass [...] cannot have a radius measured from outside smaller than: $P_o = \alpha$. For a sphere of incompressible fluid the limit will be $9\alpha/8$. (For the Sun α is equal to 3 km, for a mass of 1 gram is equal to $1.5 \cdot 10^{-28}$ cm.)

In fact at the centre of a sphere having a radius equal to the limit value, the pressure of the fluid would become infinite.

Schwarzschild's contributions to general relativity rank as some of the most important ever. Among other things, his exterior solution (34) made it possible to derive the three empirical consequences in a more satisfactory way than Einstein had been able to do. Notice, however, that these predictions are not enough in themselves to give general relativity a unique status. Those who think that it is very unlikely that two different theories may reproduce the same empirical predictions (at a certain historical time) should consider that (34) is also a solution of all field equations having at the left side an expression formed from the derivatives, of *any* degree, of the Ricci tensor $R_{\mu\nu}$; this elementary remark implies that the three famous predictions are also made by all theories postulating such field equations.[31]

Notwithstanding their rôle in fixing some of the loose ends of Einstein's arguments, Schwarzschild's results were puzzling in other ways. First of all, here was another solution (after Minkowski space–time) for the vacuum field equations, something which

---

[31]This example is of more than merely methodological interest; see [73].

seemed inconsistent with Mach's principle. Second, (34) fails for $R = \alpha$, which corresponds in Schwarzschild's coordinates to the centre of the sphere ($r = 0$). While this could be accepted as the analogue of the singular point of the Newtonian potential for a point mass, still it was not very clear what meaning could be attached to a singularity *of* space–time (rather than *in* space–time).

These perplexities could both be simply dissolved, at that time. The metric (34) does not describe the whole universe: it can be considered to be valid only for an isolated system; and Mach's principle is a topic for cosmology, rather than for the astronomy of the solar system (more will be said about it in Chap. 6). As for the second objection, Schwarzschild did not have to worry about the $R = \alpha$ singularity, because (34) was only valid for the *exterior* of a spherical mass, and he had proved that such a mass had *always* a radius bigger than α (inside the mass it is (35) which must be applied). In 1923 French physicist Marcel Brillouin gave an explicit formulation of this argument, in very clear terms [10].

Only a few decades later, after discussions in which very different opinions emerged [46,48], the exterior Schwarzschild solution (34) was taken to represent the field of an utterly collapsed star; and $R = 0$ (which, obviously, simply does not make sense in Schwarzschild's original derivation) was interpreted as a true singularity (in opposition to $R = \alpha$, which was re-interpreted as a coordinate singularity).[32] Neither Einstein nor Eddington (arguably "the most distinguished astrophysicist of his time" – [13, p. 93]) ever accepted that such stars could exist, or that general relativity implies that they should exist. Indeed Einstein suggested in 1935 how to extend (34) to obtain a solution where the region $R < \alpha$ is deleted,[33] and in 1939 he advanced a (suggestive, rather than demonstrative) argument implying that "the 'Schwarzschild singularity' does not appear for the reason that matter cannot be concentrated arbitrarily. And this is due to the fact that otherwise the constituting particles would reach the velocity of light" [39].

An important step in the achievement of a new consensus was an article published in 1960 and describing a simple construction of a maximal extension for Schwarzschild's original space–time [67]. Constructing an extension of a singular space–time is useful to understand the nature of its singularities. However, it is sobering to consider that "at least *eleven* completely inequivalent extensions" of Schwarzschild's solution have been found.[34]

The name 'black holes' was introduced in 1969. Notice that the idea that there could be bodies from which no radiation could be emitted (because their escape velocity is bigger than c) had been advanced much earlier by Newtonian physicists, such as John Michell (1783) and Laplace, who called them "dark bodies" (cf. Appendix A in [60]); the first paper to discuss this possibility in connection with Schwarzschild's solution appeared in 1929 [80]. However, the relativistic black hole is a much weirder entity than the classical one. For instance, after crossing the hypersurface $R = \alpha$ what was previously a radial coordinate becomes a *time* coordinate! This is not strange in itself as far as mathematics goes, but one may legitimately question whether it is physically reasonable, and the same may be asked for other curious features of the modern interpretation of

---

[32]For more on singularities, see Chap. 6, §18.

[33]It is the idea of the "Einstein–Rosen bridge" ([45]; see also the 1936 essay "Physics and Reality", in [41], particularly pp. 353–4).

[34][96, p. 839]; in this essay one can find a description of these extensions.

the Schwarzschild solution: for example whether they are consistent with taking it as an answer to the physical problem of determining the field of a *static* spherical mass.

In the last thirty years black holes have also been studied from a quantum-mechanical point of view, a new field of inquiry in which British theoretical physicist Stephen Hawking's contributions have been seminal. However, the whole field is still highly controversial, including its basic prediction: "black hole radiation" [62].

By now black holes (described by several other metrics, too) have become almost a byword for Einstein and relativity, and a huge literature on them is today available. However, it is not clear that, should the not overwhelming observational evidence for their existence be explained away, one should have less respect for general relativity, let alone reject it. Surely general relativity does not, in itself, dictate the customary physical interpretation of the Schwarzschild exterior solution.[35]

Many other exact solutions of the Einstein's equation have been found in the last ninety years. Their study has grown into a small academic industry, whose contributions, it must be admitted, are more often of a mathematical than of a physical nature.[36] This is one of the ways general relativity has enriched the field of differential geometry. However, a crucial difference between general relativity and Lorentzian geometry is that a Lorentzian metric, to make sense physically, must be accompanied by a realistic matter model described by a stress-energy tensor such that *together* they satisfy the field equations. Just finding a Lorentzian metric and then formally *defining* a stress energy tensor as suitably proportional to the Einstein tensor of that metric does not normally work.

16. Einstein's Dissatisfaction with His Theory

One point that deserves to be emphasized is that Einstein never considered general relativity as the last word on the subject of gravitation. In fact he started proposing changes in his field equations very soon. In 1917, as we shall see in the next chapter, he introduced a new term in them in order to allow for what then seemed to him a physically reasonable cosmological solution:

$$R_{\mu\nu} - \lambda g_{\mu\nu} = -\kappa \left(T_{\mu\nu} - \frac{1}{2} T g_{\mu\nu} \right).$$

Just two years later he suggested another, 'trace-less' version:

$$R_{\mu\nu} - \frac{1}{4} S g_{\mu\nu} = -\kappa T_{\mu\nu} \tag{36}$$

in order to deal with the possibility that the gravitational fields play "an essential part in the structure of the particles of nature" [35]. In his popular book on relativity he wrote in a footnote (never deleted in subsequent editions):

> The general theory of relativity renders it likely that the electrical masses of an electron are held together by gravitational forces [40, p. 52].

---

[35] For defenses of Schwarzschild's original interpretation of his exterior solution and criticism of black hole theory see [2,3]; cf. [96, p. 757]; for a participant history of black hole research, interspersed with portraits of the main characters see [103].

[36] For a sample, analysed in depth, see Chapter 5 of [60] (pp. 117–79), and for a virtually complete survey see the encyclopaedic work [98].

His reasons for trying to find a unified theory of gravitation and electricity, following the first proposals of Weyl and Eddington, were rooted in his perception of the field equations as fundamentally defective insofar as they relegated to the stress-energy tensor all forces that could not be, at the moment, analysed in geometrical terms.

In 1936 he compared his field equations to "a building, one wing of which is made of fine marble (left part of the equation), but the other wing of which is built of low-grade wood (right side of equation)" [41, p. 342]. He elaborated this point in his *Autobiographical Notes*:

> The second member on the left side [i.e. $-Sg_{\mu\nu}/2$] is added because of formal reasons; for the left side is written in such a way that its divergence disappears identically in the sense of the absolute differential calculus. *The right side is a formal condensation of all things whose comprehension in the sense of field-theory is still problematic.* Not for a moment, of course, did I doubt that *this formulation was merely a makeshift* in order to give the general principle of relativity a preliminary expression. For it was essentially not anything *more* than a theory of the gravitational field, which was artificially isolated from a total field of as yet unknown structure.[37]

Einstein's search for a unified theory was pursued during the last thirty years of his life. The prevailing, though not universal, opinion today on these efforts is that they were not only unsuccessful, but also that they were basically flawed because of his unwillingness to use quantum mechanics as an essential ingredient of the unification.

17. Conclusive Remarks

Up to the early 1950s, general relativity was a little-frequented subject, amongst physicists – a theory that had to be praised, but that could be safely ignored. As one of Einstein collaborators, Peter Bergmann, told Abraham Pais: "You only had to know what your six best friends were doing and you would know what was happening in general relativity" [81, p. 268]. Its supporting evidence was sparse, questionable, and unstable: essentially it reduced to the changing experimental verdicts on the three notorious tests. (In the next chapter we shall deal with a field of physics on which the impact of general relativity has been very big; of course a full appraisal of general relativity also requires considering its cosmological applications.)

The three empirical predictions were about the only ones that general relativity could refer to for about forty years and, on the whole, the confirming evidence gathered in those decades was not of the highest quality [47]. Though the deflection of light, after Eddington's and Crommelin's celebrated eclipse expeditions in 1919, instantly made Einstein a world-famous star, it was a controversial piece of evidence from the beginning, and was not even consistently confirmed by the next few eclipse observations. True, the gravitational redshift earned Einstein the Royal Society's medal in 1926, because of the authoritative data on Sirius B presented by Walter Adams of the Mount Wilson observatory; it is just a pity that in the meantime these data have dissolved into thin air, and that there is more than a suspicion that the eminent astronomer just saw what he "expected to find, even if it didn't exist" [63, pp. 65–72]. And, as to Mercury perihelion, it had never been a very sound piece of evidence in the first place.

---

[37][92, p. 75]; all italics, except for the last, are mine.

Gravitational waves might have been, and in a sense are, another great subject for empirical research.

Unfortunately, on the theoretical side, notwithstanding Einstein's two early papers on them [32,34], the very question of whether or not general relativity really predicts that such waves exist is a tricky one (cf. [69]). Eddington was against them. Einstein himself for a time, in 1937, was convinced that he had a proof that gravitational waves do not exist. Pais in his 1982 biography of Einstein abstains from taking sides on how valid the fundamental quadrupole formula is [81, p. 281].

Incidentally, those unfamiliar with the way physics develops may think that, in contrast to whether or not a given prediction has been confirmed, the issue of whether or not a given theory does make that prediction is a logical question on which there can be no serious disagreement. The rarely acknowledged truth is that the links between an ordinary scientific theory and its predictions have varying degrees of strength, and it is completely normal for different experts to differ in their estimates of how strong or loose a certain link is.

On the empirical side, the detection of gravitational waves has been announced several times since the 1960s, but it is fair to sum up the situation by saying that all positive results so far have been contradicted. Today the field is thriving, in the sense that big international projects centred on gigantic instruments like LIGO (Laser Interferometer Gravitational-Wave Observatory) and Virgo are running, but still no hard evidence of gravitational waves has been found.[38] It has been written [108]:

> The discovery and study of the formation of a black hole by means of gravitational waves would provide a stunning test of relativistic gravity.

This test, however, has still to come to fruition.

More recent tests of general relativity have been performed, with outcomes generally considered satisfactory (cf. Chap. 7). But empirical confirmations should be neither exaggerated nor over-rated; in some fields of science (like astrophysics or cosmology, as opposed to, say, medicine) they are not in themselves more sound or important than theoretical achievement, as evaluated in terms of consistency, mathematical rigour and conceptual transparency. In his later years Einstein insisted on this point on several occasions. Of course even these qualities may be difficult to evaluate and controversial. However, it is safe to say that, concerning the aesthetic appeal of general relativity, there has been a wider consensus on its beauty than on that of any other physical theory [13].

References

[1]   Abiko S. 2000: "Einstein's Kyoto address: 'How I created the theory of relativity' ", *Historical Studies in the Physical Sciences*, **31**, Part 1, pp. 1–35.

[2]   Abrams L. S. 1979: "Alternative space–time for the point mass", *Phys. Rev. D*, **20**, pp. 2474–2479; "Erratum", *Phys. Rev. D*, **21**, p. 2438.

[3]   Abrams L. S. 1989: "Black holes: the legacy of Hilbert's error", *Can. J. Phys.*, **67**, pp. 919–26.

[4]   Abraham M. 1914: "Die neue Mechanik", *Scientia*, **15**, pp. 8–27 (French translation: pp. 10–29).

[38]The fact that a sector of the scientific community can prosper for decades with no real success on record – a phenomenon which, needless to say, is far from being restricted to physics! – is in itself an interesting historical and sociological topic. An instructive, recent account of the sociology of the gravitational waves community is contained in [16].

[5] Abraham M. 1914: "Sur le problème de la relativité", *Scientia*, **15**, pp. 101–3.
[6] Assis A. K. 1999: *Relational Mechanics*, Montreal, Apeiron.
[7] Barbour J. B., Pfister H. 1995: *Mach's Principle: From Newton's Bucket to Quantum Gravity*, Boston – Basel – Berlin, Birkhäuser.
[8] Born M. 1962: *Einstein's Theory of Relativity*, New York, Dover.
[9] Bozhkov Y., Rodrigues Jr. W. A. 1995: "Mass and Energy in General Relativity", *General Relativity and Gravitation*, **27**, pp. 813–9.
[10] Brillouin M. 1923: "Les points singuliers de l'univers d'Einstein", *J. de Physique*, 6th ser., **23**, pp. 43–6 (an English translation by S. Antoci is available on "arXiv:physics/0002009").
[11] Cartan E. 1923–1924: "Sur les variétés à Connexion Affine et la Théorie de la Relativité Généralisée", *Annales Scientifiques de l'Ecole Normale Supérieure*, **40** (1923), pp. 325–412; **41** (1924), pp. 1–25.
[12] Cattani C., De Maria M. 1989: "The 1915 Epistolary Controversy between Einstein and Tullio Levi-Civita", pp. 175–200 of *Einstein and the History of general relativity* (Howard D., Stachel J., eds.), Boston, Birkhäuser.
[13] Chandrasekhar S. 1987: *Truth and Beauty. Aesthetics and Motivations in Science*, Chicago – London, University of Chicago Press.
[14] Clarke C. 1979: *Elementary General Relativity*, London, Edward Arnold.
[15] Clifford W. K. 1955: *The Common Sense of the Exact Sciences*, New York, Dover.
[16] Collins H. M. 2003: "Lead into gold: the science of finding nothing", *Stud. Hist. Phil. Sci.*, **34**, pp. 661–91.
[17] Corry L., Renn J., Stachel J. 1997: "Belated Decision in the Hilbert–Einstein Priority Dispute", *Science*, **278**, pp. 1270–3.
[18] D'Inverno R.: *Introducing Einstein's Relativity*, Cambridge University Press.
[19] Eddington A. S. 1918: *Report on the Relativity Theory of Gravitation*, London, Fleetway Press.
[20] Eddington A. S. 1920: *Space, Time and Gravitation*, Cambridge University Press.
[21] Ehlers J. 1991: "The Newtonian Limit of General Relativity", pp. 95–106 of *Classical Mechanics and Relativity: Relationship and Consistency* (ed. by Ferrarese G.), Napoli, Bibliopolis.
[22] Einstein A. 1907: "Relativitätsprinzip und die aus demselben gezogenen Folgerungen", *Jahrbuch der Radioaktivität*, **4**, pp. 411–62; **5**, pp. 98–9.
[23] Einstein A. 1911: "Über den Einfluss der Schwerkraft auf die Ausbreitung des Lichtes, *Ann. Phys.*, **35**, pp. 898–908.
[24] Einstein A. 1913: "Zum gegenwärtigen Stande des Gravitationsproblems", *Physikalische Zeitschrift*, **14**, pp. 1249–62.
[25] Einstein A. 1914: "Zum Relativitäts-problem", *Scientia*, **15**, pp. 337–48 (French translation: 139–50).
[26] Einstein A. 1914: "Die formale Grundlage der allgemeinen Relativitätstheorie", *Königlich Preussische Akademie der Wissenschaften* (Berlin), *Sitzungsberichte*, pp. 1030–85.
[27] Einstein A. 1915: "Zur allgemeinen Relativitätstheorie", *Königlich Preussische Akademie der Wissenschaften* (Berlin), *Sitzungsberichte*, pt 2, pp. 778–86.
[28] Einstein A. 1915: "Zur allgemeinen Relativitätstheorie (Nachtrag)", *Königlich Preussische Akademie der Wissenschaften* (Berlin), *Sitzungsberichte*, pt 2, pp. 799–801.
[29] Einstein A. 1915: "Erklärung der Perihelbewegung des Merkur aus der allgemeinen Relativitä tstheorie", *Königlich Preussische Akademie der Wissenschaften* (Berlin), *Sitzungsberichte*, pt 2, pp. 831–9.
[30] Einstein A. 1915: "Die Feldgleichungen der Gravitation", *Königlich Preussische Akademie der Wissenschaften* (Berlin), *Sitzungsberichte*, pt 2, pp. 844–7.
[31] Einstein A. 1916: "Grundlage der allgemeine Relativitätstheorie", *Ann. Phys.*, **49**, pp. 769–822.
[32] Einstein A. 1916: "Näeherungsweise Integration der Feldgleichungen der Gravitation", *Königlich Preussische Akademie der Wissenschaften* (Berlin), *Sitzungsberichte*, pt 1, pp. 688–96
[33] Einstein A. 1918: "Prinzipielles zur allgemeinen Relativitätstheorie", *Annalen der Physik*, **55**, pp. 241–4.
[34] Einstein A. 1918: "Über Gravitationswellen", *Königlich Preussische Akademie der Wissenschaften* (Berlin), *Sitzungsberichte*, pp. 154–67.
[35] Einstein A. 1919: "Spielen Gravitationsfelder im Aufber der materiellen Elementarteilchen eine wesentliche Rolle?", *Königlich Preussische Akademie der Wissenschaften* (Berlin), *Sitzungsberichte*, pt 1, pp. 349–56 (transl. in [44], pp. 189–98).
[36] Einstein A. 1920: "Meine Anwort. Ueber die anti-relativitätstheorie G.m.b.H.", *Berliner Tageblatt*, August 27.

[37] Einstein A. 1921: "Geometrie und Erfahrung", trans. in Einstein 1994, pp. 254–68.

[38] Einstein A. 1922: *The Meaning of Relativity*, Princeton, Princeton University Press.

[39] Einstein A. 1939: "Stationary system with spherical symmetry consisting of many gravitating masses", *Ann. Math.*, **40**, pp. 922–36.

[40] Einstein A. 1954: *Relativity. The Special and the General Theory*, 15th ed., Routledge, London and New York.

[41] Einstein A. 1994: *Ideas and Opinions* [1954], New York, The Modern Library.

[42] Einstein A., Grossmann M. 1913: *Entwurf einer verallgemeinerten Relativitätstheorie und einer Theorie der Gravitation*, Leipzig, Teubner.

[43] Einstein A., Infeld L. 1938: *The Evolution of Physics*, New York, Simon and Schuster.

[44] Einstein A., Lorentz A. A., Weyl H., Minkowski H. 1932: *The Principle of Relativity*, Dover.

[45] Einstein A., Rosen N. 1935: "The Particle Problem in the General Theory of Relativity", *Phys. Rev.*, **48**, pp. 73–7.

[46] Eisenstaedt J. 1982: "Histoire et Singularités de la Solution de Schwarzschild", *Archive for History of Exact Sciences*, **27**, pp. 157–98.

[47] Eisenstaedt J. 1986: "La relativité générale à l'étiage: 1922–1955", *Archive for History of Exact Sciences*, **32**, pp. 115–85.

[48] Eisenstaedt J. 1987: "Trajectoires et impasses de la solution de Schwartzschild, 1925–1955", *Archive for History of Exact Sciences*, **37**, pp. 275–357.

[49] Engel A., Shucking E. (eds.) 1997: *The Collected Papers of Albert Einstein, vol. 7, The Berlin Years: Writings, 1914–1917, English Translation*, Princeton–Oxford, Princeton University Press.

[50] Engel A., Shucking E. (eds.) 2002: *The Collected Papers of Albert Einstein, vol. 7, The Berlin Years: Writings, 1918–1921, English Translation*, Princeton–Oxford, Princeton University Press.

[51] Euler L. 1732: "De linea brevissima in superficie quacumque duo quaelibet puncta jungente", *Comment. Acad. Petropol.*, **3** (ad annum 1728), pp. 110–24.

[52] Fermi E. 1922: "Sopra i fenomeni che avvengono in vicinanza di una linea oraria", *Rend. Lincei*, **31**, pp. 21–3, 51–2, 101–3.

[53] Friedrichs K. 1927: "Eine invariante Formulierung des Newtonschen Gravitationsgesetzes und des Grenzübergunganges vom Einsteinschen zum Newtonschen Gesetz", *Mathematische Annalen*, **98**, pp. 566–75.

[54] Gauss C. F. 1828: *Disquisitiones generales circa superficies curvas* (in *Werke 4*, pp. 217–58).

[55] Gerber P. 1898: "Über die räumliche und zeitliche Ausbreitung der Gravitation", *Zeitschrift für Mathematik und Physik*, **43**, pp. 93–104.

[56] Graneau P., Graneau N. 2003: "Machian Inertia and the Isotropic Universe", *General Relativity and Gravitation*, **35**, pp. 751–70.

[57] Ginzburg V. L. 1983: "The Heliocentric System and the General Theory of Relativity (from Copernicus to Einstein)", pp. 254–308 in Gribanov *et al.* 1983.

[58] Giorgi G., Cabras A. 1929: "Questioni relativistiche sulle prove della rotazione terrestre", *Rend. Lincei*, **9**, pp. 513–7.

[59] Gribanov *et al.* 1983: *Einstein and the Philosophical Problems of 20th-Century Physics* [1979], Moscow, Progress Publishers.

[60] Hawking S. W., Ellis G. F. R. 1973: *The Large-Scale Structure of Space–Time*, Cambridge University Press.

[61] Held A. (ed.) 1980: *General Relativity and Gravitation. One Hundred Years After the Birth of Albert Einstein*, New York and London, Plenum Press.

[62] Helfer A. D. 2003: "Do black holes radiate?", *Rep. Prog. Phys.*, **66**, pp. 943–1008.

[63] Hetherington N. S. 1988: *Science and Objectivity*, Ames, Iowa State University Press.

[64] Howard D., Norton J. D. 1993: "Out of the Labyrinth? Einstein, Hertz, and the Göttingen Answer to the Hole Argument", pp. 30–62 of *The Attraction of Gravitation: New Studies in the History of General Relativity* (Earman J., Janssen M., Norton J. D., eds.), Boston–Basel–Berlin, Birkhäuser.

[65] Kox A. J. 1988: "Hendrik Antoon Lorentz, the Ether, and the General Theory of Relativity", *Arch. Hist. Exact Sci.*, **38**, pp. 67–78.

[66] Kretschmann E. 1914: "Über den physikalischen Sinn der Relativitätspostulate, A. Einsteins neue und seine ursprünglicher Relativitätstheorie", *Annalen der Physik*, **53**, pp. 575–614.

[67] Kruskal M. D. 1960: "Maximal Extension of Schwarzschild Metric", *Phys. Rev.*, **119**, pp. 1753–5.

[68] Lang S. 2004: *Linear Algebra*, New York – Heidelberg – Berlin, Springer-Verlag.

[69] Logunov A., Mestvirishvili M. 1989: *The Relativistic Theory of Gravitation*, Moskow, Mir Publishers.

[70] Lusanna L. 2004: "Book Review", *Found. Phys.*, **34**, pp. 1281–2.

[71] Macdonald A. 2001: "Einstein's hole argument", *Am. J. Phys.*, **69** (2), pp. 223–5.

[72] Mach E. 1960: *The Science of Mechanics. A Critical and Historical Account of its Development*, La Salle, Open Court.

[73] Mannheim P. D. 2000: "Attractive and Repulsive Gravity", *Found. Phys.*, **30**, pp. 709–46.

[74] Maxwell J. C. 1920: *Matter and Motion* [1976], ed. with notes and appendices by J. Larmor, New York, Dover 1991.

[75] Minkowski H. 1909: "Raum und Zeit", *Jahresbericht der deutschen Mathematiker-Vereinigung*, **18**, pp. 75–88, *Physikalische Zeitschrift*, **10**, pp. 104–11.

[76] Misner C. W., Thorne K. S., Wheeler J. A. 1973: *Gravitation*, New York, Freeman.

[77] North J. D. 1990: *The Measure of the Universe* [1965], New York, Dover.

[78] Norton J. 2004: "The Hole Argument", *Stanford Encyclopedia of Philosophy*, http://plato.stanford.edu/entries/space-time-holearg/.

[79] O'Neill B. 1983: *Semi-Riemannian Geometry With Applications to Relativity*, New York, Academic Press.

[80] Oppenheimer J. R., Snyder H. 1939: "On Continued Gravitational Contraction", *Phys. Rev.*, **56**, pp. 455–459.

[81] Pais A. 1982: *"Subtle Is the Lord . . .". The Science and Life of Albert Einstein*, New York, Oxford University Press.

[82] Pauli W. 1958: *Theory of Relativity*, New York, Dover.

[83] Poincaré H. 1902: *La Science et l'Hypothèse*, Paris, Flammarion.

[84] Poincaré H. 1906: "Sur la dynamique de lélectron, *Rendiconti del Circolo Matematico di Palermo*, **21**, pp. 129–75.

[85] Ricci G., Levi-Civita T. 1901: "Méthodes de calcul différentiel absolu et leurs applications, *Mathematische Annalen*, **54**, pp. 125–201.

[86] Reichenbach H. 1942: *From Copernicus to Einstein* [1928], trans. from the German, New York, Dover 1980.

[87] Robertson H. P. 1949: "Geometry as a branch of physics", in [92], pp. 315–32.

[88] Roseveare N. T. 1982: *Mercury's Perihelion from Le Verrier to Einstein*, Oxford, Clarendon Press.

[89] Rowe D. E. 2002: "Einstein's Gravitational Field Equations and the Bianchi Identities", *The Mathematical Intelligencer*, **24** (4), pp. 57–66.

[90] Sachs R. K., Wu H. 1977: *General Relativity for Mathematicians*, New York – Heidelberg – Berlin, Springer-Verlag.

[91] Salvioli S. E. 1972: "On the Theory of Geometric Objects", *J. Differential Geometry*, **7**, pp. 257–78.

[92] Schilpp P. A. (ed.) 1949: *Albert Einstein: Philosopher–Scientist*, Evanston (Ill.), Library of Living Philosophers.

[93] Schwarzschild K. 1900: "Über das zulässige Krümmungsmaass des Raumes", *Vierteljahrsschrift der astronomischen Gesellschaft*, **35**, pp. 337–47 (English transl. in *Class. Quant. Grav.*, **15** (1998), pp. 2539–44).

[94] Schwarzschild K. 1916: "Über das Gravitationsfeld einer Massenpunktes nach der Einsteinschen Theorie", *Königlich Preussische Akademie der Wissenschaften* (Berlin), *Sitzungsberichte*, pp. 189–96 (an English translation by S. Antoci and A. Loinger is available on "arXiv:physics/9905030").

[95] Schwarzschild K. 1916: "Über das Gravitationsfeld einer Kugel aus inkompressibler Flüssigkeit nach der Einsteinschen Theorie", *Königlich Preussische Akademie der Wissenschaften* (Berlin), *Sitzungsberichte*, pp. 424–34 (an English translation by S. Antoci and A. Loinger is available on "arXiv:physics/9912033").

[96] Senovilla J. M. M. 1997: "Singularity Theorems and Their Consequences", *General Relativity and Gravitation*, **29**, pp. 701–848.

[97] Stachel J. 1980: "Einstein and the Rigidly Rotating Disk", pp. 1–15 of Held 1980.

[98] Stephani H., Kramer K., MacCallum M., Hoenselaers C., Herlt E. 2003: *Exact Solutions of Einstein's Field Equations*, 2nd ed., Cambridge University Press.

[99] Struik D. 1987: *A Concise History of Mathematics*, 4th ed., New York, Dover.

[100] Synge J. L. 1966: "What is Einstein's Theory of Gravitation?", pp. 7–15 of B. Hoffmann (ed.), *Perspectives in Geometry and Relativity*, Bloomington – London, Indiana University Press.

[101] Synge J. L. 1971: *Relativity: The General Theory*, Fourth Printing, Amsterdam, North-Holland (First Edition: 1960).

[102] Tavakol R., Zalaletdinov 1998: "On the Domain of Applicability of General Relativity", *Found. Phys.*, **28**, pp. 307–31.

[103] Thorne K. S. 1994: *Black Holes & Time Warps*, New York – London, Norton.

[104] Weyl H. 1922: *Space–Time–Matter*, transl. of the 4th (1921) ed., with a new preface (1950), New York, Dover.

[105] Weyl H. 1949: "Relativity theory as a stimulus in mathematical research", *Proc. Am. Math. Soc.*, **93**, pp. 535–41.

[106] Whitrow G. J., Morduch G. E. 1965: "Relativistic theories of gravitation. A comparative analysis with particular refernce to astronomical tests", *Vistas in Astronomy*, **6**, pp. 1–67.

[107] Whittaker E. 1951–53: *A History of the Theories of Aether and Electricity*, New York, Dover 1989.

[108] Will C. M. 1999: "Gravitational Radiation and the Validity of General Relativity", *Physics Today*, Oct., pp. 38–43.

Physics Before and After Einstein
M. Mamone Capria (Ed.)
IOS Press, 2005
© 2005 The authors

Chapter 6

The Rebirth of Cosmology:
From the Static to the Expanding Universe

Marco Mamone Capria

Among the reasons for the entrance of Einstein's relativity into the scientific folklore of his and our age one of the most important has been his fresh and bold approach to the cosmological problem, and the mysterious, if not paradoxical concept of the universe as a three-dimensional sphere. Einstein is often credited with having led cosmology from philosophy to science: according to this view, he made it possible to discuss in the progressive way typical of science what had been up to his time not less "a field of endless struggles" than metaphysics in Kant's phrase.

It is interesting to remember in this connection that the German philosopher, in his *Critique of Pure Reason*, had famously argued that cosmology was beyond the scope of science (in the widest sense), being fraught with unsolvable contradictions, inherent in the very way our reason functions. Of course not everybody had been impressed by this argument, and several nineteenth-century scientists had tried to work out a viable image of the universe and its ultimate destiny, using Newtonian mechanics and the principles of thermodynamics. However, this approach did not give unambiguous answers either; for instance, the first principle of thermodynamics was invoked to deny that the universe could have been born at a certain moment in the past, and the second principle to deny that its past could be infinite.

Einstein's contribution to cosmology surely strengthened confidence in the power of human reason to investigate every conceivable subject-matter, including the most ambitious ones. It is from Einstein's cosmological writings that physicists drew the kind of boldness that decades later enabled some of them to author books with such titles as *A Brief History of Time* and *The Mind of God*. A different question is whether the road opened by Einstein led in fact to that consensus that most people consider as a trademark of the scientific endeavour. As we shall see (and as will be even more clear from Chap. 13) this is at least doubtful.

The starting point of relativistic cosmology was Einstein's 1917 paper, "Cosmological considerations on the general theory of relativity" [17] where, just one year after the apparently definitive formulation of general relativity, a novelty was introduced at the very core of the theory. Some years later, Einstein somewhat repented of this step, but by that time it had taken on a life of its own. In any case, as we shall see, the development of relativistic cosmology was an essential factor in the progress in the understanding of the power and limits of general relativity itself.

1. Einstein and Newton's Universe

In tackling in 1917 the problem of the structure of the universe, "this fundamentally important question", Einstein started by pointing out that in Newtonian physics the equivalence between the law of universal attraction and Poisson's equation is not strict, but requires suitable conditions at infinity, which in general relativity translate into the assumption that "it is possible to select a system of reference so that at spatial infinity all the gravitational potentials $g_{\mu\nu}$ become constant" [17, p. 177].

Einstein tried to show that Newtonian physics is unable to deal properly with the cosmological problem for an infinite universe. If matter is uniformly distributed in such a universe, then the gravitational potential should be *divergent* at infinity, rather than tend to a finite limit (this is a version of the so-called *gravitational paradox* of Newtonian physics). The only wayout – Einstein went on – is to assume that the mass density is not constant, but goes to zero faster than $1/r^2$ as the distance r from a given 'centre' goes to infinity. So Newton's *material* universe turns out to be essentially finite, and with a centre, though it is contained in an infinite (Euclidean) space.

Interestingly, Newton had already discussed this topic, if from a different angle, and drawn the *opposite* conclusion. In the first of his four letters to the Reverend Bentley (dated 10 December 1692), an important source for the study of his cosmological thought, Newton explained how the different celestial structures which we observe could well have arisen from an initially uniform, *infinite* distribution of matter, while from a finite distribution this development would have been very unlikely:

> As to your first Query, it seems to me that if the Matter of our Sun and Planets, and all the Matter of the Universe, were evenly scattered throughout all the Heavens, and every Particle had an innate Gravity towards all the rest, and all the whole Space, throughout which this Matter was scattered, was but finite; the Matter on the outside of this Space would by its Gravity tend towards all the Matter on the inside, and by consequence *fall down into the middle of the whole Space, and there compose one great spherical Mass.* But if the Matter was *evenly disposed throughout an infinite Space*, it could never convene into one Mass, but some of it would convene into one Mass and some into another, so as to make an infinite Number of great Masses, scattered from one to another throughout all that infinite Space [[5, pp. 281–2]; italics added].

In other words, with an infinite distribution Newton thought that it was possible to avoid a different difficulty facing cosmology, namely that of all matter collapsing into a single big mass – a destiny apparently inconsistent with astronomical observation. Thus Newton was advancing the idea that in an infinite material universe a dynamic equilibrium is possible; that is, it is possible for the gravitational field to vanish everywhere in the large, while locally what is known today as 'gravitational instability' can give rise to stars and stellar systems.

Einstein had also other reasons to reject a Newtonian finite universe. He argued, by means of statistical considerations, that one cannot have a plausible Newtonian cosmology even by assuming conditions at infinity for the gravitational potential, since in that case one has to accept the possibility that some heavenly body could escape to infinity, unless the difference of potentials between 'here' and 'infinity' was big enough. The last possibility Einstein rejected, as inconsistent with the then available observational evidence of *small values for the "stellar velocities"*. Another statistical difficulty of Newtonian cosmology arose if Boltzmann's distribution for the molecules of a gas in thermal equilibrium was applied to the stars.

To solve these difficulties within Newtonian physics, Einstein advanced a suggestion, "which does not in itself claim to be taken seriously", but was meant to introduce a change in his field equations which, on the contrary, he wished to advance quite seriously. If at the left-hand side of Poisson's equation:

$$\Delta\phi = 4\pi G\rho, \tag{1}$$

a term is added in the form:

$$\Delta\phi - \lambda\phi = 4\pi G\rho, \tag{2}$$

where λ is a "universal constant", then the new equation clearly admits a solution ϕ corresponding to a uniform mass distributions ρ_0: namely, the constant potential function $\phi = -4\pi G\rho_0/\lambda$. This change eliminates the need of enforcing conditions at infinity.

The idea of (2) was probably inspired by the modified gravitational potential ϕ_N built out of the mass-point potential $Ae^{-r\sqrt{\lambda}}/r$ (A is a constant), as proposed by the German theoretical physicist Carl Neumann in 1896. Neumann introduced this form of the potential in order to solve the gravitational paradox in the form of the impossibility to assign, *at any point*, a finite value to the gravitational potential corresponding to a uniform infinite mass distribution. In fact ϕ_N goes to zero at infinity even with such a mass distribution, and (2) is precisely the equation which it satisfies.

2. Conditions at Infinity and Mach's Principle

The hypothesis that at *spatial* infinity the $g_{\mu\nu}$ should be Minkowskian in some suitable coordinate system was the very hypothesis Einstein had adopted in his previous papers when dealing with the astronomical problems of the solar system (Chap. 5, §11), but for the universe as a whole such an hypothesis was "by no means evident *a priori*", as he said – for a number of reasons.

First, in relativity a condition at spatial infinity is necessarily dependent on the choice of a coordinate system, and so it appears to conflict with the principle of general co-variance. Second, such conditions all contradict the principle of relativity of inertia (or Mach's principle), according to which: "In a consistent theory of relativity there can be no inertia *relatively to 'space'*, but only an inertia of masses *relatively to one another*" [17, p. 180]. In fact under the stated conditions at infinity, the $g_{\mu\nu}$ would only slightly differ from the Minkowskian values, with the result that "inertia would indeed be *influenced*, but would not be *conditioned* by matter (present in finite space)" [p. 183]; and, as we know, this requirement is an essential part of Mach's principle. Third, statistical objections can be raised in relativistic cosmology just as in the Newtonian case.

Of course, another option would be to abstain from fixing any conditions at infinity for the universe as a whole, and instead to fix such conditions in a case by case manner, for space–times modelling particular physical systems (as was done by Schwarzschild when deriving his solution).[1] However, this agnostic position was not to Einstein's liking:

---

[1] Note, however, that the derivation of the Schwarzschild solution does not require that the metric is asymptotically Minkowskian, since this (as well as staticity) happens to be a *consequence* of spherical symmetry.

This is an incontestable position, which is taken up at the present time by de Sitter. But I must confess that such a complete resignation in this fundamental question is for me a difficult thing. I should not make up my mind to it until every effort to make headway toward a satisfactory view had proved to be vain [p. 182].

We shall hear again of Willem de Sitter, a professor of astronomy at the university of Leiden, who published between 1916 and 1917 three influential papers on general relativity, which, among other virtues, were responsible for awakening Arthur Eddington's fateful interest in the theory. Thus Einstein decided to *remove* the problem of imposing conditions at infinity, by using the radical remedy of postulating that space is a "self-contained continuum of finite spatial (three-dimensional) volume". In fact, if no 'infinity' (and no boundary) exists, then, obviously, no 'conditions at infinity' (or boundary conditions) are needed any more.

3. Einstein's Static Universe

The idea that an infinite, Euclidean space is not the only conceivable geometric model for the universe, largely pre-dates Einstein's cosmological paper. Indeed, it has been convincingly argued that the way Dante Alighieri in his *Divina Commedia* (fourteenth century) manages the system of "heavens" of his Paradise implies that he thought of them as if they were two-dimensional analogues of the parallels of an ordinary sphere: in other words, that he re-intepreted the Tolemaic cosmos as a three-dimensional sphere *ante litteram* (cf. [67,14]). However, it is to the nineteenth century that we owe an explicit and mathematically precise conjecture to this effect, and first of all to Riemann's generalization to higher dimension of Gauss' differential geometry of surfaces (Chap. 5, §7). Following in Riemann's footsteps, Clifford had written:

> We may postulate that the portion of space of which we are cognizant is practically homaloidal [i.e. flat], but we have clearly no right to dogmatically extend this postulate to *all* space. A constant curvature, imperceptible for that portion of space upon which we can experiment, or even a curvature which may vary in an almost imperceptible manner with the time, would seem to satisfy all that experience has taught us to be true of the space in which we dwell [4, p. 201].

Clifford had in fact a preference for the hypothesis that the universe, except for small local variations, was a space with constant positive curvature. Now this was precisely one of the hypotheses endorsed by Einstein in his 1917 paper [17]; let us list them, each one followed by a short commentary.

(I) *There is a privileged, time-orthogonal coordinate system, such that, with respect to the time coordinate matter looks "permanently at rest"* (staticity).

As we shall see, this hypothesis was weakened in subsequent theories, as regards the staticity of the matter distribution, but not in the all-important assumption of the existence of a privileged time coordinate (which will be called *cosmic time*), which was clearly a robust injection of Newtonianism into 'general relativity'.

(II) *The universe is a 3-sphere*, that is, it can be represented (except for local variations) as the set of points $(\xi^1, \xi^2, \xi^3, \xi^4)$ in 4-dimensional Euclidean space satisfying the equation:

$$\left(\xi^1\right)^2 + \left(\xi^2\right)^2 + \left(\xi^3\right)^2 + \left(\xi^4\right)^2 = R_0^2. \tag{3}$$

This can be seen as a special application of what will be named the *cosmological principle*, which assumes that the universe is both homogeneous and isotropic, as regards both geometry and mass-energy content.[2] Versions of the cosmological principle occur in most cosmological theories.

The 3-sphere is the only (up to isometry) *compact* 3-manifold with constant curvature which satisfies the further topological condition of being simply-connected (the property that every loop can be continuously deformed to a point). These conditions cannot be satisfied with a nonpositive constant curvature, though – it must be added – the condition of simply-connectedness has no obvious physical justification. It is worth pointing out that at the time the classification of 3-spaces with constant curvature was still at a very primitive stage, and topological considerations entered the discussion in a rather naive form; practically the only topological discussions had to do with the choice between the 3-sphere and the projective 3-space (also called "elliptic space": this is the simplest instance of a multiply-connected 3-space with constant positive curvature). In a letter to Weyl of June 1918 Einstein wrote that his preference for the 3-sphere was rooted in an "obscure feeling" [34, p. 78].

(III) *On a large scale, the matter content of the universe can be modelled as a pressure-less, incoherent fluid* (a dust, cf. Chap. 5, §10), so that in the privileged system the only nonvanishing component is the matter density ρ – constant both in space, because of (II), and in time, because of (I).

Hypothesis (III) has been considerably modified in other cosmologies, as different matter models have been advanced, the most popular being that of a perfect fluid, with nonvanishing pressure. What has hardly ever been modified is the fluidodynamical description itself (cf. [63, Note IX, pp. 415–6]).

From (I) and (II) it follows that the space–time metric in the privileged coordinate system has the form:

$$ds^2 = -R_0^2 \, d\sigma^2 + c^2 \, dt^2 \tag{4}$$

where $d\sigma^2$ is the metric of the unit 3-sphere S^3, that is:

$$d\sigma^2 = \sum_{a,b=1}^{3} \left(\delta_{ab} + \frac{x^a x^b}{R_0^2 - ((x^1)^2 + (x^2)^2 + (x^3)^2)}\right) dx^a \, dx^b,$$

or just

$$d\sigma^2 = \left(\delta_{ab} + \frac{x^a x^b}{R_0^2 - ((x^1)^2 + (x^2)^2 + (x^3)^2)}\right) dx^a \, dx^b, \tag{5}$$

as we may write using Einstein's convention on indices (cf. Chap. 5, §7). The coordinate system in which the metric is expressed in (5) is obtained by parallel projection along the ξ^4-axis in $\mathbb{R}^4$; it covers the upper (or the lower) hemisphere, except for the equatorial 2-sphere ($\xi^4 = 0$). It is important to remember that for 'most' manifolds (the *n*-sphere

---

[2] The first explicit statement of this principle is said to have been made in 1933 by E.A. Milne [63, pp. 156–8].

being a classic example) no single coordinate system exists covering the whole space, so all explicit coordinate representations of the metric can only describe what happens in a subspace. On the other hand, given the homogeneity of the 3-sphere, the expression (5) can be thought of as being valid in the neighbourhood of any given point.

Topologically speaking, Einstein's space–time is $S^3 \times \mathbb{R}$, and can be seen as the 4-dimensional cylinder in Minkowskian $\mathbb{R}^5$, that is, the set of all points $(x^0, x^1, x^2, x^3, x^4)$ in $\mathbb{R}^5$ satisfying:

$$\left(x^0\right)^2 + \left(x^1\right)^2 + \left(x^2\right)^2 + \left(x^3\right)^2 = R_0^2. \tag{6}$$

By applying the field equations to these data, one gets, after some computations, two equalities:

$$\frac{1}{R_0^2} g_{ab} = 0, \qquad -\frac{3}{R_0^2} = -\kappa c^2 \rho.$$

Now this result hardly makes sense: in fact, from the first equality one has that R_0 should be infinite, that is, that the 3-sphere should degenerate into ordinary Euclidean 3-space; but then the second equality implies that the matter density should be *null*, which means that the universe is *empty*! To put it in spatio-temporal terms, one is back to Minkowski space–time, with no matter in it.

4. The Cosmological Constant

Which assumption was to blame for this unseemly conclusion? Einstein's guess was that the culprit is not to be found among (I)–(III), but that it lies hidden in his very field equations. He proposed to change them into the following variant, which he claimed was "perfectly analogous to the extension of Poisson's equation given by (2)"

(IV) *Modified field equations*

$$R_{\mu\nu} - \lambda g_{\mu\nu} = -\kappa \left(T_{\mu\nu} - \frac{1}{2} T g_{\mu\nu} \right), \tag{7}$$

where λ is a constant, the so-called *cosmological constant*. Using (7) one obtains the following equations:

$$\frac{1}{R_0^2} g_{ab} = \lambda g_{ab}, \qquad -\frac{3}{R_0^2} + \lambda = -\kappa c^2 \rho,$$

whence it immediately follows $\lambda = 1/R_0^2 = \kappa c^2 \rho / 2$. These equalities establish a connection between the constant λ and the total mass M of the spherical universe. In fact, since the volume of a 3-sphere of radius R_0 is $2\pi^2 R_0^3$, the total mass of Einstein's universe is:

$$M = 2\pi^2 R_0^3 \rho = \frac{4\pi^2}{\kappa c^2} \lambda^{-1/2}.$$

As Hermann Weyl commented, "this obviously makes great demands on our credulity" [79, p. 279]. Notice, however, that now there is no absurdity. By the introduction of λ it is possible to solve the (new) field Eqs (7) with a space–time and a matter model obeying conditions (I)–(III).

It is worth pointing out that Einstein's claim that (7) is the general relativistic version of (2) is wrong. It seems that the first scientist to realize it (a quarter of century later, in 1942), or at least to publish it, was the German cosmologist Otto Heckmann [39]. In fact it is easy to prove that the Newtonian limit (Chap. 5, §12) of (7) is *not* (2), but:

$$\Delta\phi + \lambda c^2 = \frac{\kappa\rho c^4}{2},$$

which can be rewritten as

$$\Delta\phi = \frac{\kappa c^4}{2}\left(\rho - \frac{2\lambda}{\kappa c^2}\right). \tag{8}$$

This equation is clearly satisfied by a constant potential in case one puts $\lambda = \kappa c^2\rho/2$ (note that this is the same equality which must be satisfied by Einstein's solution). One interesting consequence is that by this change in Poisson's equation Newton's insight into the behaviour of matter "evenly disposed throughout an infinite Space" (§1) happens to be fully vindicated: the net gravitational field at any point is zero, and this ensures that a Newtonian infinite material universe can indeed be stationary. In other words, the way Einstein modified the field equations to solve the cosmological problem closely resembles Newton's suggestion to Bentley on how to deal with an infinite universe![3]

Even more remarkable, perhaps, is that a long sequence of eminent authors missed this basic point and blindly endorsed Einstein's stated analogy between (2) and (7) ("generations of physicists have parroted this nonsense", as is said in [39, p. 723]). Clearly most working scientists are just too anxious to publish some 'new' piece of research of theirs to spend a sufficient amount of time reviewing the foundations of their disciplines; so they frequently end up by relying on authority much more than on rational belief, in contrast with the scientific ethos as ordinarily proclaimed.

At the end of his paper Einstein pointed out that the cosmological term was needed "only for the purpose of making possible a quasi-static distribution of matter, as required *by the fact of the small velocities of the stars*" ([17, p. 188]; italics added). From the following of the story it will be clear that Einstein also expected the introduction of λ to ensure that *only* a quasi-static (i.e. static on a sufficiently large scale) distribution was possible. This conjecture received a first blow a few months later.

5. De Sitter's Space–Time (1917)

As is clear, the Eqs (7) are also verified by any *empty* (i.e. $T_{\mu\nu} = 0$) space–time with a metric $g_{\mu\nu}$ such that:

$$R_{\mu\nu} = \lambda g_{\mu\nu}. \tag{9}$$

Any space–time satisfying this relation generates a difficulty for the Machian interpretation of general relativity, because it is a space–time with no matter content and yet

---

[3]This is in agreement with the 'genuinely Newtonian' cosmology described in [59], though this author, too, considers (2) as the legitimate Newtonian limit of (7).

with a metric solving the (vacuum) field equations. We recall that the original field equations had been conceived as describing how matter determines 'inertia', which means, under the assumption of the principle of equivalence, the behaviour of particles subjected to gravitation only (this behaviour being in turn described by the geodesics of the space–time). On the other hand, the λ-term makes it easier, in principle, to have plenty of formally acceptable space–times *with no matter content at all.*[4] The Minkowski and Schwarzschild space–times satisfy this condition with $\lambda = 0$ (that is, the original vacuum field equations), but what about a nonzero λ, such as the one needed for Einstein's static space–time?

The Dutch astronomer whom Einstein had cited in his first cosmological paper, de Sitter, published a few months later a solution of the modified field equations which was to prove one of the most important in the history of cosmology, not only of *relativistic* cosmology [8].

The idea (suggested to de Sitter by Paul Ehrenfest) is to obtain an homogeneous and isotropic space–time by adapting condition (3), which defines a spherical 3-*space*, so that it defines a 'spherical' *space–time*. Consider all points in $\mathbb{R}^5$ – regarded as the 5-dimensional Minkowski space – at space-like distance R_0 from the origin:

$$-\left(x^0\right)^2 - \left(x^1\right)^2 - \left(x^2\right)^2 - \left(x^3\right)^2 + c^2\left(x^4\right)^2 = -R_0^2. \tag{10}$$

This is the four-dimensional version of a familiar quadric surface, the one-sheet hyperboloid; in order to make it into a *space–time* one has to consider the metric induced on it from the *Lorentzian* structure of $\mathbb{R}^5$. Using the coordinate system obtained by parallel projection along the x^0-axis, one obtains the following expression for the metric, which closely resembles (5) (remember that for all repeated Greek indices, sum from 1 to 4 is understood):

$$ds^2 = \left(\eta_{\mu\nu} - \frac{\eta_{\mu\rho}\eta_{\nu\sigma}x^\rho x^\sigma}{R_0^2 + \eta_{\rho\sigma}x^\rho x^\sigma}\right)dx^\mu\,dx^\nu. \tag{11}$$

where $\eta_{\mu\nu}$ is the matrix of the usual Minkowski metric in $\mathbb{R}^4$ (Chap. 5, §4). This four-dimensional manifold is like the 3-sphere (or the n-sphere, for that matter, $n \geq 2$) in also having constant curvature (equal, with our conventions, to $-1/R_0^2$). By using the formula for the Ricci tensor given in Chap. 5 and with computations very similar to those that are made for the Einstein's metric, one finds:

$$R_{\mu\nu} = \frac{3}{R_0^2}g_{\mu\nu}, \tag{12}$$

so that the left-hand side of the modified field equations is

$$R_{\mu\nu} - \frac{1}{2}Sg_{\mu\nu} + \lambda g_{\mu\nu} = \left(-\frac{3}{R_0^2} + \lambda\right)g_{\mu\nu}.$$

It follows that if $\lambda = 3/R_0^2$, then (7) is satisfied *with a vanishing stress-energy tensor.* This is one part of the trouble: the λ-term makes it possible to have new space–times

---

[4]Ironically, all pseudo-Riemannian manifolds (of any signature) satisfying (9) are today called *Einstein manifolds*. For an encyclopaedic survey see [1].

with nontrivial inertia (i.e. different from Minkowski space–time) and empty. Indeed, it can be proven (for instance by using (24) and (25) below) that the only possibility for de Sitter space–time to be filled with a perfect fluid with density ρ, is for its pressure p to verify the equation $\rho + p/c^2 = 0$, which is a condition on the signs of both density and pressure which is difficult to reconcile with a realistic physical model unless one take both ρ and p equal to zero.[5]

Here comes a further twist. Equation (10) can be rewritten, by putting $t = x^4$, in the form:

$$\left(x^0\right)^2 + \left(x^1\right)^2 + \left(x^2\right)^2 + \left(x^3\right)^2 = R_0^2 + c^2 t^2 \tag{13}$$

which is easily interpreted, geometrically, as a family of 3-spheres parametrized by t and of radius

$$R(t) = \sqrt{R_0^2 + c^2 t^2}. \tag{14}$$

Clearly, $R(t)$ has a minimum (for $t = 0$) and no maximum; this means that de Sitter space–time describes a spherical universe with no beginning and no end, with a curvature which increases until it reaches the value $1/R_0^2$ and then decreases indefinitely. One may have different opinions on how to define 'staticity', but surely such a space–time cannot be rated 'static'! By introducing a new time coordinate, namely $\tau = (R_0/c) \sinh^{-1}(ct/R_0)$, one gets the following expression for the de Sitter metric:

$$ds^2 = -R_0^2 \cosh^2\left(\frac{c\tau}{R_0}\right) d\sigma^2 + c^2 d\tau^2. \tag{15}$$

This form of the metric was first published in 1922 by Cornelius Lanczos [56], a correspondent and later collaborator of Einstein [65, p. 491]. By comparison of (15) with the Einstein static metric (4) the nonstaticity of the de Sitter universe is even more evident. Thus de Sitter space–time can be taken to describe a changing, though paradoxically empty, spherical universe satisfying the modified field Eqs (7).

6. The "Discontinuities" of de Sitter Metric

Expression (15) seemed to refute Einstein's opinion that the cosmological term was effective in ruling out non-static space–times. However, de Sitter's own view was that his space–time *was* static, although not globally so. To explain this strange-looking but in fact quite sound opinion, one must take full account of the rôle played by the choice of the coordinate system, and in particular of the time coordinate. The great historical importance of de Sitter's solution lies, first of all, in providing a spectacular demonstration of the wide-ranging consequences of such a choice. In fact, if one consider the subspace of de Sitter space–time contained in $x^3 + ct > 0$, $x^3 - ct > 0$ (this is a solid five-dimensional wedge in $\mathbb{R}^5$), then one can define on it two new coordinates θ and u:

---

[5]The function ρ can hardly make sense as a *negative* physical quantity, while a negative pressure does make sense (it would represent a cohesive force in the 'cosmic fluid'), but the value $p = -\rho c^2$ "could not be even remotely approached by any known material", Tolman wrote in 1934 [77, p. 348].

$$x^3 = R_0 \cos\theta \cosh u, \qquad t = \frac{R_0 \cos\theta}{c} \sinh\left(\frac{cu}{R_0}\right), \qquad (16)$$

and the de Sitter metric acquires the following appearance:

$$ds^2 = -R_0^2 \, d\sigma^2 + c^2 \cos^2\theta \, du^2. \qquad (17)$$

It is clear that, in this coordinate system, the metric looks static indeed, as no one of its coefficients depends on t.[6] De Sitter described his solution as "the general solution for the case of a static and isotropic gravitational field in the absence of matter" (cit. in [63, p. 88]).

A space–time satisfying the modified field equations, static and empty could not but disturb Einstein. As he commented in his public reply to de Sitter [18], in case (17) were to be shown a legitimate solution of the modified field equations, this would indicate that the λ-term "does not fulfil the purpose I intended":

> Because, in my opinion, the general theory of relativity is a satisfying system only if it shows that the physical qualities of space are *completely* determined by matter alone. Therefore, no $g_{\mu\nu}$-field must exist (that is, no space–time continuum is possible) without matter that generates it.

In other words, de Sitter's solution violates Mach's principle, and yet the cosmological constant had been introduced to circumvent, in a static background, this blemish of the original field equations. So Einstein resorted to the argument that the coordinate system used in (17) was not legitimate. His point was that if $\theta = \pi/2$, then the metric degenerates (i.e. $\det(g_{\mu\nu}) = 0$), and yet "it seems that no choice of coordinates can remove this discontinuity":

> Until the opposite is proven, we have to assume that the de Sitter solution has a genuine singularity on the surface $[\theta = \pi/2]$ in the finite domain; i.e., it does not satisfy the field Eqs (7) for any choice of coordinates.

Thus Einstein argued that de Sitter's space–time is not really "a world free of matter, but rather like a world whose matter is concentrated on the surface $[\theta = \pi/2]$", as is the case for "the immediate neighbourhood of gravitating mass points".

Einstein's refutation of de Sitter's counterexample is remarkably weak. Equation (17) is nothing but the expression in a certain coordinate system of metric (15); and metric (15) is indeed defined on *all* of de Sitter space–time, and it *does* satisfy (7) everywhere! There is no "discontinuity" to "remove" at $\theta = \pi/2$: one only need consider in the neighbourhood of any point p such that $\theta = \pi/2$ the analogue of the coordinate system giving (17) (in this new coordinate system, the angular variable will be a $\bar{\theta}$, with, say, $\bar{\theta}(p) = 0$). Such coordinate systems certainly exist, given the homogeneity of de Sitter's space–time. It follows, with no need for further inquiry, that this space–time has no "genuine singularity".[7]

As shown in private correspondence at the end of 1918, Einstein eventually recognized that de Sitter's solution was perfectly regular; nonetheless, he did not publish a retraction of his previous criticism [50].

---

[6]Notice that the first coordinate θ is just one angular coordinate of the 3-sphere, while the time coordinate u is given explicitly by $u = \tanh^{-1}(\frac{ct}{x^3}) = \frac{1}{2}\log\frac{x^3 + ct}{x^3 - ct}$.

[7]This point was well explained by Eddington in his *Report* of the same year [11, p. 88].

7. Weyl Enters the Controversy (1922–1923)

In the fourth edition of his treatise *Space–Time–Matter* Hermann Weyl discussed the contrast between the cosmological views of Einstein and de Sitter. He put it in the form as to which coordinate system – the one leading to (15) or the one leading to (17) – was more adequate to "represent the whole world in a regular manner":

> In the former case the world would not be static as a whole, and the absence of matter in it would be in agreement with physical laws; de Sitter argues from this assumption [...]. In the latter case we have a static world that cannot exist without a mass-horizon; this assumption, which we have treated more fully, is favoured by Einstein [79, p. 282].

In other words, Weyl supported Einstein's view that de Sitter's solution was not really empty of matter, but that the matter was, so to speak, 'hidden' behind the boundary of the coordinatized region.

However, in 1923 Weyl modified his views considerably ([80,82]; cf. [83]), and in so doing he posed a foundational stone of relativistic cosmology for decades to come.

In the 4-dimensional hyperboloid representing de Sitter's space–time, any of the lines obtained as the intersection with a 2-plane containing the line

$$x^0 = x^1 = x^2 = 0, \qquad x^3 + cx^4 = 0,$$

is a timelike geodesic Γ – a generating line of the hyperboloid, in fact. Now, the set of all points of space–time which can be influenced by such a worldline (its *range of influence* [*Wirkungsbereich*], as Weyl calls it)[8] is given by one 'half' of the hyperboloid: to be explicit, in the case of the line Γ_0 defined by the 2-plane $x^0 = x^1 = x^2 = 0$, this 'half' is the set M of points belonging to the half-space $x^3 + cx^4 > 0$. The important fact is that M is the range of influence not only of Γ_0, but of *the whole 3-parameters family of geodesics* Γ *obtained by the same procedure*.

Weyl's proposal was to interpret each Γ as the worldline of a "star", and, given that all these "stars" are causally connected since the "infinitely distant past" and that no influence can come from outside, to take M as the *whole* of our space–time. This assumption suggests that we select and use a coordinate system which (1) covers the whole of M (and not just the 'wedge' region); and (2) represents the Γ by parallel lines. Such a coordinate system indeed exists, and in it the metric acquires the form:

$$ds^2 = -e^{2cT/R_0}\left(dx^2 + dy^2 + dz^2\right) + c^2\, dT^2. \tag{18}$$

Note that this form of the metric is invariant for a translation of the spatial coordinates $(x, y, z) \mapsto (x + x_0,\ y + y_0,\ z + z_0)$ and also for the composite of a time coordinate translation $T \mapsto T + T_0$ with the spatial dilation $(x, y, z) \mapsto e^{-cT_0/R_0}(x, y, z)$, so that all 'fundamental observers' can be considered as equivalent. Their common origin, Weyl wrote in the German fifth edition of *Space–Time–Matter*, "makes understandable the small velocities of the stars" [80, p. 285].

Another consequence pointed out by Weyl is that even if the spatial coordinates of every star are constant, this does not mean that the distance between two stars is also constant. In fact if $D(0)$ is the distance at $T = 0$ of two stars, then the distance at time T is $D(T) = D_0 e^{cT/R_0}$.

---

[8]In modern notation [40] this is denoted as $J^+(\Gamma)$.

Weyl derived a crucial formula, giving the change of frequency of the light emitted by a star, as received by another star:

$$\frac{\Delta \nu}{\nu} \approx - \sin \frac{D}{R_0} \approx - \frac{D}{R_0}. \tag{19}$$

This means that the spectral lines of stars are shifted towards lower frequencies (*red-shifted*) when they are received by another star, and the redshift z is approximately proportional to the distance. This is one of the first appearances of what is known as "Hubble's law"; as we shall see this property of de Sitter's space–time will be important in directing Edwin Hubble's attention to this kind of relation in his data.

Note that, from the formula giving the relative distance of two stars, it follows that their relative speed is:

$$v = \frac{c}{R_0} D, \tag{20}$$

that is, *the speed is (exactly) proportional to the distance*, and therefore v turns out to be approximately proportional to the redshift. So, for all practical purposes, the relationship between redshift and distance can be interpreted as a Doppler effect.

The assumption that a cosmological space–time must admit a 3-parameter family of 'synchronized' timelike geodesics diverging from the past and hypersurface-orthogonal,[9] representing the fundamental cosmological objects ("stars" or, a few years later, galaxies and, still later, clusters of galaxies), survived the demise of de Sitter space–time as the centrepiece of relativistic cosmology, and became known as *Weyl's principle*.

This principle in fact singles out a class of privileged observers: once it is accepted, the static version of de Sitter's metric (17) cannot be considered as physically relevant, because the observers having u as their proper time *are not geodesic*. In relativistic cosmology, Weyl's fundamental observers came to play the rôle of the 'inertial systems' of Newtonian physics, just as the 'cosmic' (or 'universal') time took the place of the supposedly dead absolute time. This circuitous return of the basic Newtonian concepts, as we shall see (§16), will not please a great philosopher-scientist such as Kurt Gödel.

Moreover, the requirement of 'divergence from the past' rules out (15), which otherwise would satisfy the principle. In fact this condition implies that in the past the 'stars' were closer to each other than they are now, and this establishes a cosmological asymmetry between past and future which, as we shall see, was a critical assumption in the development of cosmology.

Note, finally, that the form (18) of the de Sitter metric describes an obviously nonstatic universe again, but with a surprise: the universe is no more a 3-sphere as in (15), but a *Euclidean 3-space*. In other words, the *space* curvature (as opposed to the *space–time* curvature, which of course is always $-1/R_0^2$, since it does not depend on the choice of the coordinate system) is zero. The moral is that, by introducing suitable coordinate systems, we can have *radically different space geometries* (Chap. 5, §5). This is one of the most striking illustrations of the fact that the general covariance principle, if not suitably restricted, leads to a radical relativization of the physical geometry of *space* (as opposed to space–time). In this sense one can say that this vindicates Poincaré's

---

[9]This means that the universe at a certain cosmic time is represented by a space-like 3-submanifold, which is crossed orthogonally by the fundamental worldlines.

'conventionalism' as regards physical geometry, though not his prediction that physicists would never abandon Euclidean geometry (Chap. 5, §9), as will be even more clear from the following.

8. Friedmann's Evolutionary Universes (1922, 1924)

While de Sitter, Einstein, Weyl and others were struggling about how to correctly interpret de Sitter's solution, a more radical development was taking place at Petrograd on the Baltic sea. The de Sitter metric, as displayed in formulas (15) and (18), had been the first example of a metric of the type

$$ds^2 = -R(t)^2 \, d\sigma^2 + c^2 \, dt^2, \tag{21}$$

which must be thought of as defined on a product manifold $\Sigma \times I$, where $(\Sigma, d\sigma^2)$ is a Riemannian 3-manifold with normalized constant curvature (i.e. the curvature K is 1 or 0 or -1), I is an open interval of the real line $\mathbb{R}$, and $R : I \to]0, +\infty[$ is a function which can be interpreted as the "radius" of the universe (or, more generally, its *scale factor*). Using formula 20 from Chap. 5, §8, we can express all these metrics in the form:

$$ds^2 = -R(t)^2 \left[\frac{dr^2}{1 - Kr^2} + r^2 \left(d\phi^2 + \sin^2 \phi \, d\chi^2 \right) \right] + c^2 \, dt^2, \tag{22}$$

where $K = 0, 1, -1$. Notice that (21) (or (22)) embodies both the cosmological principle (§3, II) and Weyl's principle (apart from the past-future asymmetry).

Of course also Einstein's metric falls into this class (for $R(t) \equiv R_0$). So it was a natural question to investigate whether other space–times of this type existed, apart from de Sitter's and Einstein's, either with $\lambda = 0$ or with $\lambda \neq 0$. Einstein had somewhat preempted this line of research, by implying that his static space–time was the *only* nonempty solution allowed by the modified field Eqs (7); in particular, he thought that all physically meaningful solutions of (7) were static.

So it took a Russian meteorologist to challenge Einstein's authority and proceed to an inquiry of the *non-stationary* solutions. His name was Aleksandr Aleksandrovich Friedmann; he made his major contributions to relativistic cosmology between 1922 and 1924.

In his 1922 paper, "On the curvature of space" [32], Friedmann considered the metrics of the form

$$ds^2 = -R(t)^2 \, d\sigma^2 + M(x) \, dt^2, \tag{23}$$

satisfying (7), under the assumption that Σ is a 3-sphere and matter a dust, as for Einstein's space–time. He showed that they divided into two classes, according to whether $\dot{R} = 0$, i.e. $R(t)$ is a constant function (stationary case), or $\dot{R} \neq 0$ (non-stationary case), the first class comprising only Einstein's and de Sitter's metrics (the latter in the form (17)).[10] In 1924 he published a second paper, "On the possibility of a world with a constant negative curvature of space" [33], dealing with the case that Σ is not a sphere but a space with $K = -1$ (a *hyperbolic* 3-space). In fact the main argument is the same

---

[10]If we start from (21), 'static' and 'stationary' can be used equivalently.

with any curvature $K = 1, 0, -1$, the non-stationary case always reducing to metrics of the form (21).

The mass density ρ is supposed to depend on t only; the stress-energy tensor is given, in the coordinate system where (21) holds, by a $T_{\mu\nu}$ having only one nonzero component: $T_{44} = \rho c^4$.

Putting all these data into (7) and indicating by dots the derivatives with respect to t, one obtains just two scalar equations:

$$\frac{\dot{R}^2 + 2R\ddot{R} + c^2 K}{c^2 R^2} - \lambda = 0, \tag{24}$$

$$\frac{3(\dot{R}^2 + c^2 K)}{c^2 R^2} - \lambda = \kappa c^2 \rho. \tag{25}$$

After a few elementary transformations on (24), one obtains the (now) famous *Friedmann's equation*:

$$\dot{R}^2 = \frac{\lambda c^2 R^2}{3} + \frac{A}{R} - c^2 K, \tag{26}$$

where A is a constant which can be expressed, by exploiting (25), as:

$$A = \frac{\kappa c^4 \rho R^3}{3}. \tag{27}$$

Equation (26) describes the behaviour of the squared time derivative of the radius of the spherical universe, and as a consequence it permits one to say how the radius itself $R(t)$ behaves in time.

Mathematically, the basic issue is to find under which conditions the right side vanishes, or equivalently what are the positive real roots (since R is necessarily positive) of the cubic equation $\lambda x^3/3 - Kx + A/c^2 = 0$, depending on λ and A.

If $K = 1$, it is easy to see that there is one critical value λ_c, defined by

$$\lambda_c = \left(\frac{2c^2}{3A}\right)^2,$$

and a critical interval J of real numbers such that the behaviour of $R(t)$ changes qualitatively according to where λ and $R(0)$ lie with respect to λ_c and J. An elementary computation shows that λ_c is equal to the cosmological constant of an Einstein static space–time containing the same total mass.

A summary of the multifarious non-stationary alternatives runs as follows (as stated above, Friedmann shows that stationary solutions exist only if $\lambda \neq 0$, thus proving that Einstein's argument for the cosmological constant is partly correct).

Suppose λ lies between 0 and λ_c; then if $R(t_0)$ lies on the right of J, the universe expands forever, either monotonically (if $\lambda = \lambda_c$) or after an initial contraction phase;[11] if $R(t_0)$ lies on the left of J, then the universe expands from $R = 0$ to a certain positive value of R, and then contracts: in other words the universe is *periodic*, or as it is more commonly known today, *oscillating* (although the whole of its story might well be in-

---

[11] Actually this last possibility, which includes the non-stationary form (15) of de Sitter metric, is not mentioned in [32].

cluded in a single cycle!).[12] Suppose λ is negative; then the universe is in all cases an oscillating one. Suppose finally that $\lambda > \lambda_c$; then the universe can only expand forever from $R = 0$.

In the case of negative constant curvature ($K = -1$) the argument follows the same pattern. Here the dividing line is whether the cosmological constant is positive (or zero), or negative; in the first case, the universe expands forever from $R = 0$; in the second case we have an oscillating universe. Friedmann did not list these different possibilities, but limited himself to pointing out that no stationary universe with a positive density can exist.

In the concluding section of [33] Friedmann tackled the problem of the finiteness of infinity of the universe, and argued, very reasonably, that "the cosmological equations, by themselves, are not sufficient to solve the problem of the finiteness of our universe [...]", and that other hypotheses, of a topological nature, were needed to reach any conclusions. In particular he emphasized that the case of negative curvature is also compatible with finiteness. In this he showed prescience, since the first examples of compact 3-manifolds with constant negative curvature were only discovered a few years later. (Today the issue of the topology of space–time is very much alive; cf. Chap. 13).

9. The Einstein–Friedmann Controversy

How did Einstein react to Friedmann's first article? It is not unreasonable to suspect that the new options crowding around his basic hypothesis of a spherical universe obeying (7) must have provoked a strong revulsion in him. To make things worse, the cosmological constant, far from being a gatekeeper stymieing cosmological anarchy, was contributing to it... Whatever the circumstantial and psychological reasons, it is a fact that he came hastily to the conclusion that Friedmann had to be wrong on a very basic point, which destroyed the main theorem of his paper, namely, that (7) admits non-stationary solutions. He wrote as much in a short note which appeared on 8 September 1922, in the same journal [21]. In it he stated that Friedmann had made a simple mathematical mistake, from which the non-constant ρ had been born. According to Einstein, from the equation div $T = 0$ (cf. Chap. 5, §13) it just followed:

$$\frac{\partial \rho}{\partial t} = 0, \tag{28}$$

implying that both ρ and $R(t)$ had to be constant. Thus Friedmann's paper was useful, after this substantial correction, only insofar as it unwittingly proved that (7) could not have non-stationary solutions. In other terms, Friedmann had provided, according to Einstein, an incorrect proof of the fact that the only solution of (7), under the given assumptions, was Einstein static space–time!

Friedmann learnt about [21] by chance, from a letter of a colleague, Y.A. Krutkov, himself quite knowledgeable on relativity, who was working in Germany at the time. Thus, on 6 December of the same year, Friedmann addressed to Einstein directly a very courteous letter, where he described in detail the computation which made him doubt

---

[12] In his semipopular book Friedmann referred to the "cycles of existence" of the Hindu religion, and also to "creation from nothing", although just as "curiosities" [34, p. 206].

Einstein's conclusion (28). The point is a quite simple one, in fact, and can be formulated in the following few lines. In general the divergence of $T_{\mu\nu}$ is:

$$(\text{div}\,T)_\mu = \frac{1}{|g|^{1/2}} \frac{\partial}{\partial x^\lambda}\left(g^{\lambda\nu}|g|^{1/2}T_{\mu\nu}\right) - g^{\lambda\sigma}\Gamma^\nu_{\lambda\mu}T_{\nu\sigma},$$

and it is easy to check that in the examined case all components of div$T$ vanish identically, except for the 4th; thus, since div$T = 0$, one must have:

$$0 = (\text{div}\,T)_4 = \frac{\partial}{\partial t}\left(|g|^{1/2}\rho\right) = \frac{\partial}{\partial t}\left(hc\rho R^3\right),$$

where $h = \det(h_{ab})$ and $d\sigma^2 = h_{ab}\,dx^a\,dx^b$. Of course this equation *does* not imply that either ρ or R is constant, but only that the product ρR^3 is (note that the factor hc is time-independent); and this is exactly what must be expected, in agreement with (27).

Friedmann was right: Einstein, not he, had made an error – and quite an elementary error at that. So he felt safe in asking Einstein to let him know whether he agreed with the computation, and suggested that Einstein published a retractation in the same journal, perhaps together with excerpts from the letter. At the time Einstein was abroad, and only returned in Berlin in March 1923. It is not known whether he read Friedmann's letter at that time, but, if he did, it must not have shaken his faith that a non-static solution for (7) simply could not exist.

By a further lucky coincidence, Krutkov was in Leiden as a visiting scholar in May, and he met Einstein, who had travelled there to attend Lorentz's last public lecture before retirement. Krutkov insisted on reading Friedmann's article with him, and succeeded in bringing Einstein to recognize his mistake. In a letter dated May 18, two days after Einstein's return to Berlin, he wrote: "I won Einstein in his argument against Friedmann. Petrograd's honour is rescued!". In fact on 21 May Einstein did submit a retractation note [22], as short as the previous one, where he mentioned Friedmann's letter and the discussions with Krutkov, admitted a computational error of his own, and stated that Friedmann's claim, that the field equations have not only stationary, but also "dynamical (i.e. time-varying) centrally-symmetric solutions", was "correct and important".

After the publication of the second note, "everybody was much impressed by my fight with Einstein and my eventual victory", as Friedmann wrote in a letter of 13 September 1923; he added that he was glad of this, since from then on his articles would have had a smoother passage to publication in the scientific journals. In 1924, the same year that his second (and last) cosmological article [33] appeared,[13] Friedmann published a popular book on relativity, entitled *The Universe as Space and Time*, and planned with a colleague a four-volume technical treatise on relativity. Only the first volume of this major work came out, however, since he suddenly fell ill with typhus and died in 1925, aged 37.

10. What Came of Friedmann's Discovery?

In his first draft of [22], which has survived, Einstein had inserted a clause to the effect that to Friedmann's evolutionary solutions "it was hardly possible to give a physical

---

[13] As is clear from Friedmann's letter to Einstein quoted above, the results of [33] had essentially been obtained by him in 1922.

meaning" [34, p. 47]. Although he ultimately decided to delete it, it is clear that he was convinced that the physical universe was not subject to any large-scale motions. In any case, he did not work on the consequences of Friedmann's articles during the next few years; as a matter of fact, he devoted to cosmology only a few dozen pages in the rest of his scientific production.[14] In itself this is a curious fact.

What is definitely strange is that no one seems to have taken notice of Friedmann's results, before or after the publication of the two Einstein notes. For instance, Weyl failed to mention Friedmann's first article in the 1923 edition of his *Raum, Zeit, Materie* [80], and this is perhaps forgivable; but he did not cite [32] and [33] even in his expository 1930 article where he remained loyal to de Sitter space–time [83].[15]

In 1927 Georges Lemaître – a Belgian mathematician, astronomer, and Catholic priest (since 1923) – discussed the evolutionary solutions for a perfect fluid [58]; what he added to Friedmann's treatment, apart from the generalization to $p \neq 0$, was an examination of recent astronomical data relevant to an assessment of the cosmological models.

In his paper Lemaître did not refer to Friedmann's articles. According to his own account, it was Einstein who at the Solvay conference in Bruxelles, of the same year, mentioned them to him. It has been surmised that the reason the articles and notes of Friedmann and Einstein, published on *Zeitschrift für Physik*, "the best known journal in theoretical physics of that epoch", had completely escaped Lemaître's attention is that he could not read German [34, p. 56], but one can doubt this explanation.[16] It is somewhat amusing that in a letter of 5 April 1930, Lemaître felt he had to inform de Sitter – who certainly knew German well! – of the very *existence* of Friedmann's articles and Einstein's notes. Nevertheless, at a meeting of the British Association in 1931, de Sitter talked of the "brilliant discovery" of "the expanding universe", which had been made ... by Professor Lemaître, of course, and "discovered by the scientific world about a year and a half ago, three [sic!] years after it had been published".

What these facts suggest is that there was a widespread resistance of some of the leading cosmologists to citing Friedmann's articles. It has been said that the reason was that Friedmann's articles were too mathematically oriented, but this is not only an inadequate explanation in itself,[17] but it is also to some extent misleading, given that Friedmann went to such lengths as to estimate the age (or, to be more precise, the "period") of the universe, in the last paragraph of his 1922 paper:

> It is left to remark that the "cosmological" quantity λ remains undetermined in our formulae, since it is an extra constant in the problem; possibly electrodynamical considerations can lead to its evaluation. If we set $\lambda = 0$ and $M = 5 \cdot 10^{21}$ solar masses, then the world period becomes

---

[14]In an appendix to the 1945 edition of [20], Einstein discussed the cosmological problem, giving Friedmann belated recognition.

[15]It is worth adding that in his 1950 preface to the reprint of the English translation of the fourth edition of [79], Weyl did not spend a single word on cosmology, not even to refer to the considerably revised section on cosmology in the fifth German, untranslated edition. A few years later, in the reprint of his famous 1921 *Encyclopädie* survey [66], Pauli devoted one long supplementary note (pp. 219–23) to Friedmann's and Lemaître's work.

[16]That it cannot be the full explanation is indicated by the circumstance that in [58] Lemaître cites two German articles, by Lanczos [56] and by Weyl [82].

[17]We shall dwell in the next section on the sparse and controversial observational evidence which was all astronomy provided at the time.

of the order of 10 billion years. But these figures can surely only serve as an illustration for our calculations.[18]

The last clause seems to be rather a ground for praising Friedmann's scientific sense and moderation, than to indict the physical relevance of his results (by the way, his estimate is close to more recent estimates, cf. [72, p. 182]). Moreover, the very mathematical nature of Friedmann's contribution would only have increased its chances of being read and utilized. A plausible conjecture has to do with the sociology of small groups: the very few people working in cosmology at the time, who were virtually all committed to general relativity and personally related to Einstein, may have been loath to advertise a recent episode which did not throw a very positive light on Einstein's mathematical skill and physical insight and showed the serious lack of uniqueness of the answers given by his acclaimed theory. Also, Friedmann had gone too far for an outsider, by showing that one of the basic assumptions shared by the leaders in the field was wrong.[19]

However, in the end, the expanding universes introduced by Friedmann, and then rediscovered by Lemaître, Howard P. Robertson and Arthur G. Walker, did find their way into recognition. How the FLRW solutions (from the initials of the four people just mentioned) became the main ingredient of modern cosmology is the subject of the next sections.

11. The Astronomical Evidence

In 1905 a well-known astronomer, Agnes Mary Clerke, opened Chapter 26 of her book *The System of the Stars* with the following words:

> The question whether nebulae are external galaxies hardly any longer needs discussion. It has been answered by the progress of research. No competent thinker, with the whole of the available evidence before him, can now, it is safe to say, maintain any single nebula to be a star system of co-ordinate rank with the Milky Way [cit. in [7, p. 214]].

How persuasive. And yet, ironically, it was the very year two revolutions in physics were slowly making a major advance; surely a better time could have been chosen to be so final on a controversial topic. To partially excuse that 1905 confidence and to put in perspective the whole observational business in cosmology, it may be useful to remember that "only one per cent of the light in the night sky comes from beyond our Galaxy" [10, p. 1128].

The view that the galaxies, or *nebulae*, as they were called, were external to the Milky Way and in fact were systems similar to it was known as the *island universe theory*. In 1914 Eddington supported this opinion as a "working hypothesis".

---

[18] [32, p. 2000]. In the "Editor's Note" to the English translation of Friedmann's articles which we have used here, one reads that "[Friedmann] considered only a non-zero cosmological constant λ [...]" [55, p. 1987]. As the preceding quotation in the text evidently shows, this is not correct.

[19] As to mathematical skill, de Sitter had been aware, since at least 1917, that "Einstein occasionally made elementary mistakes in calculus" [52, p. 453]. Unfortunately, a reluctance to expose the failings of a genius like Einstein is conspicuous in most of the scholarly literature on him. For instance the following passage is *all* Pais has to say, in his standard 552-page biography, about the Einstein–Friedmann controversy: "*1922*. Friedmann shows that Eq. (15.20) admits nonstatic solutions with isotropic, homogeneous matter distributions, corresponding to an expanding universe [F1]. Einstein first believes the *reasoning* is incorrect [E45], then *finds an error in his own objection* [E46] and calls the new results 'clarifying'" ([65, p. 288]; italics added).

The first determination of the radial speed of a spiral galaxy, Andromeda, was made in 1912 by Vesto Slipher, of the Lowell observatory, and published in 1913. It was an *approaching* velocity of 300 km/s, not a recessional one, but as Hubble was to write at the end of his life, "the first step in a new field is the great step. Once it is taken, the way is clear and all may follow" [49]. In 1915, from the examination of the spectra of 15 spiral galaxies, Slipher concluded that they all had high radial recessional speeds, from 300 to 1100 km/s. Eddington's future collaborator in the eclipse expeditions to measure the light deflection, A.C.D. Crommelin, who in 1912 had argued against the island universe theory, expressed in 1917 the "hope" that this "sublime and magnificent" conception would "stand the test of future examination".

In 1920 what will be later called the "Grand Debate" took place between Harlow Shapley and Heber Curtis before the U.S. National Academy of Sciences (and then, in amplified form, in the published texts); among other things, the first astronomer defended the traditional view and the latter the island universe theory. One of Shapley's main pieces of evidence was provided by Adrian van Maanen, of the Mount Wilson observatory, who had detected large angular rotations in spiral galaxies, indicating that they must be rather close to the solar system. In 1924 a Cambridge authority, W.M. Smart, had confirmed van Maanen's measures, of which he extolled "the extraordinary precision", uttering another of those definitive statements:

> I do not believe that anyone would be so bold as to question the authenticity of the internal motions – regarded either as rotational or as a stream motion – found by van Maanen; in fact, the more one studies the measures, the greater is the admiration which they evoke [cit. in [41, p. 101]].

Today these 'unquestionable' results have dissolved into thin air. The story has been summed up by saying that "van Maanen had read his expectations into his data",[20] which is a reasonable hypothesis – also to be applied in other cases, as we shall see in a moment.

In his autobiography Shapley recalled: "They wonder why Shapley made this blunder. The reason he made it was that van Maanen was his friend and he believed in friends!" [7, p. 354]. In fact the event had a less edifying and gentlemanly side. Another astronomer, Knut Lundmark, from Sweden, had published in 1921 and 1922 two papers undermining van Maanen's results. Shapley's reaction had been to write a distinctly intimidating letter to Lundmark, where he qualified the latter's critical remarks of van Maanen's work as "not of significance in the larger problem", and suggested that he could find "many flaws or hasty conclusions" in Lundmark's own work. Though irritated at first, Lundmark eventually did not insist on this point, and in fact van Maanen could publish in 1923, without further ado, a paper citing Lundmark's observations in support of the soundness of his own data!

Van Maanen's colleague at Mount Wilson, Edwin Hubble, thought differently, but abstained from putting his opinion in print, and also delayed the publication (until 1925) of results on the distances of galaxies M31 (Andromeda) and M33, estimated by him to be about 930,000 light-years – a huge amount, well in agreement with the island universe theory. Mount Wilson's officials discouraged him from publishing evidence contradicting the work of another member of the same institution. Eventually, in 1935, Hubble published in the *Astrophysical Journal* his remeasurements of four of the galaxies

---

[20][41, p. 110]; this well-written and scholarly small book contains a fascinating reconstruction of the case (pp. 83–110).

studied by van Maanen [48], and concluded that there was no evidence of rotations. Van Maanen, who was allowed to reply in the same issue, did not fully retract his results, but admitted that his more recent measurements "show considerably smaller values of the apparent rotational component than those first obtained", though "the persistence of the positive sign is very marked", thus requiring further investigations [78].

12. "Hubble's Law"

In 1929 Hubble stated what came to be called *Hubble's law*, namely, that redshifts and distances of galaxies are linearly related: $z = Hd$ (note that the observational counterpart of 'distance' here is 'apparent magnitude'). Several authors have pointed out that Hubble's observations suggested a quadratic rather than a linear relation (cf. [28, p. 378]). That the functional dependence had to be linear was surely inspired by de Sitter's model, and in the first place by his allegiance to general relativity, as confirmed by the following passage in his main article:

> The outstanding feature, however, is the possibility that the velocity-distance relation may represent the de Sitter effect, and hence that numerical data may be introduced into discussions of the general curvature of space [47].

Note that among the relativistic expanding universes the so-called Hubble's 'constant' is a *true* constant *only* in the case of de Sitter space–time: in the other FLRW space–times (except for the Einstein static space–time, of course, where it vanishes)[21] H is in fact a function of cosmic time. This strengthens the case for Hubble's 'reading his expectations into his data'. Nonetheless, it must be stressed that Hubble, to the end of his life [49], did not endorse the view that his discovery was decisive evidence for the expanding universe cosmology.

In the case of zero pressure and zero cosmological constant, one defines the *critical density* ρ_c as

$$\rho_c = \frac{3H^2}{\kappa c^4},$$

and so the sign of the curvature of the spatial sections is connected with the value of the Hubble's constant by the equation:

$$K = \frac{\kappa c^2 R^2}{3}(\rho - \rho_c) = \frac{\kappa c^2 R^2 \rho_c}{3}(\Omega - 1).$$

where $\Omega = \rho/\rho_c$ is called the *density parameter*. This shows the importance of evaluating H in order to determine at one stroke (under the given hypotheses) both the universe's geometry and its destiny. The positive K corresponds to $\Omega > 1$ and to a closed universe whose fate is a final collapse; the other two possibilities ($K = -1$, that is, $\Omega < 1$; and $K = 0$, that is, $\Omega = 1$) correspond to a higher or lower (respectively) rate of expansion.

It is however worth noting that the observational differences in the predictions of the three models ($K = 0$, $K = 1$, $K = -1$) are not remarkable: "all three types

---

[21] Notice that this is not true in the interpretation of Einstein's space–time by I. Segal in his chronometric cosmology; cf. Chap. 13.

make surprisingly similar predictions rather than grossly different ones as the trichotomy suggests", as the authors of a useful textbook in general relativity wrote in 1977 [72, p. 177].[22]

13. The Einstein–de Sitter Space–Time and the Demise of λ

It was in 1931 that Einstein abandoned for the first time the cosmological constant, in a paper where he endorsed an oscillatory model. In 1932 Einstein and de Sitter published a short joint paper, where they described a special case of Friedmann's solutions, the one with zero cosmological constant, zero spatial curvature and zero pressure. (The *Einstein–de Sitter metric*, as was to be named, had not been discussed by Friedmann, probably because his main concern was with non-Euclidean spatial geometries; however, it had been discussed in 1929 by Robertson.) The radius of the universe has the form: $R(t) = R(t_0)(t/t_0)^{2/3}$, and the space–time is topologically like $\mathbb{R}^4$. A computation shows that the scalar curvature of this space–time is $S(t) = 4/(3t^2)$. This implies that all the fundamental observers, for which t is the proper time, have a finite past: t cannot go past zero, because at zero the scalar curvature blows up to infinity. This is an example of a genuine space–time singularity, which can be physically interpreted as describing a universe expanding from a state of infinite density. In Sec. 17 we shall see how it was discovered that this property is not the awkward outcome of too special symmetry assumptions.

In their paper Einstein and de Sitter put $\lambda = 0$, but they pointed out that "an increase in the precision of data derived from observations will enable us in the future to fix its [i.e. of λ] sign and to determine its value" [26]. *This is precisely the same as Friedmann's position* (§10). According to George Gamow, Einstein once told him that the introduction of the cosmological constant had been "the greatest blunder of my life"; perhaps, talking with Gamow, who had been Friedmann's student, he was thinking also of the mathematical mistake connected with his temporary rejection of Friedmann's work (§9). It is true, however, that by 1919 Einstein had written that the λ-term was "gravely detrimental to the formal beauty of the theory" [19].

These facts may perhaps serve to explain, in part, the following amusing story by Eddington:

> Einstein came to stay with me shortly afterwards, and I took him to task about [the dropping of the Λ in [26]]. He replied: "I did not think the paper was important myself, but de Sitter was keen on it". Just after Einstein had gone, de Sitter wrote to me announcing to visit. He added: "You will have seen the paper by Einstein and myself. I do not myself consider the result of much importance, but Einstein seemed to think that it was" [cit. in [3, p. 123]].

In any case, the question of the cosmological constant cannot be dismissed so easily, as the following remarks show (more will be said in §18).

The modified field Eqs (7) can be re-written, equivalently, in the form (note the sign of λ):

---

[22]A sustained argument in favour of an open universe (with $\Omega \approx 0.2$) was presented 20 years later by P. Coles and G.F.R. Ellis; the authors admit that "the case [they] have made [...] though strong is by no means watertight" [6, p. 204].

$$R_{\mu\nu} - \frac{1}{2}Sg_{\mu\nu} + \lambda g_{\mu\nu} = -\kappa T_{\mu\nu}.$$

Now there is a theorem, proved by Hermann Vermeil in 1917 and included by Hermann Weyl in Appendix II of [79], that states that the only rank 2 symmetric tensor, containing the $g_{\mu\nu}$ and its first and second derivatives, and furthermore linear in the second derivatives of the $g_{\mu\nu}$, is of the form:

$$E_{\mu\nu} = c_1 R_{\mu\nu} + c_2 Sg_{\mu\nu} + c_3 g_{\mu\nu}.$$

If one requires, in addition, that $E_{\mu\nu}$ has zero divergence (that is, $g^{\lambda\nu}\nabla_\lambda E_{\mu\nu} = 0$), then

$$E_{\mu\nu} = c_1\left(R_{\mu\nu} - \frac{1}{2}Sg_{\mu\nu} + \lambda g_{\mu\nu}\right).$$

From this point of view it is clearly the *a priori* decision of neglecting λ, rather than that of including it, that needs a justification. The curious thing is that Einstein referred to a 'Vermeil-like' theorem in his 1916 outline [15, p. 144] – he did it before the theorem had been proved, and his statement there is mistaken precisely because it omits the c_3 term! It is true that this omission can be justified *a posteriori* by appeal to the Newtonian limit (§4); nonetheless the popular perception of the λ-term as a 'blunder' is certainly off the mark.[23]

It is interesting that Eddington and Lemaître both felt very strongly about the cosmological constant; the former wrote: "I would as soon think of reverting to Newtonian theory as of dropping the cosmical constant"; and the latter, when asked what had been the most important conceptual novelty introduced by general relativity, answered "without a moment's hesitation [...] 'the introduction of the cosmical constant' " [3, p. 122].

Eddington commented in 1933 that he had difficulty in discussing the proposal of Einstein and de Sitter, since he did

> not see what are 'the rules of the game'. These proposals are left as mathematical formulations, all doubtless compatible with what we observe; but there seems nothing to prevent such formulations being indefinitely multiplied [13, p. 63].

As to the singular initial state, Eddington found it unacceptable: as he wrote in 1928: "As a scientist I simply do not believe that the Universe began with a bang" – little or big, we are tempted to add [53, p. 46]. He showed in 1930 [12] that the Einstein space–time is unstable, because if it happens that ρ at some instant is different from $2\lambda/\kappa c^2$ (see §4), then from that moment on the universe will either be expanding or contracting (this can be seen as an easy consequence of (24)–(25)). Eddington wrote:

> I may mention that the proof of the instability of the Einstein configuration was the turning point in my own outlook. Previously the expanding universe (as it appeared in de Sitter's theory) had appealed to me as a highly interesting possibility, but I had no particular preference for it [13, p. 60n].

His preference was in fact for a nonsingular, spatially spherical universe asymptotically arising from Einstein static space–time and developing into de Sitter space–time (the so-called *Lemaître-Eddington model*: $K = 1$, $\lambda = \lambda_c$, with pressure). Eddington

---

[23]For a short history of the cosmological constant, and its recent resurrection, the note by G.F.R. Ellis [29] should be consulted.

was convinced that it was the only reasonable cosmological model available, particularly because it did not conflict with geological time estimates.[24]

Lemaître, on the other hand, gave his preference in 1931 to a singular model starting with a "primeval atom" – a first version of what is today called the "standard" cosmology.

14. Neo-Newtonian Cosmology

As we have seen, the cosmological discourse has been framed into Newtonian categories since the beginning of relativistic cosmology.

Apparently Lemaître was the first to notice, in 1931, that elementary considerations of Newtonian mechanics made it possible to derive Friedmann's Eq. (26) [34, pp. 221–2].

Three years later, the British mathematical physicists Edward Milne and William McCrea showed in detail that the Friedmann solutions are the relativistic counterpart of the possible evolutions of an homogeneous and isotropic fluid (solid) sphere kept together by gravitation only. In particular the Friedmann equations can be derived from the dynamical and the continuity equations for this fluid.

Since Newtonian space is Euclidean, the curvature of Friedmann's solutions must undergo a radical reinterpretation: the three alternatives correspond to the sphere expanding forever with exactly ($K = 0$) or more ($K = -1$) than the escape velocity, or collapsing on itself ($K = 1$). The cosmological constant can be interpreted in Newtonian terms as an additional "repulsive force proportional to distance", and Hubble's law is also valid.

Milne [60] wrote:

> It seems to have escaped previous notice that whereas the theory of the expanding universe is generally held to be one of the fruits of the theory of relativity, actually all the at-present-observable phenomena could have been predicted by the founders of mathematical hydrodynamics in the eighteenth century, or even by Newton himself.

Of course the fundamental observers in the Newtonian version cannot be fully equivalent, since they are accelerated observers, with different accelerations, so Milne's claim does not exactly hold, unless we correct Newton's physics by introducing the equivalence principle; moreover, a finite sphere has a centre and a boundary, so its points do not enjoy exactly the same view! The main interest in this revival of Newtonianism lies, apart from its paedagogical value, in the way it showed how close general relativity had come, in dealing with the cosmological problem, to classical mechanics.[25]

15. The Steady-State Theory

In 1948 two papers appeared advancing a cosmological theory which deviated from the orthodox line of using general relativity as the basis for cosmological speculation. Their authors, Hermann Bondi and Thomas Gold on one side, Fred Hoyle on the other, had

---

[24]"In Eddington's time most cosmological models except his own implied an 'age of the universe' less than 2×10^9 years. This was less than half the age of the oldest rocks on Earth!" (W. McCrea, in [13, p. xviii]).

[25]For a recent treatment see [74, pp. 18–28], where it is explained how one can circumvent the difficulties mentioned in the text.

extensively discussed the issue previously, so their convergence was not a case of independent discovery.[26]

Bondi and Gold adopted a simple deductive approach, assuming the relativistic space–time but not the field equations – that is, the *geometric* component of general relativity without the *dynamical* component. The main feature was the postulation of the so-called *perfect cosmological principle*, stating that the universe has to look the same not only at each point (homogeneity) and in each direction (isotropy), as in the FLRW models, but *also for each time*. Of course this principle would also be satisfied by Einstein's static universe, but the strength of the new theory was that it took into account, convincingly, those observational data that had defeated Einstein's universe, first of all the accepted redshift-distance relationship for galaxies.

Starting from (21), which is needed in order to comply with the ordinary cosmological principle, the further determination of $R(t)$ and $d\sigma^2$ is effected by exploiting the 'perfect' clause. In fact the curvature of the universe at time t is $K(t) = k/R(t)^2$, where $k = -1, 0, 1$, so for this (observable) quantity to be independent of time the only possibility (if we reject Einstein space–time) is $k = 0$, that is, the universe must be Euclidean:

$$ds^2 = -R(t)^2\left(\left(dx^1\right)^2 + \left(dx^2\right)^2 + \left(dx^3\right)^2\right) + c^2\,dt^2.$$

Moreover, the proportionality factor between velocity (observable as redshift) and distance of galaxies is given in such a universe by the Hubble constant $H(t) = \dot{R}(t)/R(t)$. The perfect cosmological principle requires that $H(t)$ be constant, say $H(t) \equiv H_0$, giving

$$R(t) = R(0)e^{H_0 t}.$$

So, if we posit $R(0) = 1$, the final expression of the metric is:

$$ds^2 = -e^{2H_0 t}\left(\left(dx^1\right)^2 + \left(dx^2\right)^2 + \left(dx^3\right)^2\right) + c^2\,dt^2, \tag{29}$$

which coincides with the metric of de Sitter space–time as displayed in (18) if $R_0 = c/H_0$. In fact this last condition is not completely a matter of notation, since it can only hold if $H_0 > 0$. A further attractive feature of the steady-state approach is that it enables one to derive this inequality – that is, that the universe is expanding – from two established astronomical facts of the most elementary nature, namely: (1) that the universe is not in thermodynamic equilibrium, and (2) that the sky at night is dark.[27]

Of course an infinite and uniform universe which is forever expanding must have one disturbing peculiarity: new matter must be created continuously, in order for the matter density to remain constant (this is not required in FLRW space–times, where the matter density is time-dependent, cf. (27)). In other words, the steady-state cosmology had to introduce a systematic violation of the local energy-mass conservation (cf. Chap. 5, §13). It is true that the quantity of created matter needed is very small, nonetheless to many physicists this seemed rather hard to swallow.

As we have seen, the derivation of de Sitter metric in the steady-state approach is obtained without using Einstein's field equations; as a consequence the steady-state uni-

---

[26] It seems that Gold was the originator of the basic idea [53, pp. 173–9].

[27] Point (2) is enough to dismiss a static and eternal *Euclidean* universe if, furthermore, stars are assumed to be uniformly distributed (*Olbers paradox*).

verse must not be thought as empty (or filled with exotic matter), and in fact it can accommodate *any* constant mass density ρ_0. The weak side of this freedom is that, contrary to the FLRW models, there is no link between H and ρ (§11). This also depends on the fact that no alternative field equations describing the matter creation process were presented by Bondi and Gold.

It was Hoyle's paper which filled this gap. Hoyle introduced a scalar field C (that is, a real-valued function on space–time) depending only on the time coordinate, and suggested the following modification of Einstein's equations:

$$R_{\mu\nu} - \frac{1}{2}Sg_{\mu\nu} = -\kappa(T_{\mu\nu} + C_{\mu\nu}), \tag{30}$$

where $C_{\mu\nu}$ is the covariant second derivative of C (i.e. $C_{\mu\nu} = \nabla_\nu C_\mu$). It can be verified that if $T_{\mu\nu}$ is a matter tensor for a dust of density ρ and C is proportional to t, then (30) has the de Sitter metric as a solution, with $T_{\mu\nu} \neq 0$.

The steady-state theory was on several occasions defended by Bondi on methodological grounds, since it was allegedly *more* open to falsification, thus more scientific (in Popperian terms), than its rivals. The implication was that one had to prefer, at least provisionally, the steady-state theory because it was more at risk of being in conflict with observations. Given the scarcity of observational evidence at the time, the argument looked disquietingly close to claiming support for the theory from non-existent data.[28] Interestingly, Popper himself never endorsed the steady-state theory, of which he disliked the very cornerstone, the perfect cosmological principle; he also came to a rather poor opinion of the scientific status of physical cosmology as a whole [53, pp. 224–6].

Hoyle's line was different. He stressed the way the steady-state theory, with his field equations, allowed the avoidance of ad hoc initial conditions to satisfy Mach's principle; he and his collaborator Jayant V. Narlikar [44] argued that *any* solution of (30) tended asymptotically to the de Sitter universe, thus guaranteeing the link between the space–time metric and matter which is at the core of Mach's principle, and which is lost in the FLRW cosmology, as will be even more clear from the next section.

16. The Gödel's Universes

At the end of the 1940s the great logician Kurt Gödel started working on general relativity. The results of this inquiry were outlined in three short papers [36–38], which are among the most important in the history of general relativity.

As we have seen, the cosmological solutions of the field equations so far examined share two properties which are clearly inspired by the 'old' Newtonian setting: they have a privileged time coordinate $x^4 = t$, and matter defines a class of privileged, freely-falling observers (modelling the galaxies); the worldlines of these observers have a velocity vector $u^\mu = (0, 0, 0, 1)$ which is orthogonal to the level hypersurfaces $t = const.$ In order for these conditions to hold for a metric $ds^2 = g_{\mu\nu}\, dx^\mu\, dx^\nu$ one must have

---

[28]The worthy side of the argument, that is, that the bolder theories should be given more attention since their very failures are likely to "throw light upon the state of contemporary science and may indicate where it requires supplementing", was of course not a very original idea (the quotation is by mathematician H.T.H. Piaggio, quoted approvingly by Eddington [13, p. 124]).

$$g_{a4} = 0, \qquad g_{44} = const.$$

The first condition ensures that matter is not rotating. A universe where matter has large-scale rotation clearly violates Mach's principle (cf. Chap. 5, §1).

In his first article Gödel presented explicitly a new exact solution of the modified field equations, with matter in the form of a dust, which does not admit any cosmic time and where matter (in the form of a dust with constant density) rotates. The metric, which can be considered as defined on all of $\mathbb{R}^4$, has a simple form:

$$ds^2 = a^2 \left[-\left(dx^1\right)^2 - \frac{e^{2x^1}}{2}\left(dx^2\right)^2 - \left(dx^3\right)^2 + \left(e^{x^1}dx^2 + dx^4\right)^2 \right], \tag{31}$$

and is a solution of (7) with a negative cosmological constant. It is easy to see that the fourth coordinate vector field is timelike; however, the fourth coordinate is not a cosmic time, i.e. a scalar function on space–time which is increasing on every future-pointing timelike curve. In fact this space–time does not admit *any* cosmic time; we shall see why after a short detour.

Gödel's model enjoys a curious property, which may be introduced by quoting a passage from Weyl's treatise [79, p. 274]:

> [...] it is not impossible for a world-line (in particular, that of my body), although it has a time-like direction at every point, to return to the neighbourhood of a point which it has already once passed through. The result would be a spectral image of the world, more fearful than anything the weird fantasy of E.T.A. Hoffmann has ever conjured up. In actual fact the very considerable fluctuations of the $[g_{\mu\nu}]$ that would be necessary to produce this effect do not occur in the region of world in which we live.

Weyl went on to remark that "there is a certain amount of interest in speculating on these possibilities inasmuch as they shed light on the philosophical problem of cosmic and phenomenal time". This is in fact what had prompted Gödel to work out his solution, and in his paper he proved that his space–time does contain closed timelike curves; indeed there are closed time-like curves through *every* point![29] This is interesting both in itself, and because it clearly shows that in such a a space–time there can be no cosmic time function. Moreover, this confirms that the mere form of the stress-energy tensor does not dispense in general relativity from appeal to specifically cosmological assumptions if one is to arrive at the FLRW space–times.

A further reason of interest is that Gödel space–time is simply-connected; this is important because it is rather easy to construct space–times with closed timelike curves (or even geodesics) if non-simply-connected topologies are allowed (e.g. one can identify points in Minkowski space–time).

Finally, as Gödel writes, his solution shows that, according to general relativity, "it is theoretically possible [...] to travel into the past, or otherwise influence the past" [36, p. 447]. Today speculation on 'time travel' is thriving, and the boundary between science and science fiction is becoming increasingly difficult to discern [76,62].

---

[29]Though not closed *geodesics*. Incidentally, in his short article Gödel never claimed that, but two authors, one being S. Chandrasekhar, 'corrected' him in 1954 of this supposed 'mistake' ... What is more and worse, it seems that it took *nine* years before someone pointed out in print that Gödel was innocent of the charge! See [64].

The second paper, a very dense one, extracted a philosophical moral from the existence of such solutions. That time is not an objective entity out there has been argued by idealistic philosophers; the relativity of the time order in special relativity seemed to confirm this claim, but now general-relativistic models of the universe have reinstated a sort of absolute time under the name of cosmic time.[30] So one might argue that *even according to general relativity*, time cannot be denied an objective existence, at least on a cosmological scale. But Gödel's solutions show that a cosmology *without* cosmic time is perfectly compatible with general relativity. So after all there is still a remarkable agreement between the idealistic concept of time and modern physics: the inexorable, unique, one-way flowing of time may still be considered as a subjective illusion.[31]

Note that (31) is a stationary space–time, thus there are no cosmological redshifts in it. In his third paper Gödel stated, with little or no proofs, that rotating and non-static universes, spatially homogeneous and with closed timelike curves, exist. This seminal paper [38] was delivered as a communication at the 1950 International Congress of Mathematics, and over the next two decades it was a major impulse to research into exact solutions and topological and causality problems in general relativity.

17. The Big-Bang Cosmology 'Wins'

The discussion on the relative virtues of the steady-state cosmology and the evolutionary models – the "Big Bang" cosmology, as it was named after a scornful remark in 1950 by Fred Hoyle – during the decade starting in the mid-Fifties increasingly concentrated on astronomical issues. The most relevant results, which seemed to favour the evolutionary models, were the radio-sources counts, the redshifts of the quasars, and, perhaps most importantly, the black-body microwave radiation accidentally discovered by Arno Penzias and Robert Wilson in 1964, and interpreted as a relic of the 'big explosion' by Robert Dicke and collaborators.

One of the reasons of the success of this theory – favourably mentioned by Pope Pius XII in 1951 [53, p. 256] – was the possibility of marrying particle physics with cosmology when dealing with the first few instants of the universe after the Big Bang. The first paper discussing nucleosynthesis in the initial 'hot' stage of the universe was published by Georg Gamow in 1946.[32]

Not all competent scientists, however, have been convinced by these results that the Big Bang is the only acceptable picture of the universe (cf. also Chap. 13). In a lecture at the International Centre for Theoretical Physics (Trieste, Italy), in June 1968, Paul A.M. Dirac had this to say on cosmology [9, pp. 127–8]:

---

[30]Gödel quotes a lecture by James Jeans in 1935. In 1943 Jeans still wrote: "The hypothesis that absolute time and space do not exist brings order into man-sized physics, but seems so far to have brought something like chaos into astronomy. Thus there is some chance that the hypothesis may not be true. [...] It may be that before it can make sense, the new astronomy must find a way of determining an absolute time, which will then describe as cosmical time. The space–time unity will then be divided into space and time separately by nature itself" [51, pp. 67, 68].

[31]Einstein's reply ended in dubitative form: "It will be interesting to weigh whether these [Gödel's solutions] are not to be excluded on physical grounds" [73, p. 688].

[32]See Gamow's later survey article [35].

One field of work in which there has been too much speculation is cosmology. There are very few hard facts to go on, but theoretical workers have been busy constructing various models for the universe, based on any assumptions that they fancy. These models are probably all wrong. It is usually assumed that the laws of nature have always been the same as they are now. There is no justification for this. The laws may be changing, and in particular quantities which are considered to be constants of nature may be varying with cosmological time. Such variations would completely upset the model makers.

As to the supposedly 'clinching' evidence of the cosmic background radiation, this was Fred Hoyle's comment thirty years after the discovery:

How, in big-bang cosmology, is the microwave backgroud explained? Despite what supporters of big-bang cosmology claim, it is not explained. The supposed explanation is nothing but an entry in the gardener's catalogue of hypotheses that constitutes the theory. Had observation given 27 kelvins instead of 2.7 kelvins for the temperature, then 27 kelvins would have been entered in the catalogue. Or 0.27 kelvin. Or anything at all [43, p. 413].

In fact, apart from these methodological criticisms, the big-bang cosmology has to face a number of serious difficulties; some of them will be briefly mentioned in the next section.

18. Singularities and Other Open Problems

Classically, if in a theory some physical quantity becomes infinite for some value of a given parameter, this is considered a symptom that the theory reaches its limits of application near that value. In fact, since no real-world physical quantity is infinite, there must be something defective in the theory. Now one common feature of the expanding cosmological models with $\lambda = 0$ is that they all have one or two finite values of the cosmic time for which the metric degenerates. As we have seen (§6), Einstein's criterion for a singularity was the vanishing of $g = \det(g_{\mu\nu})$. In fact the problem of how to define a singularity in general relativity proved to be more tricky.

A common view about why singularities arise in the FLRW cosmological models was that they embodied unrealistic symmetry assumptions. A first step to remove this doubt was made in 1955, when an Indian theoretical physicist, Amalkumar Raychaudhuri succeeded (not very easily, in fact) in having a paper published in *Physical Review* in which he questioned this opinion [68]. Suppose that in a relativistic space–time, with matter modelled as a dust of density ρ, we have a coordinate system where the x^4-coordinate lines are geodesics which coincide with the worldlines of the particles of matter, for which $x^4 = t$ is the proper time. Then along any such worldline, say Γ, the following equation holds:

$$\frac{1}{G}\frac{\partial^2 G}{\partial t^2} = \frac{1}{3}\left(c^2\lambda - \frac{\kappa c^4 \rho}{2} - \phi^2 + 2\omega^2\right), \tag{32}$$

where $G = |g|^{1/6}$, and ω and ϕ are functions such that ω vanishes if and only if the matter worldlines are hypersurface-orthogonal, and ϕ vanishes if matter expands isotropically. From this equation it follows that if $\omega = 0$ and $\lambda \leq 0$, then G has a negative second derivative with respect to t, and this implies that at some time in the past G, and therefore g, has to vanish. Raychaudhuri's conclusion was that there was a "singularity at a finite time in the past as in isotropic models. Thus a simple change-over to anisotropy does not solve any of the difficulties" [68, p. 1125].

The argument is ingenious and correct insofar as it suggests that anisotropy is not the key, but it is inconclusive. The vanishing of g may well depend on the limits of validity of the coordinate system; for instance, it may be easily shown to occur for a suitable family of geodesic observers in (empty!) Minkowski space–time. What should be proven is that something *really* wrong, that is, something not coordinate-dependent, occurs on some timelike (or lightlike) geodesic. The simplest criterion for saying that something wrong is occurring on a timelike or lightlike geodesic is when its natural parameter (proper time, in the timelike case) cannot be extended beyond a certain finite value, either in the future or in the past. If not all inextendible timelike (resp. lightlike) geodesics are *complete* (i.e. their natural parameter ranges over all the real line), then space–time is said to be *timelike* (resp. *lightlike) incomplete*. A space–time which is incomplete in this sense and is not just a subspace of another, bigger space–time, is certainly a good candidate to be classified as genuinely *singular*.

In the mid-1960s a number of theorems were proven – by Roger Penrose, Stephen W. Hawking, Robert Geroch and others – stating that a space–time satisfying some rather broad global conditions (implying the non-existence of closed timelike curves, for instance) is singular in the sense just explained. Versions of (32) were a basic ingredient of the proofs. It is interesting to remember that for Einstein (cf. §5) the existence of a singularity in *his* sense was enough to rule out a space–time as a legitimate solution of his equations. What the singularity theorems proved was, roughly, that in relativistic cosmology singular space–times were *not* exceptional.[33]

The natural interpretation of the singularity theorem was endorsed by Hawking and George F.R. Ellis in their classic treatise *The Large-Scale Structure of Space-Time*, published in 1973:

> It seems to be a good principle that the prediction of a singularity by a physical theory indicates that the theory has broken down, i.e. it no longer provides a correct description of observations. The question is: when does General Relativity break down? [40, pp. 362–3].

Hawking and Ellis's guess is that "a breakdown occurs for lengths between 10^{-15} and 10^{-33} cm". They go on to connect the theorems with the big-bang picture:

> In any case, the singularity theorems indicate that the General Theory of Relativity predicts that gravitational fields should become extremely large. That this happened in the past is supported by the existence and black body character of the microwave background radiation, since this suggests that the universe had a very hot dense early phase.

Here "the General Theory of Relativity" is to be understood with the proviso that the cosmological term is forbidden; otherwise the singularity theorems might be seen as offering a good reason to preserve it, in the spirit of the Eddington–Lemaître space–time (§13). It is true that there are reasons suggesting that the reintroduction of the λ-term would not be enough to avoid a singularity, at least if the accepted theory of nucleosynthesis is maintained [29].

In any case, the cosmological term has other chances for a resurrection. In 1967 Russiam physicist Yakov Zel'dovich showed that in quantum field theory the stress-energy

---

[33]The survey [75] gives an interesting and useful discussion (pp. 796–808) of the scope and limits of the singularity theorems.

tensor of the vacuum behaves like a cosmological constant. The idea reappeared in 1980 in an influential variant of big-bang cosmology at the heart of which is the postulate that, very shortly after the big-bang (about 10^{-30} seconds later), there was a period of "inflation", that is, of very fast expansion. There are about 60 versions of "inflationary comology" on offer, to date; what is worse, the theoretical energy density of the quantum vacuum corresponds to a λ which is different from the one compatible with astronomical observations by 120 orders of magnitude (according to one estimate): this is "one of the major unsolved problems of theoretical physics" [29].

In 1990 Hawking summed up the state of the art in cosmology as regards another big open problem of standard cosmology, that of "dark matter":

> So there are two possibilities; either our understanding of the very early Universe is completely wrong, or there is some other form of matter in the Universe that we have failed to detect. The second possibility seems more likely, *but the required amount of missing 'dark' matter is enormous; it is about a hundred times the matter we can directly observe* [70, p. viii].

A rather curious situation, which will not fail to provide in the future years food for thought both for scientists and lay observers.

19. Conclusive Remarks

Cosmology is a peculiar kind of science (if it is a science at all) for the most obvious of reasons, namely that it is the study of a class of objects of which there is just one instance. In a sense it can be compared to geology or natural history, except for the worrisome circumstance that its object – space–time – is a theoretical construct in a stronger sense than can be said of the Earth or of the extinct life forms which we try to reconstruct from their relics. Though by now many more observations have been gathered than Einstein and his contemporaries could have dreamt of, cosmology cannot be but an observational, rather than an experimental, science, and moreover one whose empirical basis cannot be made sense of "without some admixture of ideology" [40, p. 134]. *"There are no purely observational facts about the heavenly bodies"*, as Eddington wisely wrote in 1933 ([13, p. 17]; the italics are his).

One example is given by the cosmological principles. The standard one (§3) is often defended on grounds of Copernican humility; here is how the author of one of the classic treatises on cosmology, Richard C. Tolman, put it: "[...] it avoids the anthopocentric assignment of a unique importance to our own location in the universe", though he immediately added the somewhat inconsistent remark that it enables us to regard "the observations that we obtain as fairly representing the character of those which would be obtained from similar locations in other portions of the universe" [77, p. 363]. In other words, our position in the universe lacks "importance" only in some indefinite sense of the word, but is otherwise *very* good, since it allegedly permits us to use local observations for global purposes. The net outcome is clearly one of ('Copernican'?) scientific optimism.[34]

General relativity has surely enlarged the conceptual tool-box of the cosmologists, enabling them to discuss in mathematical terms a bewildering variety of more or less

---

[34] In fairness to Tolman, it must be added that a few lines below he writes that it is "most important" to emphasize that "the assumption is to be regarded merely as a working hypothesis".

well-behaved or weird possible worlds. However, it must be emphasized that even in this largest of the physical systems – the universe – relativistic cosmology has consistently relied upon Newtonian analogies. Gödel's contributions arose as a reaction to the Newtonian dominance which he saw, correctly, as entrenching upon the philosophical interpretation of cosmological models. It still remains to be seen whether the strange possibilities opened by general relativity and revealed by Gödel and others after him should be rated as a strength or a weakness of the theory.

Of all fields in which an orthodoxy can coagulate, cosmology would seem to be one of the most unlikely, given the fragility of both theory and evidence; nevertheless, even here this has happened. (Religion might be judged, on the same grounds, an even more unlikely candidate, and yet . . .). The phenomenon has many reasons, like the strongly hierarchical structure of the scientific community, and (not least) the sorry tendency of scientific professionals in most fields today to overstate their case in order to seduce the public into supporting increasingly expensive research programmes. However, inflated claims may eventually backfire and cause discredit both to the individuals and to their disciplines [10].

In the history of cosmology, as in other scientific fields, we see time and again that what receives the most attention is not necessarily what is correct or even promising, but what is in agreement with the opinions, and even errors, of the scientific leaders. 'Authoritative' ideas can resist correction for decades, the more so – obviously – if there are very few people who do most of the fundamental work in the area.[35] Moreover, even competent outsiders (like Friedmann) may simply be ignored unless they succeed in connecting (like Lemaître) with the establishment. Fortunately, if we count arguments rather than heads, as we always should in all matters scientific, cosmology has never stopped being an healthily controversial field, and strictures on dissidents have never succeeded in completely stifling the debate even on the most fundamental points (cf. Chap. 13).

One last remark is in order. Substantial progress in cosmology, more than elsewhere, is a long term business. It will ultimately depend on how long humankind will resist the destructive forces which threaten to put an untimely end to its adventure. This is a most telling instance to show that scientific research – if it is meant to be more than a way of channeling one's competitive drive or of gaining material assets – cannot be separated from humanitarian concerns and participation in general cultural and political issues.

References

[1] Besse A. 1987: *Einstein Manifolds*, New York – Heidelberg – Berlin, Springer Verlag.

[2] Bondi H., Gold T. 1948: "The Steady State Theory of the Expanding Universe", *Monthly Notices of the Royal Astronomical Society*, **108**, pp. 252–70.

[3] Chandrasekhar S. 1987: *Truth and Beauty*, Chicago – London, The University of Chicago Press.

[4] Clifford W. K. 1955: *The Common Sense of the Exact Sciences*, New York, Dover.

---

[35]Cf.: "On various occasions in the history of cosmology the subject has been dominated by the bandwagon effect, that is, strongly held beliefs have been widely held because they were unquestioned or fashionable, rather than because they were supported by evidence. As a result, particular theories have sometimes dominated the discussion while more convincing explanations were missed or neglected for a substantial time, even though the basis for their understanding was already present" [28, p. 367]. There is plenty of evidence to show that cosmology is not exceptional in this respect.

[5] Cohen I. B. (ed.) 1978: *Isaac Newton's papers & Letters on Natural Philosophy and related documents*, 2nd edn, Cambridge (Mass.) and London, Harvard University Press.

[6] Coles P., Ellis G. F. R. 1997: *Is the Universe Open or Closed*, Cambridge University Press.

[7] Crowe M. 1994: *Modern Theories of the Universe from Herschel to Hubble*, New York, Dover.

[8] de Sitter W. 1917: "On the relativity of inertia", *Proc. Kon. Akad. Wetensch. Amst.*, **19**, pp. 1217–25.

[9] Dirac P. A. M. 1968: "Methods in theoretical physics", pp. 125–43 of *Unification of Fundamental Forces*, Cambridge University Press, 1990.

[10] Disney M. J. 2000: "The Case Against Cosmology", *General Relativity and Gravitation*, **32**, pp. 1125–34.

[11] Eddington A. S. 1918: *Report on the Relativity Theory of Gravitation*, London, Fleetway Press.

[12] Eddington A. S. 1930: "On the instability of Einstein's spherical world", *Monthly Not. Roy. Astr. Soc.*, **91**, pp. 412–6.

[13] Eddington A. S. 1933: *The Expanding Universe*, Reprint with a preface (1987) by W. McCrea, Cambridge University Press.

[14] Egginton W. 1999: "On Dante, Hyperspheres, and the Curvature of the Medieval Cosmos", *J. Hist. Ideas*, **60**, pp. 195–216.

[15] Einstein A. 1916: "Grundlage der allgemeine Relativitätstheorie", *Ann. Phys.*, **49**, pp. 769–822.

[16] Einstein A. 1917: *Über die spezielle und die allgemeine Relativitätstheorie, Gemeinverständlich*, Braunschweig, Vieweg.

[17] Einstein A. 1917: "Kosmologische Betrachtungen zur allgemeinen Relativitätstheorie", *Preussische Akademie der Wissenschaften, Sitzungsberichte*, P. 1, pp. 142–52.

[18] Einstein A. 1918: "Kritisches zu einer von Hrn. De Sitter gegebenen Lösung der Gravitationsgleichungen", *Preussische Akademie der Wissenschaften, Sitzungsberichte*, pp. 270–2.

[19] Einstein A. 1919: "Spielen Gravitationsfelder im Aufber der materiellen Elementarteilchen eine wesentliche Rolle?", *Königlich Preussische Akademie der Wissenschaften* (Berlin), *Sitzungsberichte*, pt 1, pp. 349–56 (transl. in Einstein et al. 1932, pp. 189–98).

[20] Einstein A. 1922: *The Meaning of Relativity*, Princeton, Princeton University Press.

[21] Einstein A. 1922: "Bemerkung zu der Arbeit von A. Friedmann 'Über die Krümmung des Raumes' ", *Zeitschrift für Physik*, **21**, p. 326.

[22] Einstein A. 1923: "Notiz zu der Arbeit von A. Friedmann 'Über die Krümmung des Raumes' ", *Zeitschrift für Physik*, p. 228.

[23] Einstein A. 1931: "Zum kosmologischen Problem der allgemeinen Relativitätstheorie", *Preussische Akademie der Wissenschaften, Sitzungsberichte*, pp. 235–7.

[24] Einstein A. 1954: *Relativity. The Special and the General Theory*, 15th ed., Routledge, London and New York.

[25] Einstein A. 1994: *Ideas and Opinions* [1954], New York, The Modern Library.

[26] Einstein A., De Sitter W. 1932: "On the relation between the expansion and the mean density of the universe", *Proc. Nat. Acad. Sci.*, **18**, pp. 213–4.

[27] Einstein A., Lorentz A. A., Weyl H., Minkowski H. 1932: *The Principle of Relativity*, New York, Dover.

[28] Ellis G. F. R. 1989: "The Expanding Universe: A History of Cosmology from 1917 to 1960", pp. 367–431 of *Einstein and the History of General Relativity*, ed. by D. Howard and J. Stachel, Basel, Birkhäuser.

[29] Ellis G. F. R. 2003: "A historical review of how the cosmological constant has fared in general relativity and cosmology", *Chaos, Solitons and Fractals*, **16**, pp. 505–12.

[30] Engel A., Shucking E. (eds.) 1997: *The Collected Papers of Albert Einstein, vol. 7, The Berlin Years: Writings, 1914–1917*, English Translation, Princeton-Oxford, Princeton University Press.

[31] Engel A., Shucking E. (eds.) 2002: *The Collected Papers of Albert Einstein, vol. 7, The Berlin Years: Writings, 1918–1921*, English Translation, Princeton-Oxford, Princeton University Press.

[32] Friedmann A. A. 1922: "Über die Krümmung des Raumes", *Zeitschrift für Physik*, **10**, pp. 377–87. English translation in *General Relativity and Gravitation*, **31** (1999), pp. 1991–2000.

[33] Friedmann A. A. 1923: "Über die Möglichkeit einer Welt mit konstanter negativer Krümmung des Raumes", *Zeitschrift für Physik*, **21**, pp. 326–33. English translation in *General Relativity and Gravitation*, **31** (1999), pp. 2001–8.

[34] Friedmann A. A., Lemaître G. 1997: *Essais de Cosmologie*, with an introduction by J.-P. Luminet, Paris, Seuil.

[35] Gamow G. 1949: "On Relativistic Cosmogony", *Rev. Mod. Phys.*, **21**, pp. 367–73.

[36] Gödel K. 1949: "An Example of a New Type of Cosmological Solutions of Einstein's Field Equations of Gravitation", *review of Modern Physics*, **21**, pp. 447–50.

[37] Gödel K. 1949: "A remark about the relationship between relativity theory and idealistic philosophy", pp. 557–62 of [73].

[38] Gödel K. 1952: "Rotating Universes in General Relativity Theory", in L. M. Graves (ed.), *Proceedings of the International Congress of Mathematicians*, Cambridge (Mass.), vol. 1, pp. 175–81.

[39] Harvey A., Schucking E. 2000: "Einstein's mistake and the cosmological constant", *Am. J. Phys.*, **68** (8), pp. 723–7.

[40] Hawking S. W., Ellis G. F. R. 1973: *The Large-Scale Structure of Space–Time*, Cambridge University Press.

[41] Hetherington N. S. 1988: *Science and Objectivity*, Ames, Iowa State University Press.

[42] Hoyle F. 1948: "A New Model for the Expanding Universe", *Monthly Notices of the Royal Astronomical Society*, **108**, pp. 372–82.

[43] Hoyle F. 1997: *Home Is Where the Wind Blows*, Oxford University Press.

[44] Hoyle F., Narlikar J. V. 1962: "Mach's principle and the creation of matter", *Proc. Roy. Soc. London*, **A270**, pp. 334–41.

[45] Hubble E. P. 1925: "NGC 6822, a remote Stellar System", *Astrophys. J.*, **62**, pp. 409–33.

[46] Hubble E. P. 1926: "Extra-Galactic Nebulae", *Astrophys. J.*, **64**, p. 321.

[47] Hubble E. P. 1929: "A Relation between Distance and Radial Velocity among Extra-Galactic Nebulae", *Proc. Nat. Acad. of Sciences*, **15**, pp. 168–73 (reproduced in Crowe 1994, pp. 338–45).

[48] Hubble E. P. 1935: "Angular Rotations of Spiral Nebulae", *Astrophys. J.*, **81**, pp. 334–5.

[49] Hubble E. P. 1953: "The Law of Red-Shifts", *George Darwin Lecture*, May 8, 1953, edited by A. R. Sandage.

[50] Janssen M., "The Einstein-De Sitter Debate and Its Aftermath", www.tc.umn.edu/~janss011/pdf%20files/Einstein-De%20Sitter.pdf.

[51] Jeans J. 1943: *Physics and Philosophy*, New York, Dover 1981.

[52] Kahn C. & F. 1975: "Letters from Einstein to de Sitter on the nature of the Universe", *Nature*, **257**, pp. 451–4.

[53] Kerzsberg P. 1989: *The Invented Universe*, Oxford University Press.

[54] Kragh H. 1996: *Cosmology and Controversy*, Princeton, Princeton University Press.

[55] Krasiński A., Ellis G. F. R. 1999: "Editor's Note", *General Relativity and Gravitation*, **31**, pp. 1985–90.

[56] Lanczos C. 1922: "Bemerkung zur De Sitterschen Welt", *Physikalische Zeitschrift*, **23**, pp. 539–43.

[57] Lemaître G. 1925: "Note on De Sitter's Universe", *J. Math. Phys.*, **4**, pp. 37–41.

[58] Lemaître G. 1927: "Un universe homogène de masse constante et de rayon croissant, rendant compte de la vitesse radiale des nébuleuses extragalactiques", *Annales de la Société scientifique de Bruxelles*, series A, **47**, pp. 29–39.

[59] Lemons D. S. 1988: "A Newtonian cosmology that Newton would understand", *Am. J. Phys.*, **44**, pp. 502–4.

[60] Milne E. A. 1934: "A Newtonian Expanding Universe", *Quart. J. Math. Oxford*, **5**, pp. 64–72.

[61] Milne E. A., McCrea W. 1934: "A Newtonian Expanding Universe", *Quart. J. Math. Oxford*, **5**, pp. 73–80.

[62] Nahin P. J. 1993: *Time Machines*, New York, American Institute of Physics.

[63] North J. D. 1990: *The Measure of the Universe* [1965], New York, Dover.

[64] Ozsváth I., Schucking E. 2003: "Gödel's trip", *Am. J. Phys.*, **71**, pp. 801–5.

[65] Pais A. 1982: *Subtle Is the Lord… The Science and Life of Albert Einstein*, New York, Oxford University Press.

[66] Pauli W. 1958: *Theory of Relativity*, New York, Dover.

[67] Peterson M. A. 1979: "Dante and the 3-sphere", *Am. J. Phys.*, **47**, pp. 1031–5.

[68] Raychaudhuri A. 1955: "Relativistic Cosmology. I", *Physical Review*, **98**, pp. 1123–6.

[69] Raychaudhuri A. 1957: "Singular State in Relativistic Cosmology", *Physical Review*, pp. 172–3.

[70] Riordan M., Schramm D. 1993: *The Shadows of Creation*, Foreword by Stephen Hawking, Oxford University Press.

[71] Robertson H. P. 1933: "Relativistic Cosmology", *Reviews of Modern Phsycis*, **5**, pp. 62–90.

[72] Sachs R. K., Wu H. 1977: *General Relativity for Mathematicians*, New York – Heildelberg – Berlin, Springer-Verlag.

[73] Schilpp P. A. (ed.) 1949: *Albert Einstein: Philosopher-Scientist*, Evanston (Ill.), Library of Living Philosophers.

[74] Sciama D. W. 1993: *Modern Cosmology and the Dark Matter Problem*, Cambridge University Press.

[75] Senovilla J. M. M. 1997: "Singularity Theorems and Their Consequences", *General Relativity and Gravitation*, **29**, pp. 701–848.

[76] Thorne K. S. 1994: *Black Holes & Time Warps*, New York – London, Norton.

[77] Tolman R. C. 1934: *Relativity Thermodynamics and Cosmology*, New York, Dover 1987.

[78] van Maanen A. 1935: "Internal Motions in Spiral Nebulae", *Astrophys. J.*, **81**, pp. 336–7.

[79] Weyl H. 1922: *Space–Time–Matter*, transl. of the 4th (1921) ed., with a new preface (1950), New York, Dover.

[80] Weyl H. 1923: *Raum, Zeit, Materie. Vorlesungen über allgemeine Relativitätstheorie*, 5th ed., Springer.

[81] Weyl H. 1923: "Entgegnung auf die Bemerkungen von Herrn Lanczos über die de Sittersche Welt", *Physikalische Zeitschrift*, **24**, pp. 130–1.

[82] Weyl H. 1923: "Zur allgemeinen Relativitätstheorie", *Physikalische Zeitschrift*, **24**, pp. 230–2.

[83] Weyl H. 1930: "Redshift and relativistic cosmology", The Lon. Edin. Dubl. Phil. Mag. and J. of Sci., **9**, pp. 936–43.

Physics Before and After Einstein
M. Mamone Capria (Ed.)
IOS Press, 2005
© 2005 The authors

Chapter 7

Testing Relativity

Klaus Hentschel

University of Berne, Switzerland

1. Introduction: Einstein's Attitude towards Experiments

This chapter covers various experimental tests of special and general relativity, with an emphasis on those conducted during Einstein's lifetime.[1] But what was Einstein's own attitude toward this activity? As the popular image would have it, Einstein would dismiss any empirical result contradicting his predictions, preferring to follow his own intuition.

One origin of this myth is the apocryphal story reported by a neo-Kantian philosopher, Ilse Rosenthal-Schneider.[2] Reminiscing in a memoir dated 23 July 1957 (which finally appeared in 1981, i.e. decades after Einstein's death in 1955) about meetings as a student in Berlin in 1919 with Einstein, von Laue and Planck, she recounted witnessing how Einstein received the first telegram from Lorentz[3] about the provisional results of his light deflection measurements. She purportedly asked Einstein what he would have done if the results had turned out otherwise and his alleged reply was: 'Then too bad for the observations; my theory is right anyway.'

If Rosenthal-Schneider's story were construed as implying that Einstein was indifferent to empirical verifications, then this would contradict what we know from several other sources. Letters to his mother and his colleagues from the time reveal that he was quite excited about Eddington's results as well as about his earlier success in calculating the previously unexplained slight motion of Mercury's perihelion, which agreed so well with the anomalies recorded since the mid-nineteenth century.

Einstein – at least until his epistemological shift in the early 1920s – was very keen on empirical verification of the three testable predictions of general relativity. In 1917 he even went so far as to hire an astronomer, Erwin Finlay Freundlich (1885–1964) to pursue this issue as the first full-time staff member of the Kaiser Wilhelm Institute for

---

[1] Because of space limitations, my footnotes will generally only refer to a few key papers and secondary sources. Surveys of the vast primary literature are given in Tonnelat [90], Will [94,99] and Hentschel [56].

[2] On the following, see Hentschel [50] and further references to primary documents given therein.

[3] Actually Rosenthal-Schneider refers to *Eddington*'s telegram even though Eddington did not send any telegrams to Einstein. Einstein was first informed by a telegram sent by H.A. Lorentz from The Hague on Sept. 22, 1919, then by a more detailed letter dated Oct., 7, 1919, and then by another Lorentz-telegram dated Nov. 14, 1919 soon after the famous joint meeting of the Royal Society and the Royal Astronomical Society on Nov. 6, 1919. Eddington first sent a letter to Einstein on Dec. 1, 1919 expressing his great pleasure to be now in personal communication with Einstein. The fact that Rosenthal-Schneider does not know the name of the telegram's sender puts her account further into question.

Physical Sciences, which Einstein headed as the founding director.[4] And in December 1919, right after congratulating and thanking Eddington for his attempts to test the deflection of light on the 1919 solar eclipse expedition (see below), he went even further, emphasizing the still open issue of gravitational redshift: 'If it were proven that this effect did not exist in nature, then the whole theory [of gravitation and relativity] would have to be abandoned.'[5]

However, as Einstein saw it, such 'proof' would not result from any single experiment or observation, but rather from a cluster of experiments all pointing in one direction and barring feasible alternatives.[6] As we shall see in the following, testing the theories of relativity was no easy matter and in most cases his predictions were only satisfactorily confirmed after his death in 1955.

2. Experimental Tests of Special Relativity

All testable deviations of Einstein's special theory of relativity from classical mechanics and classical electrodynamics are very small. They are usually second-order effects, that is $\sim v^2/c^2$, with v being the velocity of the observer (for instance, c. 30 km/sec for aether–wind experiments relating to the Earth's motion around the Sun), and c, the velocity of light, which is c. 30 million km/sec in vacuo. So $v^2/c^2 \sim 10^{-8}$ is a very small ratio indeed.

Einstein's 1905 paper on the electrodynamics of moving bodies was *not* written with the Michelson–Morley experiment in mind but with entirely different motives.[7] Nevertheless, the negative outcome of that experiment, of which Einstein was certainly aware from Lorentz's 1895 essay or Wien's 1898 survey, for instance, was also naturally explained in terms of the constancy of light. The velocity of light was independent of the velocity of its emitter (which eliminated all emission theories of light such as Ritz's),[8] and also independent of the direction in which the two arms of the Michelson interferometer were turned (isotropy).[9] Thus experiments such as the Michelson–Morley ones could not detect any relative motion of the Earth and the aether, not just because of the experimental accuracy then obtainable, but in principle. The Fizeau drag coefficient, known empirically since Fresnel's interpretation of Fizeau's experiment of 1851 involving a relative motion of light and water, could also be derived by inserting c/n and the

---

[4]On Freundlich's efforts to find avenues for testing general relativity, see Hentschel [53] and [55] with references to Freundlich's many articles regarding particularly gravitational redshift and light deflection.

[5]A. Einstein to A.S. Eddington, 15. Dec. 1919, Collected Papers of Albert Einstein, call no. 8 263–2: 'Wenn bewiesen wäre, dass dieser Effekt in der Natur nicht existiert, so müsste die ganze Theorie verlassen werden.'

[6]Einstein's quasi-holistic attitude towards experiments was most likely inspired by Pierre Duhem's philosophy of science and explains why Einstein could very well reject *singular*, isolated experiments in conflict with his theories such as Kaufmann's 1906 data that seemed to contradict Einstein's formula for relativistic mass increase: see Einstein [25, p. 439].

[7]On this point, cf. Holton [58]. For a detailed analysis of Einstein's path to special relativity and a contrast to Lorentz's and Poincaré's approaches, see Miller [68]. On the Michelson experiment of 1881 and its many repetitions by Michelson and Morley in 1887, and then by Morley and others, see Swenson [87].

[8]On this point see de Sitter [13]; on other early experiments and observations relevant for special relativity, even though many of them had been conducted before 1905, see Laub [62] and Lenard [63, pp. 498 ff].

[9]For the latest limit on the anisotropy of the velocity of light $\delta c/c_0 < 3 \times 10^{-15}$ and for the independence of c from the velocity of the emitter $\Delta c/c_0 < 6 \times 10^{-12}$, obtained by comparing the resonance frequency of two cyrogenic resonators over a period of more than a year, see Peters and Müller [71].

water's velocity w into Einstein's formula for the relativistic addition of velocities. The slight correction factor $1/(1 + v_1 v_2/c^2)$ to the classical superposition rule automatically led to the Fizeau drag coefficient $1 - 1/n^2$ without any need for special assumptions:

$$v_{rel} = \frac{v_1 + v_2}{1 + v_1 v_2/c^2} \rightarrow v = c/n + w\left(1 - \frac{1}{n^2}\right) + \text{negligible terms} \sim v^2/c^2.$$

Einstein's reinterpretation of the Lorentz transformation also led to testable predictions such as the contraction of bodies moving relative to the observer's frame of reference ($\sim 1/\gamma$) and the corresponding dilation of time intervals ($\sim \gamma$). Because of the closeness to 1 of the relativistic factor $\gamma = 1/(1 - v^2/c^2)^{1/2}$, however, any practical testing of these second-order effects was difficult.

The time dilation was confirmed after the discovery of unstable particles, so-called muons or μ-leptons, among cosmic rays reaching Earth. Their decay time of 2.2×10^{-6} sec is very well determined from laboratory experiments. It was also known that they are generated by impacts of protons from extraterrestrial cosmic rays with atoms in the outer regions of the terrestrial atmosphere. Even at the maximum speed, the velocity of light c, this would only allow them to travel about 0.66 km before decaying, far less than the c. 16 km needed to reach sea level. The only explanation for their abundant presence on the Earth's surface is that their decay time, invariant in their own frame of reference, had been stretched by the γ-factor of 20 or 30 relative to our frame of reference because of their extremely high speed of about $0.99c$. Similar effects of the lengthening of decay times of artificially created unstable particles could later also be observed in high-energy physics experiments, where muons were accelerated in a circular accelerator ring to 99.7% of c. The observed 12-fold increase in their lifetimes agreed with relativistic predictions to an accuracy of 2% in the first measurement of this kind at CERN in 1966. The reliable measurements of particle velocities and decay times make these tests today among the best means to test the special theory of relativity.[10]

3. On $E = mc^2$

Einstein's mass–energy relationship, $E = mc^2$, has probably become the most famous equation in physics. When he derived it in 1905, Einstein suggested that it might be testable by measuring changes in mass after the decay of radioactive substances, such as radium salts.[11] Two years later, Max Planck pointed out that the loss of mass of one gram mole of radium within one year of decay would be too small to be measured by the limited techniques then available. It took another 25 years until the relation could be directly confirmed experimentally.[12]

In 1937, a German review article on the issue could conclude that the equivalence of mass and energy had turned into 'an empirically-based fundamental law of physics'.[13]

---

[10] On the preceding see Rossi *et al.* [77,78] and further references in Marder [65, Sec. 2.7 and Chap. 5].

[11] See Einstein [24], especially towards the end of the short note (p. 71 of the English translation).

[12] On the rather intricate story about Bainbridge's realization that Cockroft and Walton's experiment actually afforded a direct test of Einstein's formula, see the excellent survey article of Stuewer [86] and refs therein.

[13] See Braunbek [6, p. 11]; on Fritz Hasenöhrl's 1904 paper and on alternative interpretations of mass–energy equivalence within classical physics see e.g. Lenard [63, pp. 510 ff., 549 ff] and the editor's critical comments on pp. 366–369.

Even though Einstein's name was not once mentioned in this article, the validity of his formula could no longer be denied, even in Nazi Germany. Further tests of the special theory of relativity arose from combining it with quantum theory in what is called quantum electrodynamics (which is beyond the scope of this chapter). As far as the mass–energy equivalence is concerned, special relativity can be regarded as one of the most reliably confirmed theories of modern physics.

4. General Remarks about Experimental Tests of General Relativity

Even more so than special relativity, general relativity is exceedingly difficult to verify, as all relativistic effects deviating from Newtonian gravitational theory are proportional to GM/c^2R, with G the gravitational constant, c again the velocity of light, and M the mass of the body massive enough to create curvature effects in space–time, R the distance from it. On the surface of the Sun, this factor equals 10^{-6}, on the surface of the Earth it is just 10^{-9}, and on the surface of a ten-ton block of aluminium it is 10^{-22}. Only for exotic astronomical objects, such as black holes, does this factor approach unity, but experimenters too close to black holes would have other problems. For all terrestrial experiments and astronomical observations within our solar system, the relevant solution of Einstein's gravitational field equations is the one published in 1916 by the astronomer Karl Schwarzschild (1873–1916), valid for the spherically symmetric space around a mass point or a spherically shaped fluid in space such as the Sun (cf. Chap. 5, §15).

$$ds^2 = -\frac{dr^2}{1 - 2m/r} - r^2\left(d\Theta^2 + \sin^2\Theta\, d\phi^2\right) + \left(1 - \frac{2m}{r}\right)dt^2.$$

The main features of this solution in terms of experimentally testable effects (all of which will be dealt with in detail below) are: (i) a redshift of spectral lines, (ii) a light deflection of twice the amount as derived from Einstein's 1911 Prague theory (cf. below), (iii) a perihelion motion of Mercury of about 43", and (iv) a time delay for signals passing close by the Sun. Also note that the system has a coordinate singularity at $2m/r = 1$, i.e. for the radius r equalling the so-called 'Schwarzschild radius' $r \cong 2m$ in natural units of measure ($c = 1$) and a true singularity at $r = 0$. For a long time this was regarded as a mere mathematical artifact, but the work of Subrahmanyah Chandrasekhar, Robert Oppenheimer and Hartland Snyder and (later) Stephen Hawking (all beyond the scope of this chapter) has shown that such singularities should be interpreted as black holes: collapsed systems of extremely compressed mass cause the formation of singularities in space–time from which no material body, not even a light ray, can escape.

Aside from Einstein's general theory of relativity and gravitation, about 25 alternative theories of gravitation were proposed between 1905 and 1960. In order to relate the many different types of experiments to these theories, the so-called parametrized post-Newtonian (PPN) formalism was developed. It contains a set of about 10 parameters to describe any conceivable metric theory of gravitation in its respective post-Newtonian approximations. With most of them already excluded by experiment, Einstein's theory has so far withstood every test.[14]

---

[14]For surveys of the PPN formalism see Thorne and Will [89], Will [93], Misner *et al.* [69, pp. 1068 ff], Will [95, pp. 86 ff]. For updates on the limits on the ten PPN parameters set by various experiments, see Will

5. Equality of Inertial and Gravitational Mass

In Newton's *Principia*, inertial and gravitational mass appear as separate entities, one of them endowed with the strange property to resist all changes of motion (inertia), the other responsible for the mutual attraction of two masses according to Newton's inverse square law. Newton was aware of this duality and actually tested the equality of these two types of mass in experiments involving pendulums consisting of empty containers into which various materials of identical weight could be mounted. The resulting periods seemed not to depend on the material composition, thus confirming his assumption of the equality of mass and weight to one per mille. Friedrich Wilhelm Bessel (1784–1846) improved upon Newton's accuracy by a factor 10 in 1832, and in 1923 this rose to a factor 100.

Further advances had to await a different type of instrument: a special torsion balance designed by the Hungarian geophysicist Baron Roland von Eötvös (1848–1919).[15] Basically this consisted of two weights suspended at the two ends of a horizontally hung rod, isolated from vibrations. Because of the Earth's rotation, two different forces act on these two masses: gravitation, pulling both of them downward, and a centrifugal force pulling the masses away from the axis of the Earth's rotation. Even though this centrifugal force is about 470 times smaller than gravity, it leads to a considerable tilting of the angle of the suspended masses against the vertical. If inertial and gravitational mass were not exactly equal, the centrifugal force would act differently on these two masses and a small torsion would emerge, tilting the horizontal bar along the vertical axis. Turning the whole set-up by 180 degrees would make the torsion act in the other direction, thus enabling a highly sensitive null experiment. By comparing the torsion in both orientations, the torsion should cancel out only if the equality of inertial and gravitational mass is valid.

In several series of measurements between 1889 and 1920, Eötvös and his colleagues Desidirius Pekár and Eugene Fekete carried out this test using various materials, including copper, water, asbestos, aluminium, etc., and did not find any anomalous torsion. Thus they concluded that both types of masses were identical at least up to a few parts at 10^8. Even though Einstein only learned about these experiments in 1912, he often cited them as confirmation of his principle of equivalence which he first expounded in 1907.

The Eötvös experiments set the standard for this type of test for roughly 50 years. It was only in the early 1960s and 1970s that two groups of experimentalists managed to achieve even higher accuracy.[16] They replaced terrestrial gravitation and the centrifugal force exerted by the Earth's rotation around its axis by its attraction towards the Sun and by the force induced by the Earth's motion around the Sun. This had the advantage that experimenters did not have to turn the balance since that was automatically achieved by the Earth's axial rotation, thus allowing the whole apparatus to be mounted in a vacuum with sensitive temperature and optical control systems in place. The group led by Robert H. Dicke (1916–1997) at Princeton University achieved an accuracy of $1 : 3 \times 10^{11}$, while a Russian team headed by Vladimir Braginski (*1931) claimed to have reached $1 : 10^{12}$.

[95, pp. 204 ff], [99,100], [101, p. 546]. On measurements of the Newtonian gravitational constant, see Gillies [45].

[15]On the following, see Eötvös *et al.* [37] and the earlier literature cited therein.

[16]See Dicke [15], Braginski in Bertotti (ed.) [3], Will [95, pp. 24 ff] and [97, Chap. 2 with clear diagrams].

In the second half of the 1980s, a reanalysis of Eötvös's data led Ephraim Fishbach and a few others to claim that certain residuals in these data could be understood as indicating a fifth fundamental force in nature, having a strength in between the long-range gravitational and electromagnetic forces and the ultra-short-range weak and nuclear strong forces, and a range of a kilometre or less.[17] Fishbach's reanalysis of Eötvös's data seemed to indicate that there was a correlation between differences in acceleration of various test bodies on Eötvös's balance and their respective baryon-number-to-mass ratio. Unlike gravity, which acts similarly on all masses regardless of their (sub)atomic composition, a repulsive fifth force would cause an acceleration depending upon baryon number or isospin, that would vary with the material and thus violate the principle of equivalence. But subsequent experiments carried out at various locations with very sophisticated torsion balances showed no evidence for the fifth force, down to a level of $1/1000$ and less. Nor did such a fifth force show up in Galilean-type experiments with falling masses of different compositions in a high-vacuum chamber. Ultimately, the preponderance of negative evidence with increasing constraints on the strength of such a fifth force at one hundred-thousandth of the strength of gravity led to the rejection of such an additional force of nature by the scientific community.

The Einstein equivalence principle (EEP) has been differentiated into a number of closely related assumptions.[18] One of them, the weak equivalence principle (WEP), states that test bodies fall with the same acceleration independently of their internal composition and is thus verified by the Eötvös-type experiments just discussed. It can also be put as follows:

The outcome of any local non-gravitational experiment is (i) independent of the velocity of the freely falling reference frame in which it is performed (that is local Lorentz invariance, LLI), or: the outcome is (ii) independent of where and when in the universe it is performed (local position invariance, LPI).

The EEP has been shown to be at the core of general relativity in the sense that, once valid, it implies that gravitation must be described by *metric* theories of gravitation, hence ruling out many competing, non-metric theories. Another consequence of Einstein's EEP is gravitational redshift, discussed below.

6. Gravitational Redshift

Minute shifts of solar spectrum lines, as compared with the position of these same lines in terrestrial emission spectra, had already been observed by Lewis E. Jewell, the personal assistant of Henry A. Rowland (1848–1901) at Johns Hopkins University in Baltimore around 1890. While Rowland himself tried to explain them away as a mysterious artefact of his newly invented concave gratings, Jewell persevered and proved they were real.[19]

Even though gravitational redshift is one of the simplest consequences and also, historically speaking, the first that Einstein derived, in 1907, just from the principle of equivalence as opposed to the full field equations of 1915, it remained one of the most difficult effects to test empirically. The reason is that in the solar spectrum this gravitational shift of spectral lines is superimposed by many other physical effects also leading

---

[17] On the following see e.g. Will [97, 2nd ed., 1993, Chap. 11], and Franklin [41].

[18] For a survey of the various versions of EEP see e.g. Haugan [47].

[19] See Hentschel [52], [56, Chaps 2–3], where I also discuss the early interpretations of these spectrum shifts.

to shifts in varying amounts, such as Doppler shifts induced by convection currents in the convective layer of the solar atmosphere, scattering and pressure effects. With other astronomical objects it is hard to discriminate between redshifts induced by gravitational fields and those originating from a motion of the object away from the observer, such as cosmological redshift (proportional to the object's distance according to Edwin Hubble's law of 1929).

When Einstein's astronomer Freundlich tried to find redshift in stellar spectra based on statistics in 1915/16, this cosmic expansion (another effect of general relativity dealt with in Chapter 6 of this volume) was not yet known, so he was bound to fail. The first efforts to find gravitational redshift in the solar spectrum were fraught with problems. They were expected to amount to a relative shift of 2×10^{-6} (i.e., a mere 0.01 Å difference for spectral lines in the visible range of the spectrum when compared against terrestrial spectra).[20] From 1919 to 1922, Einstein supported the efforts of two physicists from Bonn, Albert Bachem (1888–1957) and Leonard Grebe (1883–1967), to explain away the discrepancies with his predictions as due to unresolved fine structure or to stronger neighboring spectrum lines, but their photometric work remained inconclusive.[21] In the US, the spectroscopist Charles Edward St John (1857–1935) invested several years of work in a thorough search for redshift in the solar spectrum. His results from 1917 until 1921 were mostly negative (as was his predisposition against relativity theory), but in 1923 he announced that the more extensive data sets he had subsequently analysed seemed to speak in favour of gravitational redshift in the amount predicted by Einstein since 1907.[22]

The first non-controversial confirmation of gravitational redshift eventually came from Robert V. Pound (*1919) and his assistant Glen A. Rebka from a terrestrial experiment performed in 1959. They utilized the recently discovered Mössbauer effect, a recoil-less emission of low-energy gamma rays, to measure the minute frequency shift of 2.45×10^{-15} of a 14.4 keV gamma ray falling in the Earth's gravitational field at the 22.5 m elevation of the Harvard Jefferson Laboratory.[23] Soon afterwards, James William Brault (*1932) also managed to validate gravitational redshift of the strong D_1-line of sodium in the solar spectrum using a newly developed photoelectric spectrometer with its slit vibrating mechanically back and forth across a narrow region of the spectrum. That way, the position of the line peak was defined independently of subjective judgment and with a precision improved by a factor of 10 over conventional visual methods. He chose the sodium line because it was known to be emitted high in the Sun's atmosphere, above the regions strongly disturbed by pressure and convective shifts, yet deeper than the chromosphere.[24]

When Josef Carl Hafele (*1933) and R.E. Keating made a comparison of caesium atomic clocks at different altitudes in 1971, they actually measured a complex com-

---

[20] On the following, cf. Hentschel [56] for a detailed analysis of seven decades of work toward understanding the slight shifts in the solar spectrum when compared against terrestrial emission spectra.

[21] For a study on this photometric analysis see Hentschel [1992a].

[22] On this spectacular about-face of an experimenter against his own predisposition see Earman and Glymour [17], Hentschel [51].

[23] See Pound and Rebka [73]; in a refined experiment of similar type, Pound and Snider [74] verified gravitational redshift with an accuracy of less than 1%; for details cf. also Pound [72], Misner *et al.* [69, pp. 1056–1058], Hentschel [54, pp. 270 ff], [56, pp. 682–701] (also on parallel endeavours by the British team of J.P. Schiffer and W.C. Marshall and others).

[24] See Brault [5], Misner *et al.* [69, pp. 1058 ff], Hentschel [54, pp. 281 ff], and [56, Sec. 11.3, pp. 703 ff].

posite effect of gravitational redshift (proportional to the differences in altitude ranging up to the 10,000 m reachable by the commercial jet aircraft of the time) and the kinematic time dilation of special relativity (proportional to the velocity of the jet relative to our terrestrial inertial frame). Because of the Earth's rotation, the velocity differs for west-bound and east-bound flights, so they decided to take two round trips, one in each direction, and compare the resulting time gains of the flying clocks against an identical one stationed on the ground.[25] For both directions, the observed time gains, e.g., $(273 \pm 7) \times 10^{-9}$ sec for the westward direction, could be fully accounted for by the kinematic correction in the special theory of relativity $(96 \pm 10) \times 10^{-9}$ plus the gravitational redshift $(179 \pm 18) \times 10^{-9}$. As an aside: given the simple instrumentation needed for this experiment, this was likely the cheapest high-precision test of relativity theory ever conducted. An improved variation with atomic clocks carried in aircraft tracked by radar and with laser pulse time comparison was performed between May 1975 and January 1976 by Carrol O. Alley and collaborators from the University of Maryland together with Hewlett-Packard and support from the US Navy.[26] The measured effect related to the calculated effect as 0.987 ± 0.016. Further improvement beyond this 15% confidence level was only possible by altering the experimental design. Robert F.C. Vessot and Martin W. Levine (both from the Center for Astrophysics at Harvard College Observatory) used a space-bound portable version of a hydrogen maser as a clock, to be compared with two identical stationary oscillators on Merrit Island, Florida, at a precision of 1 to 1 billion. The maser was mounted on a Scout D rocket, propelled to an altitude of 10,000 km and then decoupled from the rocket and subjected to free fall. By 1976 – after two years of data reduction and systematic elimination of Doppler effects of the clock on its ballistic trajectory from Wallops Island to a spot in the Atlantic Ocean east of the Bermuda islands – the two Cambridge astrophysicists were able to say that the observed redshift was equal to that predicted by the equivalence principle $\pm 2 \times 10^{-4}$. By 1980 they improved their data analysis to an agreement of 70×10^{-6}, including the final parts of the trajectory.[27]

If gravitational redshift is accepted as a real effect, it can serve as a tool to determine the mass of the system emitting radiation. In the earliest such study, Walter Adams (1876–1956) from Mt Wilson Solar Observatory, measured the redshift of the H_β line of hydrogen in the spectrum of Sirius B, a strange companion of Sirius with very low luminosity. Its faint spectrum indicated that it was a white-hot star, and Eddington came to believe that Sirius B was a very dense star, a so-called white dwarf. Adams's measurement of the redshift made it possible to calculate its mass. At the time this was highly controversial because gravitational redshift still remained to be verified.[28] While Einstein's formula for the dependency of gravitational redshift on the differences of gravitational potential remained unscathed, Adams's spectroscopic measurements – indicating an average redshift equivalent to a Doppler shift of 19 to 21 km/s – did not. In the late 1960s, new measurements of Sirius B's gravitational redshift yielded over 80 km/s, with an estimated error of no more than 16 km/s. Retrodictive adaptation of the relativity theory to the new finding was easy: one just had to lower the estimated radius of Sirius

[25] See Hafele and Keating [46].

[26] On the following see Alley [2, pp. 17–26].

[27] See Vessot and Levine [91], Vessot et al. [92] and Alley [2, pp. 26 ff].

[28] See Adams [1] and Hetherington [57] for a historical analysis of this interesting case.

B and thus increase its mass density accordingly, but what went wrong in Adams's very careful earlier measurements could never be clarified satisfactorily.

7. Light Deflection

In 1911 Einstein also predicted the deflection of starlight by heavy masses like the Sun. Because of light scattering, this minute effect could only be observed during an eclipse when the brilliant solar disk is obstructed by the Moon and stellar images near the solar rim become visible. Einstein's Prague theory of 1911 predicted a shift of $0.85''$ for light rays grazing the solar rim, which decreased by $1/r$ as the distance r from the Sun increased.[29] This effect was derived from the assumption of equality of inertial and gravitational mass according to which light with energy E should also have a mass $m = E/c^2$ and thus be attracted to the Sun like any other small test body. As a matter of fact, in the late eighteenth and early nineteenth centuries similar quasi-ballistic calculations had already been carried out by Henry Cavendish and Georg von Soldner on the basis of Newton's particle theory of light, but both had only verified that the resulting effect would be too small to observe with contemporary visual observation techniques. When Einstein's papers drew attention to this issue again, rabid anti-relativists intent on charging Einstein with plagiarism dug up the much older papers from 1921 – overlooking Einstein's totally new approach to the question.[30] Einstein's Prague theory still comprehended a pseudo-Euclidean space–time, with the gravitational potential $\Phi = -GM/R$ replaced by the velocity of light as a new scalar potential $c = c_0 + \Phi/c^2$, so the deflection of the light rays near the Sun could also be interpreted as a minute change in the refractive index $n = c/c_0 = 1/(1+\Phi/c^2) \sim 1-\Phi/c^2$, with $\Phi < 0$. But efforts to find this minute shift in stellar positions on existing eclipse photographs were in vain. A new chance to take more photographs with this specific purpose in mind came with the eclipse of 1914. The Berlin astronomer Erwin Finlay Freundlich prepared to go on an expedition to the Crimea that summer, but their team unfortunately got detained as potential spies at the outbreak of World War I. By November 1915 Einstein had found the correct field equations for his general theory of relativity and gravitation, which yielded a light deflection of $1.75''$ for stellar rays just grazing the solar rim, decreasing as $1/r$ with increasing distance r of the star position from the solar rim. This new prediction is approximately twice that of the Prague theory because the old prediction was augmented by a contribution from the curvature of space induced by the solar mass, an effect not accounted for in the earlier theory.

Even though England had been cut off from official academic exchange with the Continent since 1914, news of the interesting effect reached the Dutch astronomer Willem de Sitter (1872–1934), who privately passed on copies of Einstein's papers to his British colleagues and then wrote a widely read survey article, 'On Einstein's theory of gravitation and its astronomical consequences'. In 1917, the Astronomer Royal Frank Watson Dyson (1868–1939) drew attention to the fact that 29 May 1919 would be an exceptionally good opportunity for testing Einstein's theory, as the image of the Sun would be in the richly populated region of the Hyades full of bright stars which would be visible

---

[29] See Einstein [26, p. 908]; cf. also Earman and Glymour [18] and Hentschel [55] on the early reception.

[30] For the classical derivations see Jaki [59], Will [98] and Eisenstaedt [36]; on the later anti-relativist plagiarism charges issued foremost by the anti-Semite Philipp Lenard, see also Hentschel [48, pp. 155–161, 570].

during an eclipse even against the solar corona. Arthur Stanley Eddington (1882–1944) was keen to lead an expedition to one of the two remote places expected to have good visibility conditions: Principe Island in the Gulf of Guinea on the West African coast, or Sobral in Brazil. Being a Quaker, Eddington was a conscientious objector, so the option to send this distinguished scientist on an expedition provided the authorities with a face-saving alternative to internment, which would otherwise have been unavoidable.[31] But the day of the eclipse was overcast on Principe where Eddington and his assistant Cottingham went, so several of his photographs did not show any stellar images at all. One, however, did permit tracing about a dozen star images through the clouds. A comparison of this photograph with one of the same region taken without the Sun supported Eddington's claim of a systematic deflection of the stellar positions with the extrapolated value at Principe being $1.61'' \pm 0.30''$. Thus the quasi-Newtonian scalar-potential models were excluded. The observational estimate agreed with Einstein's modified prediction of late 1915.

Another British expedition headed by Andrew de la Chérois Crommelin (1865–1939) and Charles Rundle Davidson (1875–1970) had been sent to Sobral, where the weather was perfect. It was found, though, that the focal length of the astrograph had changed due to the rapid temperature differences during the eclipse. Thus the stellar images on all 16 photographs were blurred and it was only with great difficulty that they could arrive at an estimate for the light deflection of $0.93'' \pm 0.05''$. The seven photographs from another coelostat in Sobral yielded $1.98'' \pm 0.12''$, hence slightly larger than Einstein's prediction but of the right order of magnitude.[32]

Despite considerable difficulties in the reduction of the data and the problematic weather conditions under which these data were collected, they were advertised as a decisive confirmation of Einstein's theory of gravitation. On 6 November 1919, immediately after the official presentation of these findings at an overcrowded joint meeting of the Royal Society and the Royal Astronomical Society, *The Times* reported: 'Revolution in Science. Newtonian Ideas Overthrown', and other major newspapers throughout the world struck the same exuberant note. Overnight Einstein became a world celebrity and relativity a household word, which – incidentally – did not mean that people knew what they were talking about.[33]

Contrary to this public perception of a clear confirmation, the scientific debate continued. Eddington's observations remained controversial: for some, they became a model of how *not* to do an experiment. For others, 'the quality and the utility of the photographs were very carefully considered.'[34] Later eclipse expeditions likewise yielded somewhat inconclusive evidence.[35] Several expeditions were launched at the next opportunity on 21 September 1922. William Wallace Campbell (1862–1938) and Robert Julius Trumpler (1886–1956) from Lick Observatory obtained $1.72'' \pm 0.15''$ with a double astrograph

---

[31] For more material on this political and religious background see Stanley [85]; on the contemporary press campaign see Sponsel [83]; on earlier (failed) expeditions cf. also Earman and Glymour [18]; Brush [7] deals with the issue of prediction *vs.* retrodiction.

[32] The most detailed analysis of Eddington's data was presented in Dyson, Eddington and Davidson [16]; cf., however, also Freundlich [44] and Hentschel [53] for a critical reanalysis of Eddington's data, claiming that Eddington should rather have reported a result of $2.2''$.

[33] On some aberrations and misinterpretations of relativity theory, cf. Hentschel [48].

[34] These are quotes from Everitt [39, p. 533] and Stanley [85, p. 89], respectively.

[35] On the following see the detailed surveys provided in Mikhailov [67], *vs.* Klüber [60], Hentschel [53], [55, Chaps 9, 12], and the extensive primary literature listed there.

and $1.82'' \pm 0.2''$ with a quadruple astrograph, both set up in Wallal, Western Australia; G.F. Dodwell and C.R. Davidson from Adelaide and Greenwich obtained $1.77'' \pm 0.4''$. C.A. Chant and R.K. Young from Victoria Observatory preferred to quote separately the different results from their three plates: $1.75''$, $1.42''$ and $2.16''$. At the eclipse on 9 May 1929 in Sumatra, Erwin Finlay Freundlich was finally successful in taking careful measurements with a double coelostat, which incorporated the first independent scale checks by means of a grid projected onto the field of view. Strictly speaking, though, his result of $2.24'' \pm 0.10''$ ruled out Einstein's prediction as beyond his margin of error, and so did various other later eclipse photographs, such as Mikhailov's in 1936, yielding $2.73'' \pm 0.31''$, or van Biesbroeck from the Yerkes Observatory on a Brazil expedition in 1947, yielding $2.01'' \pm 0.27''$. In 1974, a reanalysis of all known measurements of the deflection of light using modern computing methods[36] yielded satisfactory agreement for those stars further away from the solar image, with deviations shrinking from $+0.139 \pm 0.033''$ for an average distance of 3.43 solar radii $R_{\circ}$, to just $+0.013 \pm 0.029''$ for $11.6 R_{\circ}$.

Better agreement with Einstein's general theory of relativity and gravitation could thus far only be obtained by resorting to other frequencies. Using long baseline radio-interferometric techniques for the precise localization of powerful sources of radio waves, so-called quasars, it is now possible to trace their apparent change due to the proximity of the Sun along their paths.[37] In the first such measurement in 1968, Richard Anthony Sramek (*1943) obtained an agreement of $\pm 7\%$, and in the mid-1970s, Fomalont and Sramek even obtained agreement up to $\pm 1.5\%$. By the mid-1990s, agreement had been improved to 0.9996 ± 0.0017, using Very Long Baseline Interferometry (VLBI) between the Haystack observatory and the Owens Valley Observatory in California.[38]

Furthermore, various good examples of gravitational lensing have been found. They are called such in analogy to optical lenses, with multiple, semicircular or even completely circular images of an emitter deflected by a very heavy mass positioned somewhere in the optical path between the emitter and us. While the dark heavy mass absorbs all the direct light rays, it deflects those grazing by in our direction so we see a smeared image of the same cosmic object from other directions very nearby.[39] Thus finally, gravitational deflection can also count as fully confirmed, albeit not for light, but for other forms of electromagnetic radiation only.

8. Mercury's Perihelion Motion

Kepler's laws describe the path of the planets around the Sun as perfect ellipses with the Sun at one of the two focal points. Newton's theory of gravitation with its $1/r^2$ law of attraction allowed incorporation of these Keplerian ellipses into the theory, but only for a single planet. Other bodies present in the solar system would disturb this perfect path and lead to deviations from closed ellipses, as would any deviation from the strict $1/r^2$-law. Because there are other planets nearby, Mercury's orbit deviates from a perfect

---

[36] See Merat *et al.* [66].

[37] On the following see e.g. Sramek [84], Fomalont and Sramek [40] and Will [95, p. 172].

[38] See Maddox [64] for references and commentary.

[39] On the earliest discovery of gravitational lensing which Einstein [34, p. 507] himself had considered to be very unlikely ever to be observed, see Chaffee [8]; for more recent examples of such gravitational lenses including a nearly complete Einstein ring, see Ehlers [23, pp. 44 ff].

ellipse, with the perihelion, i.e., the point of the orbit closest to the Sun, moving by as much as 531 arc-seconds/century around the Sun. In 1859, the French astronomer Jean Joseph LeVerrier (1811–1877) announced that there was an unresolved discrepancy of *c.* 38″/*c* between Newtonian theory and increasingly precise observations. By 1882, Simon Newcomb (1835–1909) had recalculated and rechecked the observations, obtaining an unresolved difference between Newtonian theory and observation of 43″/*c*, which was very close to today's value of $43.11'' \pm 0.45''$.

Various alternative explanations were put forward, such as an increase by 10% in the mass of Venus, the existence of intramercurial planets or asteroids or an undetected moon of Mercury, or finally, deviations of the Newtonian $1/r^2$-law. But none of these hypotheses were fully satisfactory, in particular since each would have led to new problems and discrepancies with observations elsewhere. So the advance of Mercury's perihelion remained an unresolved, if innocuous, anomaly in the classical scheme.[40]

Einstein realized early on in his endeavour to generalize relativity and gravitation that this issue would also be affected. In 1913 one of Einstein's candidates for the gravitational field equations yielded a mercurial perihelion motion of $\sim -50''$/century, that is, even in the opposite direction from the known discrepancies between observations and Newtonian theory. In October 1915, another candidate came to 18″, i.e. only one-third of the amount needed. But a month later his modified field equations finally produced a perihelion advance of $45'' \pm 5''$/century, just the amount needed to explain the anomaly known since half a century. (Today's calculations yield 43.03″/century, i.e., an even better fit between theory and observation!)

In a letter to Arnold Sommerfeld dated 28 November 1915, Einstein called this surprisingly good match of his calculation with the known anomaly 'the most magnificent thing that ever happened to me' ('das Herrlichste, was ich erlebte'). Soon afterwards he wrote enthusiastically to his friend Michele Besso: 'The boldest of dreams have now been fulfilled. *General* covariance. Mercury's perihelion motion wonderfully precise.'[41] To his colleague Wander J. de Haas he even confessed that he was so excited that he was unable to work for days. Both light deflection and mercurial perihelion advance are direct effects of the slight changes of space–time metrics near large masses, but the derivation of the perihelion advance depends on the fully-fledged gravitational field equations. In this sense, it was the first test of the fully developed general theory of relativity – which might well explain Einstein's exaltation.

Later recalculations and improved observations have not substantially altered this excellent match. In the 1960s, there was some discussion about a possible discrepancy of 10% due to the effects of solar oblateness, suggested by Robert Dicke. Optical studies of the shape of the Sun by Dicke and Goldenberg seemed to confirm this in 1968, but others came to the conclusion that the optical shape of the Sun fluctuates and cannot be taken as a reliable indicator of the shape of its mass.[42] Considering this issue of solar

---

[40]For a survey of the literature on Mercury's perihelion advance, see Roseveare [76].

[41]Einstein to Besso, 10 December 1915, as quoted from the *Collected Papers of Albert Einstein*, Vol. 8, English translation volume, Princeton: Princeton University Press, 1998, p. 160.

[42]For surveys of these later discussions see Everitt [39, p. 531 ff], Will [95, pp. 176 ff], [97, Chap. 5]. The editor, Marco Mamone Capria, pointed out to me that the standard textbook argument depends essentially on using, on the one hand, Newtonian physics for the *n*-body perturbative calculation of the 531″ observed secular advance, and on the other hand, the relativistic one-body spherically symmetric Schwarzschild solution for the residual 43″. Some theoreticians, such as J.L. Synge [88, pp. 296 ff], find this schizophrenic approach to save the phenomena 'intellectually repellent', but the scientific community at large has no problem with it.

oblateness still remains to be resolved definitely, one of the leading experts, Clifford Will from the McDonnell Center for Space Science at George Washington University in St Louis summarized the situation with respect to Mercury's perihelion motion: 'It is ironic that after seventy years, Einstein's first great success remains an open question, a source of controversy and debate.'[43]

9. Other Non-standard Tests of General Relativity after 1960

9.1. The Time-Delay Measurement

In 1964, Irwin I. Shapiro (*1929) pointed out that light passing by a very heavy body such as the Sun is not just deflected (see above) but also slightly delayed. This can best be understood on the basis of Einstein's Prague theory with its Ansatz for the change of the velocity of light (slowing down in the proximity of large masses, hence taking more time). Measurement of this time delay became feasible with sophisticated radar ranging techniques that have been available since the early 1960s. Shapiro's paper inspired a series of very successful measurements of time delays in radar echos from Mercury, Venus and Mars, all showing a clear maximum when the Sun's orbit is closest to the line connecting Earth and Venus.[44] Because the delay time amounts to no more than about 200 microseconds for echo signals from Venus travelling for about half an hour before reaching Earth, these determinations have to be exact to 10^{-7}: an impressive achievement. With the landing of the Viking probes on Mars in 1976, a new climax was reached for time-delay measurements because the unmanned stations, designed to work for 90 days on Mars, actually functioned for several years. The most recent test of this time delay used multi-frequency radio links with the Cassini spacecraft during a solar conjunction in 2002. By overcoming the solar plasma noise, Bertotti and his Italian co-workers reached a sensitivity approaching the level sensitive to deviations from some non-standard cosmological models inspired by string theory. The prediction of general relativity is now confirmed up to a factor $1 + (2.1 \pm 2.3) \times 10^{-5}$.

9.2. The Gyroscope Experiment

This experiment was devised in 1959 by the theoretical physicist Leonard Isaac Schiff (1915–1971) and implemented by some of his experimentalist colleagues at Stanford University.[45] The underlying idea is to measure the precession of a gyroscope's spinning axis relative to the distant stars as the gyroscope orbits the Earth. General relativity implies two minute effects pertinent to such an orbiting gyroscope: a geodetic precession as a consequence of the curvature of space near gravitating bodies, and the so-called Lense–Thirring effect, a kind of dragging of inertial frames by rotating masses. Both these consequences of general relativity had been known about for a long time. The first was pointed out by Willem de Sitter in 1916 in one of the earliest surveys of experimental consequences of general relativity. The second dates back to a paper by Austrian physicists Josef Lense and Bruno Thirring in 1918. But measuring a tiny perturbation of the

---

[43] Will [97, p. 107].

[44] See e.g. Shapiro [81], Shapiro *et al.* [82], Reasenberg *et al.* [75], Will [97, Chap. 6] and Bertotti *et al.* [4] for the latest measurement.

[45] See e.g. Schiff [80], Misner *et al.* [69, pp. 1117 ff], Fairbank *et al.* in Bertotti (ed.) [3], Everitt [38], [39, pp. 535 ff], Will [95, pp. 208 ff], [97, Chap. 11], Everitt *et al.* in Lämmerzahl *et al.* (eds) [61].

orbit caused by the spin of the attracting body only became feasible with the availability of satellites and sophisticated techniques of stabilizing gyroscopes (pushing the limiting drift rate below 0.001″/year) in the wake of their use in guidance systems for rockets and submarines after World War II. In the terrestrial gravitational field, the disturbing influences on the gyroscope would be far greater than the few arc-seconds precession per year that have to be measured. In outer space, further from gravitating masses, it is barely feasible to measure the net effect of 6.9″ geodetic precession per year of a gyroscope orbiting the Earth about 5000 times a year at an altitude of a few hundred kilometres. Even the smaller inertial dragging effect, amounting to no more than 0.1″ or even only half that per year might be possible, depending on the orientation of the gyroscope and its orbit relative to the Earth's spinning axis.[46] The Stanford implementation of this idea consists of a set of four gyroscopes, each constituting a perfectly spherical ball about 4 cm in diameter made of optically selected fused quartz and coated with a thin film of superconducting niobium. The ball is electrically suspended and initially spun up to a speed of about 12,000 revolutions per minute by gas jets, then the gas is pumped out and the ball is allowed to spin freely in a vacuum. The direction of its spinning axis is read out by a very sensitive SQUID (superconducting quantum interface device) magnetometer, and any unforeseen sources of error are cross-checked by four identical gyroscopes side by side. This experiment, called Gravity Probe B, is currently undergoing final testing prior to launch: its weekly progress can be followed at http://einstein.stanford.edu.

9.3. The Nordtvedt Effect

Like all planets, the Earth is held together by gravitational forces. From the equation $E = mc^2$, it is possible to calculate a gravitational self-energy of the Earth, corresponding to 1 part in 10^9 of its mass. In 1968, Kenneth L. Nordtvedt (*1939) realized that in certain gravitational theories satisfying the principle of equivalence in the ordinary sense, the Earth–Moon system would violate this principle. An empirically testable consequence would be a bimonthly oscillation in the distance between the Earth and Moon that might be as large as 10 m according to the Brans–Dicke theory. Laser ranging measurements between the Earth and the Moon would make such oscillations traceable, but thus far no such effects have been found.[47] This implies the equivalence of gravitational and inertial mass for the different materials of the Earth and Moon to a few parts in 10^{11}, a level of accuracy nearly as high as that reached by Dicke, Roll and Krotkov in the 1960s in laboratory experiments testing the principle of equivalence for gold and aluminium (see §5 above).

10. Gravitational Waves

Like all other signals and forces, gravitational action can only propagate with the velocity of light. According to Einstein's theory, gravitational waves as emitted by rapidly changing mass–energy distributions in space–time – for example, by collapsing stars – would propagate as disturbances in the space–time metric with properties somewhat similar to

---

[46]For a clear exposition of the Lense–Thirring effect, see Lämmerzahl and Neugebauer in Lämmerzahl *et al.* (eds) [61]; on its recent confirmation within 10% of what is predicted by Einstein's general theory of relativity (±20% total error), by means of two laser-ranged satellites, LAGEOS I and II, see Ciufolini *et al.* [9].

[47]See e.g. Nordtvedt [70], Nordtvedt in Bertotti (ed.) [3] and Will [95, pp. 185 ff], [97, Chap. 7].

radio waves emitted by oscillating charges. Einstein's first paper on gravitational waves dates from 1918,[48] yet no efforts were made to detect these feeble ripples in space–time before the pioneering investigations of Joseph Weber (1919–2000) began in 1959.[49]

Weber's idea was to suspend a large aluminium bar, about five feet long and over two feet in diameter, and measure any mechanical oscillations of this bar by attached pyroelectric strain gauges capable of detecting changes in length as small as 10^{-14} cm. These piezocrystals would translate any deformation into an electric signal. Just as electric oscillations are induced in a radio antenna, such an aluminium bar would exhibit mechanical vibrations if – and that is a big if – its eigenfrequencies happened to resonate with the frequency of the incoming gravitational wave. Since gravity waves emitted from terrestrial objects would be far too weak to be detectable, Weber's only hope were large signals from huge cosmic objects undergoing catastrophic changes in shape and mass radius, because such events would emit the most intense gravitational waves. Besides anti-vibrational mountings for these devices to reduce even further the likelihood of accidental signals, Weber installed two gravitational wave antennas 1000 km apart, one at the Argonne National Laboratory near Chicago and the other at the University of Maryland, limiting his search to simultaneous oscillations. Against all odds, Weber was able to announce statistically significant events in 1968: pulses reportedly occurring several times a day and originating from near the centre of the galaxy.

However, many attempts to replicate Weber's findings, even with improved designs that ought to have had a much higher sensitivity, carried out between 1970 and 1975, could not confirm his findings. More troublesome still is the fact that the other teams did not even manage to interpret these events in Weber's own data. Their computer programs for data analysis apparently differed on significant assumptions about the shapes of the signal and background and on the algorithmic strategies to isolate the signal. When Weber analysed data provided to him by other teams, he again found significant correlated events, only to learn later that the assumptions about the time zones in which these two sets of data were taken were incorrect: his candidates for correlated events were actually measurements taken hours apart – thus revealing a certain in-built tendency in his programs to 'find' events too easily. The sociologist of science Harry Collins has claimed that this stalemate exemplifies what he calls the 'experimenter's regress', i.e. a kind of vicious circle between theoretical assumptions built into an experiment and the instruments used in it, and the results you get with that set-up. But other historians and philosophers of experiments, such as Allan Franklin, have argued that there are strong neutral strategies available for testing the reliability of such results, and that these were effective in Weber's case, as in others.[50]

Even with the second generation of gravitational-wave antennas similar to Weber's aluminium bars, but more sensitive by a factor of 1000 to 10,000, no one has yet succeeded in finding direct evidence for gravitational waves. Detectors (such as GEO600) working with laser detectors and laser interferometry had problems with ground vibrations but are now in the sensitivity range of 10^{-22}–10^{-23} Hz$^{-1/2}$. Other projects searching for very low-frequency gravity waves by radar ranging to spacecraft have thus far

---

[48] See Einstein [33] and Einstein and Rosen [35].

[49] For surveys of gravitational wave physics, see Misner *et al.* [69], Will [95, pp. 221 ff], [97, Chap. 12].

[50] On this controversy see Collins [10, Chap. 4] (mostly based on interviews) [11] and Franklin [42], also quoting the pertinent primary literature by Weber and the many teams that set out to check his results.

also yet to report positive results.[51] However, we do have good indirect evidence for the existence of gravitational waves originating from the energy loss of a certain binary pulsar labelled PSR 1913+16, which was discovered by Joseph H. Taylor (*1941) and Russell A. Hulse (*1950) in 1974. Soon after their discovery of this strong radiowave emitter, they realized that its pulsation period changed periodically. This could be interpreted as Doppler shifts due to orbital motion of the pulsar about a dark, but heavy, companion. Continuous radio tracking of this binary star allowed its orbital parameters to be calculated so precisely that it became clear that this system was continually losing energy. In December 1978 Taylor published a measurement of the rate of change of the orbital period of $-(2.425 \pm 0.010) \times 10^{-12}$ sec, which was fully consistent with the prediction of gravitational radiation damping in general relativity. This finding constitutes excellent indirect evidence for the existence of such gravity waves; the best we have to date. The Nobel Prize Committee must have thought so, too, since they awarded the 1993 Nobel prize to Taylor and Hulse:[52] the latest physics Nobel prize awarded in the field of experimental tests of general relativity, which has long since ceased to be 'a theorist's paradise but an experimenter's hell'.

References

[1] Adams W.S. 1925: 'The Relativity Displacement of the Spectral Lines in the Companion of Sirius', *PNAS* **11**, pp. 382–387.

[2] Alley C.O. 1979: 'Relativity and Clocks', *Proceedings of the 33rd Annual Symposium on Frequency Control*, Washington, D.C., pp. 4–39.

[3] Bertotti B. (ed.) 1974: *Experimental Gravitation: Proceedings of Course 56 of the International Summer School of Physics 'Enrico Fermi'*, New York: Academic.

[4] Bertotti B., Iess L. & Tortora P. 2003: 'A test of general relativity using radio links with the Cassini spacecraft', *Nature* **425**, pp. 374–376.

[5] Brault J.W. 1962: *The Gravitational Redshift in the Solar Spectrum*, Ph.D. Diss., Princeton University.

[6] Braunbek W. 1937: 'Die empirische Genauigkeit des Mass-Energie-Verhältnisses', *Zeitschrift für Physik* **107**, pp. 1–11.

[7] Brush S.G. 1989: 'Prediction and theory evaluation: the case of light bending', *Science* **246**, pp. 1124–1129.

[8] Chaffee Jr. F.H. 1980: 'The discovery of a gravitational lens', *Scientific American* **243**, Nov., pp. 60–68.

[9] Ciufolini I. *et al.* 1998: 'Test of general relativity and measurement of the Lense-Thirring effect with two Earth satellites', *Science* **279**, pp. 2100–2103.

[10] Collins H.M. 1985: *Changing Order. Replication and Induction in Scientific Practice*, London: Sage Publ.

[11] Collins H.M. 1994: 'A strong confirmation of the experimenter's regress', *Studies in the History and Philosophy of Science* **25**, pp. 493–503.

[12] Danzmann K. & Ruder H. 1993: 'Gravitationswellen', *Physikalische Blätter* **49**, pp. 103–108.

[13] de Sitter W. 1913: 'Ein astronomischer Beweis für die Konstanz der Lichtgeschwindigkeit', *Physikalische Zeitschrift* **14**, p. 429 sowie ibid. p. 1267: 'Über die Genauigkeit, innerhalb welcher die Unabhängigkeit der Lichtgeschwindigkeit von der Bewegung der Quellen behauptet werden kann'.

[14] de Sitter W. 1916/17: 'On Einstein's theory of gravitation and its astronomical consequences', *Monthly Notices of the Royal Astronomical Society* **76**, pp. 699–728, **77**, pp. 155–184, **78**, pp. 3–28.

[15] Dicke R.H. 1960: 'Eötvös experiment and the gravitational redshift', *American Journal of Physics* **28**, pp. 344–347.

---

[51] For surveys see Everitt [39, pp. 548 ff], Danzmann and Ruder [12], Rüdiger [79], Rüdiger *et al.* in Lämmerzahl *et al.* (eds) [79, pp. 131 ff], Ehlers [23, pp. 46 ff].

[52] See www.nobel.se/physics/1993; for the underlying physics see also Will [95, pp. 283 ff].

[16] Dyson F.W., Eddington A.S. & Davidson C. 1920: 'A determination of the deflection of light by the sun's gravitational field, from observations made at the total eclipse of May 29, 1919', *Philosophical Transactions of the Royal Society* London, *Ser. A* **220**, pp. 291–334.

[17] Earman J. & Glymour C. 1980: 'The gravitational redshift as a test of general relativity: History and analysis', *Studies in the History and Philosophy of Science* **11**, pp. 251–278.

[18] Earman J. & Glymour C. 1980: 'Relativity and eclipses: The British eclipse expeditions of 1919 and their predecessors', *Historical Studies in the Physical Sciences* **11**, pp. 49–85.

[19] Eddington A.S. 1919: 'The deflection of light during a solar eclipse', *Nature* **104**, pp. 372, 468 (see also Andrew C.D. Crommelin, ibid. **102**, pp. 444–446; **104**, pp. 280–281, 372–373; **105**, p. 8).

[20] Eddington A.S. 1919: 'The total eclipse of 1919 May 29 and the influence of gravitation on light', *Observatory* **42**, pp. 119–122.

[21] Eddington A.S. 1919: 'Astronomers on Einstein. A new geometry wanted. Dr. Eddington and relativity', *Times*, 13 Dec., No. 42,282, p. 9, col. 1 (= report on Eddington's talk at the Royal Astronomical Society, London on 12 Dec. 1919).

[22] Eddington A.S. *et al.* 1919: 'Discussion on the theory of relativity', *Monthly Notices of the Royal Astronomical Society* **80**, pp. 96–119.

[23] Ehlers J. 2002: 'Aktuelle Probleme der Gravitationsphysik', *Physik Journal* **1**, No. 7/8, pp. 43–47.

[24] Einstein A. 1905: 'Ist die Trägheit eines Körpers von seinem Energiegehalt abhängig?', *Annalen der Physik* (4th series) **17**, pp. 639–641, translated by W. Perrett and G.B. Jeffery as 'Does the inertia of a body depend upon its energy-content?', in: *The Principle of Relativity*, ed. by A. Sommerfeld, New York: Dodd, pp. 69–71.

[25] Einstein A. 1907: 'Über das Relativitätsprinzip und die aus demselben gezogenen Folgerungen', *Jahrbuch der Radioaktivität und Elektronik* **4**, pp. 411–462.

[26] Einstein A. 1911: 'Über den Einfluß der Schwerkraft auf die Ausbreitung des Lichtes', *Annalen der Physik* (4th series) **35**, pp. 898–908.

[27] Einstein A. 1915: 'Zur allgemeinen Relativitätstheorie', *Sitzungsberichte der Preußischen Akademie der Wissenschaften*, math. phys. Kl., 4 Nov. 1915, pp. 778–786; 'Nachtrag', ibid., pp. 799–801.

[28] Einstein A. 1915: 'Erklärung der Perihelbewegung des Merkur aus der allgemeinen Relativitätstheorie', ibid., pp. 831–839.

[29] Einstein A. 1915: 'Die Feldgleichungen der Gravitation', ibid., pp. 844–847.

[30] Einstein A. 1915/25: 'Die Relativitätstheorie', in *Kultur der Gegenwart. Die Physik*, Leipzig/Berlin, 1st ed. 1915, pp. 703–713; 2nd ed. 1925, pp. 783–797.

[31] Einstein A. 1916: Die Grundlagen der allgemeinen Relativitätstheorie, in *Annalen der Physik* (4th series) **49**, pp. 769–822; (b) as a book: Leipzig, 1916.

[32] Einstein A. 1917: *Über die spezielle und die allgemeine Relativitätstheorie*, Braunschweig: Vieweg; 1st ed. 1917; (b) 10th ed. 1920; (c) reprint of the 21st ed. 1969: 1979; (d) in Engl. transl.: *Relativity, the Special and General Theory*, New York: Crown, 1961.

[33] Einstein A. 1918: 'Über Gravitationswellen', *Sitzungsberichte der Preussischen Akademie der Wissenschaften, Berlin, 1918, mathem.-physikalische Klasse*, pp. 154–167.

[34] Einstein A. 1936: 'Lens-like action of a star by deviation of light in the gravitational field', *Science* **84**, pp. 506–507.

[35] Einstein A. & Rosen N. 1937: 'On gravitational waves', *Journal of the Franklin Institute* **223**, pp. 43–54.

[36] Eisenstaedt J. 1991: 'De l'influence de la gravitation sur la propagation de la lumière en théorie newtonienne. L'archéologie des trous noirs', *Archive for History of Exact Sciences* **42**, pp. 315–386.

[37] Eötvös R.V., Pekár V. & Fekete E. 1922: 'Beiträge zum Gesetze der Proportionalität von Trägheit und Gravitation', *Annalen der Physik* **68**, pp. 11–66.

[38] Everitt C.W.F. 1977: 'Gravitation, relativity, and precise experimentation', *Proceedings of the First Marcel Grossmann Meeting on General Relativity*, ed. by R. Ruffini, Amsterdam: North Holland Publ., pp. 545–625.

[39] Everitt C.W.F. 1979: 'Experimental tests of general relativity: Past, present and future', in: Riazuddin (ed.) *Physics and Contemporary Needs*, New York: Plenum, 1979, Vol. **4**, pp. 529–555.

[40] Fomalont E.B. & Sramek R.A. 1976: 'Measurements of the solar gravitational deflection of radio waves in agreement with general relativity', *Physical Review Letters* **36**, pp. 1475–1478.

[41] Franklin A. 1993: *The Rise and Fall ot the Fifth Force*, New York: AIP.

[42] Franklin A. 1994: 'How to avoid the experimenter's regress', *Studies in the History and Philosophy of Science* **25**, pp. 463–491.

[43] Freundlich E.F., von Brunn A. & Brück H. 1931: 'Die Ablenkung des Lichtes im Schwerefeld der Sonne', *Abhandlungen der Preußischen Akademie der Wissenschaften*, Berlin, issue no. 1, pp. 1–61 with plates.

[44] Freundlich E.F., von Brunn A. & Brück H. 1933: 'Weitere Untersuchungen über die Bestimmung der Lichtablenkung im Schwerefeld der Sonne', *Annalen von de Bosscha-Sterrenwacht Lembang* **5**, pp. B1–B64.

[45] Gillies G.T. 1997: 'The Newtonian gravitational constant: recent measurements and related studies', *Reports of Progress in Physics* **60**, pp. 151–225.

[46] Hafele J.C. & Keating R.E. 1972: 'Around the world atomic clocks', *Science* **177**, pp. 166–170.

[47] Haugan M.P. 1979: 'Energy conservation and the principle of equivalence', *Annals of Physics* **118**, pp. 156–186.

[48] Hentschel K. 1990: *Interpretationen und Fehlinterpretationen der speziellen und allgemeinen Relativitätstheorie durch Zeitgenossen Albert Einsteins*, Basel, Birkhäuser (= Science Networks, 6).

[49] Hentschel K. 1992: 'Grebe/Bachems photometrische Analyse der Linienprofile und die Gravitations-Rotverschiebung: 1919 bis 1922', *Annals of Science* **49**, pp. 21–46.

[50] Hentschel K. 1992: 'Einstein's attitude towards experiments', *Studies in History and Philosophy of Science* **23**, pp. 593–624.

[51] Hentschel K. 1993: 'The conversion of St. John – A case study on the interplay of theory and experiment', *Science in Context* **6**(1), pp. 137–194.

[52] Hentschel K. 1993: 'The discovery of the redshift of solar Fraunhofer lines by Rowland and Jewell in Baltimore around 1890', *Historical Studies in the Physical Sciences* **23**(2), pp. 219–277.

[53] Hentschel K. 1994: 'Erwin Finlay Freundlich, Albert Einstein, and experimental tests of the general theory of relativity', *Archive for History of Exact Sciences* **47**(2), pp. 143–201.

[54] Hentschel K. 1996: 'Measurements of gravitational redshift between 1959 and 1971', *Annals of Science* **53**, pp. 269–295.

[55] Hentschel K. 1997: *The Einstein Tower. An Intertexture of Dynamic Construction, Relativity Theory, and Astronomy*, Stanford: Stanford Univ. Press (German orig. 1992).

[56] Hentschel K. 1998: *Zum Zusammenspiel von Instrument, Experiment und Theorie am Beispiel der Rotverschiebung im Gravitationsfeld und ergänzender spektraler Verschiebungseffekte 1880–1960*, Hamburg: Kovač.

[57] Hetherington N.S. 1980: 'Sirius B and the gravitational redshift: An historical review', *Quarterly Journal of the Royal Astronomical Society* **21**, pp. 246–252, expanded as chapter 6-7 in Hetherington's book *Science and Objectivity: Essays in the History of Astronomy*, Ames: Iowa State Univ. Press, 1984.

[58] Holton G. 1969: 'Einstein, Michelson, and the crucial experiment', *Isis* **60**, pp. 133–197.

[59] Jaki S. 1978: 'Johann Georg v. Sondner and the gravitational bending of light, with an English translation of his essay on it published in 1801', *Foundations of Physics* **8**, pp. 927–950.

[60] von Klüber H. 1960: 'The determination of Einstein's light-deflection in the gravitational field of the sun', *Vistas in Astronomy* **3**, pp. 47–77.

[61] Lämmerzahl C. *et al.* (eds.) 2001: *Gyros, clocks, Interferometers..: Testing Relativistic Gravity in Space*, Berlin: Springer (= Springer Lecture Notes 562).

[62] Laub J.J. 1910: 'Über die experimentellen Grundlagen des Relativitätsprinzips', *Jahrbuch der Radioaktivität und Elektronik* **7**, pp. 405–463.

[63] Lenard P. 1920: 'Über Äther und Uräther', (a) *Jahrbuch für Radioaktivität und Elektronik* **17** (1920), pp. 307–355, (b) 2. Aufl., Leipzig: Hirzel (1922); (c) reprinted and commented upon in: Philipp L.: *Wissenschaftliche Abhandlungen*, Bd. 4, hrsg u. kommentiert v. Charlotte Schönbeck, Berlin: Verlag f. Geschichte der Naturwissenschaften und der Technik (2003).

[64] Maddox J. 1995: 'More precise solar-limb light-bending', *Nature* **377**, p. 11.

[65] Marder L. 1971: *Time and the Space Traveller*, London: Allen & Unwin, 1971.

[66] Merat P., Pecker J.-C., Vigier J.-P. & Yourgrau W. 1974: 'Observed deflection of light by the sun as a function of solar distance', *Astronomy & Astrophysics* **32**, pp. 471–475.

[67] Mikhailov A.A. 1959: 'The deflection of light by the gravitational field of the sun', *Monthly Notices of the Royal Astronomical Society* **119**, pp. 593–608.

[68] Miller A.I. 1981: *Einstein's Special Theory of Relativity: Emergence (1905) and Early Reception (1905–1911)*, Reading, Mass., Addison Wesley, 1981.

[69] Misner C.W., Thorne K. & Wheeler J.A. 1973: *Gravitation*, San Francisco: Freeman.

[70] Nordtvedt K.L. 1972: 'Gravitation theory: Empirical status from solar system experiments', *Science* **178**, pp. 1157–1164.

[71] Peters A. & Müller H. 2004: 'Einsteins Theorie auf dem optischen Prüfstand', *Physik in unserer Zeit* **35**, pp. 70–75.

[72] Pound R.V.: 'Weighing Photons', *Physics in Perspective* **2** (2000), pp. 224–268, **3** (2001), pp. 4–51.

[73] Pound R.V. & Rebka G.A.: 'Gravitational red-shift in nuclear resonances', *Physical Review Letters* **3** (1959), pp. 439–441 and ibid. 555–556. **4** (1960), pp. 274–275, and especially pp. 337–341.

[74] Pound R.V. & Snider J.L. 1965: 'Effect of gravity on gamma radiation', *Physical Review B* **140**, pp. 788–803.

[75] Reasenberg R.D. *et al.* 1979: 'Viking relativity experiment: Verification of signal retardation by solar gravity', *Astrophysical Journal* (Section Letters), pp. L219–221.

[76] Roseveare N.T. 1983: *Mercury's Perihelion. From Le Verrier to Einstein*, Oxford: Oxford Univ. Press.

[77] Rossi B. & Hilberry N. & Hoag J.B. 1940: 'The variation of the hard component of cosmic rays with height and the disintegration of mesotrons', *Physical Review* **57**, pp. 461–469.

[78] Rossi B. & Hall D.B. 1941: 'Variation of the rate of decay of mesotrons with momentum', *Physical Review* **59**, pp. 223–228.

[79] Rüdiger A. 2001: 'Auf der Lauer nach Gravitationswellen', *Physikalische Blätter* **17**(2), pp. 15–16.

[80] Schiff L.I. 1960: 'Motion of a gyroscope according to Einstein's theory of gravitation', *Proceedings of the National Academy of Sciences* **46**, pp. 871–882.

[81] Shapiro I.I. 1964: 'Fourth test of general relativity', *Physical Review Letters* **13**, pp. 789–791.

[82] Shapiro I.I. *et al.* 1977: 'The Viking relativity experiment', *Journal of Geophysical Research* **82**, pp. 4329–4334.

[83] Sponsel A. 2002: 'Constructing a 'revolution in science': The campaign to promote a favorable reception for the 1919 solar eclipse expedition', *British Journal for the History of Science* **35**, S.439–469.

[84] Sramek R.A. 1971: 'A measurement of the gravitational deflection of microwave radiation near the sun', *Astrophysical Journal* **167**, pp. L55–60.

[85] Stanley M. 2003: ''An expedition to heal the wounds of war': The 1919 eclipse and Eddington as Quaker adventurer', *Isis* **94**, pp. 57–89.

[86] Stuewer R.H. 1993: 'Mass–energy and the neutron in the early thirties', *Science in Context* **6**, pp. 195–238.

[87] Swenson L. 1972: *The Ethereal Aether*, Austin: Univ. of Texas Press.

[88] Synge J.L. 1960/71: *Relativity: The General Theory*, Amsterdam: North Holland, 1st ed. (1960), 4th printing (1971).

[89] Thorne K.S. & Will C.M. 1971: 'Theoretical frameworks for testing relativistic gravity. I. Foundations', *Astrophysical Journal* **163**, pp. 595–610 (see also Will [93]).

[90] Tonnelat M.-A. 1964: *Les vérifications expérimentales de la relativité générale*, Paris: Masson.

[91] Vessot R.F.C. & Levine M.W. 1979: 'A test of the equivalence principle using a space-borne clock', *General Relativity and Gravitation* **10**, pp. 181–204.

[92] Vessot *et al.* 1980: 'Test of relativistic gravitation with a speca borne hydrogen maser', *Physical Review Letters* **45**, pp. 2081–2084.

[93] Will C.M. 1971: 'Theoretical frameworks for testing relativistic gravity. II. Parametrised post-Newtonian hydrodynamics and the Nordtvedt-effect', *Astrophysical Journal* **163**, pp. 611–628 (= continuation of Thorne & Will [89]).

[94] Will C.M. 1979: 'The confrontation between gravitation theory and experiment', in: S.W. Hawking & W. Israel (eds.), *General Relativity. An Einstein Centenary Survey*, Cambridge: Cambridge Univ. Press, pp. 24–89.

[95] Will C.M. 1981: *Theory and Experiment in Gravitational Physics*, Cambridge: CUP (2nd ed. 1993, with a new chapter 14 including updates on experimental tests).

[96] Will C.M. 1984: 'The confrontation between general relativity and experiment: an update', *Physics Reports* **113**, pp. 345–422.

[97] Will C.M. 1986: *Was Einstein right? – Putting General Relativity to the Test*, New York: Basic Books (1986) 2nd ed. with a new chapter on the fifth force issue (1993).

[98] Will C.M. 1988: 'Henry Cavendish, Johann v. Soldner and the deflection of light', *American Journal of Physics* **56**, pp. 413–415.

[99] Will C.M. 1992: 'The confrontation between general relativity and experiment: a 1992 update', *Int. Journal of Modern Physics* **D1**, pp. 13–68.

[100] Will C.M. 2001: 'Living review in relativity 4, 2001–2004', www.livingreviews.org/lrr-2001-4 (accessed June 18, 2003).

[101] Will C.M. 2003: 'The confrontation between general relativity and experiment', *Astrophysics and Space Science* **283**, pp. 543–552.

Physics Before and After Einstein
M. Mamone Capria (Ed.)
IOS Press, 2005
© 2005 The authors

Chapter 8

Einstein and Quantum Theory

Seiya Abiko

Seirei Christopher College, 3453 Mikatahara-Town, Hamamatsu-City, 433-8558, Japan

1. Prelude: Einstein's Encounter with Planck's Theory

Albert Einstein wrote, in 1946, when he was 67, his autobiography, "Autobiographical Notes" (abbreviated as "Notes" below), which was published three years later in a book entitled *Albert Einstein: Philosopher–Scientist*, edited by Paul Arthur Schilpp [21]. According to Schilpp, initially Einstein was not willing to write it but, after much persuasion by Schilpp, Einstein changed his mind, because he realized that it was his obligation to mankind to inform them how and why he reached his conclusions and interpretations in each of his works [56].

At the start of "Notes", he states that he wrote it as his own 'obituary' and, therefore, one may believe that he tried to write it as accurately as possible. Moreover, for such a versatile writer as Einstein, we have every reason to believe that he was as lively at that time as he had been in his prime; in fact, he was to live for another, active, nine years.

In "Notes", Einstein indicates his two points of view according to which it is possible to assess physical theories in general. They are 'external confirmation' (the theory must not contradict empirical facts) and 'internal perfection' (the 'naturalness' or 'logical simplicity' of its premises). These two seem to constitute for him the invariable prerequisites of a physical theory. From both viewpoints, he criticizes mechanics as the foundation of physics. Then, he describes the two revolutionary events in physics: the introduction of the field concept by Faraday–Maxwell's electromagnetic theory, and Max Planck's investigation into thermal radiation (1900). This chapter starts from this second revolutionary event.

> In his youth, Planck studied Rudolf Clausius' *Mechanical Theory of Heat* (*Mechanische Wärmelehre*) in detail, and prepared his doctoral dissertation on the second law of thermodynamics. He continued his studies on thermodynamics and applied their results successfully to physical chemistry. On this basis, in 1888 he was appointed successor of Gustav Kirchhoff, as assistant professor at the University of Berlin and director of the Institute for Theoretical Physics [42].

In the vicinity of the university, there was the newly founded Physical-Technological Imperial Institute (Physikalische-Technische Reichsanstalt), where precise measurements of thermal-radiation spectra, needed for the technological purpose of high-temperature measuring, were pursued. To combine the thermal-radiation spectra with the temperature of an emitting body, a precise radiation distribution formula for black-body radiation[1] was necessary. After a number of trials, Planck finally found out the radiation formula (which was later named after him), which exactly reproduced the experimental data.

His next step was to derive this radiation formula from the three fundamental theories of classical physics: i.e. mechanics, electromagnetism, and thermodynamics. After considerable efforts, he finally adopted Ludwig Boltzmann's statistical method. Boltzmann, in his gas theory, had assumed that the amount of kinetic energy each molecule could take is only an integral multiple of a minimum value ε. After the statistical calculation utilizing the probability theory, Boltzmann set $\varepsilon = 0$ and thus succeeded in deducing the equilibrium energy distribution formula of gas molecules he was aiming at.

Planck had already derived, utilizing electromagnetic theory and thermodynamics, the relationship between black-body radiation energy-density $\rho(v, T)$,[2] and the average energy value of the radiation-emitting elements, named as 'resonators' (i.e. minute antennas modelling atoms and molecules) with proper frequency v at equilibrium with radiation. In order to obtain the latter average energy value of 'resonators', he applied Boltzmann's method to calculate the number of ways in which the total amount of energy is divided among 'resonators'. In a similar manner to Boltzmann's method, Planck deduced his radiation formula successfully, except for one crucial difference. In order to obtain his radiation formula, Planck had to set $\varepsilon = hv$ instead of $\varepsilon = 0$, where h is a proportionality constant which was later named Planck's constant.

On this matter, Einstein stated in "Notes" [19]:

> Planck got his radiation-formula if he chose his energy-elements ε of the magnitude $\varepsilon = hv$. The decisive element in doing this lies in the fact that the result depends on the taking for ε a definite finite value, i.e. that one does not go to the limit $\varepsilon = 0$. This form of reasoning does not make obvious the fact that it contradicts the mechanical and electrodynamic basis, upon which the derivation otherwise depends. Actually, however, the derivation presupposes implicitly that energy can be absorbed and emitted by the individual resonator only in 'quanta' of magnitude hv, i.e. that the energy of a mechanical structure capable of oscillation as well as the energy of radiation can be transferred only in such quanta – in contradiction to the laws of mechanics and electrodynamics. [...] All of this was quite clear to me shortly after the appearance of Planck's fundamental work; so that, without having a substitute for classical mechanics, I could nevertheless see to what kind of consequences this law of temperature-radiation[3] leads for the photo-electric effect[4] and for other related phenomena of transformation of radiation-energy, as well as for the specific heat of (especially) solid bodies. All my attempts, however, to adapt the theoretical foundation of physics to this new type of knowledge failed completely. It was as if the ground had been pulled out from under one, with no firm foundation to be seen anywhere, upon which one could have built.

If we believe that this statement is accurate, Einstein was shocked by Planck's theory "shortly after" its appearance in 1900, and delved deeply into the foundations of physics even before he published his first paper in 1901. The excerpt above from "Notes" tells us that his concern with black-body theory began at the same time as his scientific career.

In conflict with the above, Thomas S. Kuhn stated in his book *Blackbody* of 1978 [47]:

---

[1] 'Black-body radiation' is a kind of thermal radiation which is at equilibrium with the emitting body kept at a constant temperature.

[2] The radiation energy-density coincides with the radiation distribution formula apart from a numerical factor.

[3] 'Temperature-radiation' is another name for 'thermal radiation'.

[4] Photo-electric effect' is a phenomenon of the emission of electrons (or cathode rays) from a body irradiated by (usually ultraviolet) light, discovered by Heinrich Hertz in 1887.

Einstein's paper [i.e. the 1903 paper of the statistical trio from 1902 to 1904], written *before he showed any signs of concern with black-body theory*, bridged these gaps. [Italics added]

The "gaps" Kuhn referred to are Planck's uncertainty about the probability of states and his lack of direct statistical expressions for entropy[5] and temperature. In my view, however, it was in order to bridge "these gaps" in Planck's theory that Einstein constructed his theory of the statistical trio papers from 1902 to 1904, as will be explained below.

Usually, this kind of 'Autobiography' or 'Recollections' written in later years is regarded as unreliable due to the time that has elapsed between the events and writing about them. But, as will be shown below, we can confirm the accuracy of his description by comparing the text with numerous other documents Einstein left, e.g. his letters, papers and lectures.

In fact, the love letters he wrote in 1901 to his fiancée Mileva Marič contain the following remarkable statements [57]:

> I've begun to have reservations of a fundamental nature about Max Planck's studies of radiation, so much so that I'm reading his paper with mixed feeling. [4 April 1901]
>
> Planck assumes that a specific kind of resonator (fixed period and damping) causes the conversion of the radiation energy, an assumption that I have difficulty accepting. [10 April 1901]

Here, Einstein astutely discerned the problem in Planck's reasoning based on the classical theories of resonators.

We can also find in his love letters descriptions corresponding to the passage in "Notes" quoted above, "for the photo-electric effect and for other related phenomena of transformation of radiation-energy, as well as for the specific heat of (especially) solid bodies":

> It seems to me that it is not out of question that the latent kinetic energy of heat in solids and fluids can be thought of as the electric energy of resonators. If this is the case, then the specific heat and the absorption spectrum of solids would have to be related. [23 March 1901] [58].
>
> I was recently struck by the idea that when light is generated, a direct transformation from energy of motion into light might occur because of the parallel: the kinetic energy of the molecules – the absolute temperature – spectrum (spatial radiation energy in the state of equilibrium). [30 April 1901] [59].
>
> I have just read a wonderful paper by Lenard on the generation of cathode rays by ultraviolet light. Under the influence of this beautiful piece I am filled with such happiness and joy. [28 May 1901] [60].

These quotations from Einstein's letters prove to us that the quoted description from his "Notes" is accurate.

2. Did Planck Introduce a Physical Energy Quantum?

With regard to Einstein's comment quoted in the preceding section, "[Planck's] form of reasoning does not make obvious the fact that it contradicts the mechanical and electrodynamic basis", there arises the question: didn't Planck introduce the 'quantum postu-

---

[5] 'Entropy' is a physical quantity describing the state of a system (a substance or a collection of substances) introduced by Rudolf Clausius in 1865. He introduced it in order to express the second law of thermodynamics mathematically by describing the irreversibility of an isolated system as the increase of entropy.

late' that showed explicit contradiction with the mechanical and electrodynamic basis? This was also the point raised by Kuhn in *Blackbody*, quoted earlier. Kuhn's answer was negative [49].

In order to assess Kuhn's judgment, let us consult the portion of Planck's 1900 paper in which he is usually regarded as having introduced his 'quantum postulate' [54]:

> This constant [i.e. Planck's constant h] multiplied by the common frequency v of the resonators gives us the energy element ε in ergs, and, dividing E by ε, we get the number P of energy elements which must be divided over the N resonators. *If the ratio thus calculated is not an integer, we take for P an integer in the neighbourhood.* [my italics]

What matters to us here is the italicized sentence, where Planck clearly indicates that the total energy E of resonators is not necessarily an integral multiple of the energy element ε. Therefore, the so-called 'quantum postulate' Planck is said to have introduced should be regarded merely as:

(1) the introduction of a natural constant h,
(2) the division of the energy scale into intervals composed of 'energy elements' $\varepsilon = hv$.

The first half (the introduction of h) was an inescapable consequence of Wien's displacement law.[6] In fact, the value of h had already been introduced by Planck into Wien's radiation formula (under the notation 'b') in his 1899 paper [53], that is, *before* the construction of his own radiation formula. Also, it was only in 1906, i.e. after the introduction of the term 'light-quantum' (Lichtquantum) by Einstein, that Planck designated h as the 'quantum of action' (Wirkungsquantum) [55]. On the other hand, the second half (the division into intervals of ε) was merely a mathematical device in order to deduce a finite value for the number of divisions. Therefore, the "energy element" Planck referred to had no physical basis.

That Einstein also shared this interpretation of Planck's text can be ascertained by consulting Einstein's 1906 review [32] of Planck's book, *Lectures on Thermal Radiation* (*Vorlesung über die Theorie der Wärmestralung*, Leipzig, 1906). Here, Einstein used the term 'elementary regions' (Elementargebiete) instead of 'energy elements' (Energieelemente), the term used by Planck, which confirms that Einstein denied Planck's introduction of the discreteness of energy or energy quantization.

3. Einstein's Research Programme: Construction of Statistical Thermodynamics

As I pointed out in Chapter 4, Section 7, there was an error in "Notes" in the first and the second editions of *Albert Einstein: Philosopher-Scientist*. The relevant portion after the correction reads [22]:

> Reflections of this type made it clear to me as long ago as shortly after 1900, i.e., shortly after Planck's trailblazing work, that neither mechanics nor *electrodynamics* (except in limiting cases) claim exact validity. By and by I despaired of the possibility of discovering the true laws by means of constructive efforts based on known facts. [Italics added]

---

[6]Wien's displacement law states that v and T always appear in radiation formulas in the form v/T. This law was discovered and derived by Wilhelm Wien in 1893.

The above quotation tells us that, "shortly after Planck's trailblazing work", he realized "that neither mechanics nor electrodynamics can claim exact validity". Thus, of the three fundamental theories of classical physics (i.e. mechanics, electromagnetism, and thermodynamics), Einstein regarded thermodynamics as "the only physical theory of universal content which, within the framework of the applicability of its basic concepts, will never be overthrown" [18]. Therefore, it is certain that the first core ingredient of Einstein's research programme was 'thermodynamics'.

But, as is well known, the applicability of the basic concepts of thermodynamics is much restricted. Therefore, he should have felt it necessary to broaden its applicability. His love letters quoted in Section 1, which show that he was studying closely the microscopic mechanisms of energy transformation, could be read as indicating the direction of his efforts to broaden the applicability.

After the quoted part from his "Notes" in Section 1, the following lines read [20]:

> My own interest in these years was less concerned with the detailed consequences of Planck's results, [...] My major question was: What general conclusions can be drawn from the radiation-formula concerning the structure of radiation and even more generally concerning the electromagnetic foundation of physics?

This paragraph can be interpreted as the specification of his research programme in his early years, i.e. his aim was to draw conclusions concerning "the structure of radiation" and "the electromagnetic foundation of physics" from Planck's radiation formula.

Several other portions of his love letters (e.g. those dated 13 September 1900 [61], 3 October 1900 [63] and 15 April 1901 [62]) reveal the actual development of his research programme. According to them, he was much attracted to Boltzmann's kinetic theory of gases, the ionic theory of physical chemistry in conflict with Maxwell's electromagnetic theory, and the influence of intermolecular forces on the dissipative processes. As a matter of fact, these themes correspond to the five papers he wrote before 1905. As I pointed out elsewhere [2], in those early years Einstein was concerned with the connection between macroscopic and microscopic aspects of matter. In fact, his first and second papers try to apply the expression of intermolecular forces, which he conjectured modelling the law of universal gravitation, to the macroscopic phenomena of capillarity and of the contact electric-potential difference.

Following the above two papers on concrete physical phenomena, between 1902 and 1904 he published three papers of a more general scope on statistical physics (referred to below as 'the statistical trio') [14]. Being unsatisfied with Planck's theory, Einstein's motive for composing the statistical trio must have been somehow to construct a general thermodynamics, on which the theories, not only of fluids and solids, but also of thermal radiation could be based.

Since Planck's theory was modelled on Boltzmann's theory, Einstein's aim in the first paper of the statistical trio was to attempt to close the gap in Boltzmann's kinetic theory of heat. The gap Einstein referred to is the lack of a derivation of the laws of thermal equilibrium and the second law of thermodynamics using only the equations of mechanics and the probability calculus. More precisely, he constructed a statistical theory which treats an ensemble of systems enclosed by an environment kept at constant temperature T, i.e. the case later called the 'canonical ensemble'.

In this first paper of the statistical trio, he also derived the mathematical expression for entropy from the standpoint of mechanics, upon which he made an important comment [14, p. 47]:

This expression for the entropy is remarkable, because it depends only on E and T, while the special form of E as a sum of potential energy and kinetic energy no longer appears. This fact suggests that the result we obtained is more general than the mechanical representation here adopted.

In accordance with the above, the second statistical paper, published in 1903, begins by asking the question [14, p. 48]:

Whether the kinetic theory is really necessary for the derivation of the foundations of the theory of heat obtained in the 1902 paper, or assumptions of a more general nature may suffice?

Then he showed that, in fact, the latter is the case. He derived, without resorting to the kinetic theory, the foundations of the theory of heat with regard to the most general system expressed, not by coordinates and momenta, but by 'state-variables' (*Zustandsvariabeln*) $p_1, \ldots, p_n$.

The generality of the system treated is restricted only by two conditions: first, the existence of causal equations for the state-variables:

$$\frac{dp_i}{dt} = \phi_i(p_1 \cdots p_n) \quad (i = 1, \ldots, n) \tag{1}$$

and, second, the existence of the energy equation for these causal equations

$$E(p_1 \cdots p_n) = const. \tag{2}$$

Accordingly, his theory of statistical physics underwent a transformation from 'statistical mechanics' into 'statistical thermodynamics' at this point.

The last paper of the trio, published in 1904, was written as an addendum to the second. Here Einstein made the following important remark [14, p. 68]:

I derive the expression of the entropy of a system, which is completely analogous to that found by Boltzmann[7] for ideal gases and assumed by Planck in his theory of radiation.

Einstein derived an expression for the entropy S of the general thermodynamical system treated in the second paper, as follows (using the modern notation of Boltzmann's constant k instead of 2κ in Einstein's paper):

$$S = k \log[\omega(E)], \tag{3}$$

where

$$\omega(E)\delta E = \int_E^{E+\delta E} dp_1 \cdots dp_n. \tag{4}$$

Though he did not give this any name, we will call $\omega(E)$ the 'state-density' (as a function of energy).

As the system under consideration is enclosed by an environment kept at constant temperature T, the system can exchange energy with the environment. As a result, the

---

[7]Boltzmann expressed the entropy of a state of the system under consideration by a logarithm of the realization-probability of the state. This expression, which Planck utilized in the derivation of his radiation formula, was later named 'Boltzmann's principle' by Einstein.

energy E of the system fluctuates around its average value $\overline{E}$.[8] If we put $E = \overline{E} + \varepsilon$, then $\overline{\varepsilon^2} = \overline{E^2} - \overline{E}\,\overline{E}$ represents a measure of the thermal stability of the system, in the sense that the smaller its value, the greater the stability. With regard to this quantity, he derived, utilizing Eqs (3) and (4), the expression [14, p. 75]:

$$kT^2 \frac{d\overline{E}}{dT} = \overline{E^2} - \overline{E}\,\overline{E}. \tag{5}$$

Thus, he concluded that the constant k determines "the thermal stability of the system".

Next, he looked for a system for which it was possible to determine $\overline{\varepsilon^2}$, and stated [14, p. 76]:

> In fact, there is only a single kind of physical system for which we can surmise from experience that it possesses energy fluctuation: this is empty space filled with temperature radiation.

Thus, he finally applied his theory to black-body radiation. Applying Stefan–Boltzmann's law[9] and the value of k obtained from the kinetic theory of gases to Eq. (5), he could successfully obtain a version of Wien's displacement law $\lambda_m \propto 1/T$, i.e. the correct dependence of the wavelength λ_m of the maximum radiation energy on the temperature T. What is more, the proportionality coefficient deduced was in order of magnitude agreement with the observed value.

We see here that Einstein went one step further to his final aim. What enabled him to do so was his departure from 'statistical mechanics' to the more general 'statistical thermodynamics'. Statistical thermodynamics, and the theory of fluctuation thus constructed, underlay his doctoral dissertation, the theory of Brownian movement, the theory of light quantum (so far 1905), the first quantum theory of matter (1906), and the theory of specific heat (1907). Furthermore, as I discussed in Chapter 4, there was a close relationship between his special theory of relativity and the statistical thermodynamics.

4. Investigation into the Constitution of Radiation: The Light-Quantum Theory

In the last quarter of the nineteenth century, the phenomenon of Brownian movement was becoming widely known. While a minority of scientists still attributed its cause to such effects as electrical effects, osmosis, or surface tension, most seemed to think that it must be connected with thermal molecular motions. However, on account of the immense difference in size between small particles and molecules, those who were familiar with the kinetic theory of gases were suspicious of this view. Moreover, there was still no quantitative theory to be tested against experiments [8].

Not only did Einstein's theory of Brownian movement [25] provide a breakthrough in the understanding of this phenomena, it also did, to borrow Max Born's wording, "more than any other work to convince physicists of the reality of atoms and molecules, of the kinetic theory of heat, and of the fundamental part of probability in the natural laws" [7].

---

[8]The horizontal bar designates the average over the ensemble of systems, which is assumed to coincide with the temporal average of one of the systems.

[9]Stefan–Boltzmann's law states that the total radiation energy emitted per second from a black-body kept at temperature T takes a finite value proportional to T^4. This law was discovered by Joseph Stefan in 1879 and derived by Boltzmann in 1884.

"Notes" proceeds, after a description of the statistical trio and their successful application to Brownian movement, to a thought-experiment[10] on the Brownian movement of a small mirror suspended in a cavity radiation.[11] This thought-experiment required the assumption that "radiation energy consists of indivisible point-like localized quanta of energy $h\nu$", as follows [23]:

> Upon this recognition [i.e. the success of the theory of Brownian movement], a relatively direct method can be based which permits us to learn something concerning the constitution of radiation from Planck's formula. One may conclude in fact, in a space filled with radiation, a freely moving, quasi monochromatically reflecting [two-sided] mirror would have to go through a kind of Brownian movement, the average kinetic energy of which equals $kT/2$.[12] If radiation were not subject to local fluctuations, the mirror would gradually come to rest, because, due to its motion, it reflects more radiation on its front than on its reverse side.[13] However, the mirror must experience certain random fluctuations of the pressure exerted upon it due to the fact that the wave-packets, constituting the radiation, interfere with one another. These can be computed from Maxwell's theory. This calculation, then, shows that these pressure variations are by no means sufficient to impart to the mirror the average kinetic energy $kT/2$. In order to get this result, one has to assume rather that there exists a second type of pressure variations, which cannot be derived from Maxwell's theory, which corresponds to the assumption that radiation energy consists of indivisible point-like localized quanta of energy $h\nu$, which are reflected undivided.

The above description continues with the passage cited at the start of Section 3 (i.e. the passage in which the correction was made) and then ends with remarks on special and general relativity. Therefore, we can see that the analysis of the Brownian movement of a suspended mirror was made well before 1905 ("shortly after 1900"), i.e. before his famous light-quantum theory was constructed. Therefore, let us take up this thought-experiment of a suspended mirror, before entering into the light-quantum theory.

The thread of his thought seems to have been the following. After reading Planck's 1900 paper, Einstein thought that, in order to derive Planck's formula, the resonator energy should be restricted to discrete values. But this means, from the viewpoint of energy conservation, that the energy of thermal radiation itself should also take only a discrete set of values. Therefore, he needed a more direct way to corroborate this inference. This requirement subsequently led him to turn to the thought-experiment of Brownian movement of a suspended small mirror, which was what he meant by "a relatively direct method [. . .] to learn something concerning the constitution of radiation from Planck's formula". As we will show shortly, it was by this thought-experiment that he introduced the so-called wave-particle duality for the first time.

The concrete procedure of the calculation is given in papers written later in 1909 and 1910. According to the equipartition law of statistical thermodynamics, in the environment kept at constant temperature T, the mean-square value $\overline{v^2}$ of the mirror velocity

---

[10] A thought-experiment is an imaginative experiment considered in order to make inferences theoretically.

[11] Black-body radiation can be realized experimentally by the radiation confined within a cavity surrounded by walls kept at constant temperature. This type of radiation is called 'cavity radiation'.

[12] According to the equipartition law of statistical thermodynamics, in a system kept at constant temperature T, the average kinetic energy of a body per a degree of freedom (i.e. one of possible modes) of its motion should be $kT/2$.

[13] Maxwell's electromagnetic theory states that there is radiation pressure in the direction normal to the electromagnetic waves, and numerically equal to the radiation energy in unit volume.

v in the Brownian movement must take the value (using k instead of R/N in Einstein's paper):

$$\overline{v^2} = \frac{k}{m} T. \tag{6}$$

In the 1909 paper [26], Einstein introduced the increment Δ of the velocity v of the two-sided small mirror during a short time-interval τ. This increment of the velocity is brought about by the random fluctuation of the pressure of radiation exerted on the mirror. The condition that the value $\overline{v^2}$ should remain unchanged during τ leads to

$$\overline{\Delta^2} = \frac{2P\tau}{m}\overline{v^2}, \tag{7}$$

where m is the mass of the mirror, and P is the constant of radiation friction occurring due to the excess reflection on the front than the reverse side of the mirror (because of its motion). An expression of P is given, without its derivation in this paper, in terms of radiation energy-density ρ (as a function of v and T per unit volume and per unit frequency range).

It is in the 1910 paper [40] that the derivation of the expression for P in terms of ρ was shown, where Maxwell's equation as well as the results of the special theory of relativity of 1905 were utilized. As stated in Chapter 4 of this book, it might be in order to fulfil this urgent requirement that the special theory of relativity was constructed at that time. This expression for P was then utilized, in combination with the calculated results for $\overline{\Delta^2}$ from Maxwell's equation, to yield the expression for ρ, the result of which was Rayleigh–Jeans' radiation formula:

$$\rho = \frac{8\pi v^2}{c^3}kT. \tag{8}$$

Returning to the 1909 paper, Einstein examined the consequence of the substitution, as ρ in the expression of P, of Planck's radiation formula:

$$\rho = \frac{8\pi h v^3}{c^3}\frac{1}{e^{(hv)/(kT)} - 1}. \tag{9}$$

The resultant $\overline{\Delta^2}$ in the frequency interval $v \sim v + dv$ with surface area f of the mirror was:

$$\frac{\overline{\Delta^2}}{\tau} = \frac{1}{c}\left[hv\rho + \frac{c^3}{8\pi}\frac{\rho^2}{v^2}\right]f\,dv. \tag{10}$$

In the above, the second term on the right-hand side is the same form as that derived in the 1910 paper utilizing Maxwell's theory. The first term on the right-hand side, however, indicates the same fluctuation as that exhibited in a collection of gas molecules each having energy hv. In other words, *the first term exhibits particle-like fluctuation, while the second term exhibits wave-like fluctuation.*

Thus, wave-particle duality, the most peculiar feature of quantum theory, was first introduced by Einstein in this 1909 paper [43]. If one inserts $\rho V\,dv$ as $\overline{E}$ into Eq. (5), with ρ substituted by Planck's formula Eq. (9), the same type of result as in Eq. (10) is obtained for the energy fluctuation of the radiation component ν. According to "Notes", all these considerations were carried out well before 1905.

Encouraged by the success of these investigations into radiation, Einstein proceeded to the full consideration of the consequences of Planck's formula, the result of which was his famous light-quantum theory of 1905 [24]. In this paper, he first pointed out that, if we take the standpoint of Maxwell's electromagnetic theory and Lorentz's electron theory, the resulting temperature-radiation formula cannot be other than Rayleigh–Jeans' Eq. (8), which contradicted experience. Specifically, according to this formula, the total energy of black-body radiation in a cavity becomes infinite, in contradiction to Stefan–Boltzmann's law.

Then, he proceeded to examine the opposite limit (i.e. large ν/T) to the Rayleigh–Jeans' of the Planck's formula, viz. Wien's radiation formula:

$$\rho = \frac{8\pi h \nu^3}{c^3} e^{-(h\nu)/(kT)}. \tag{9}$$

Utilizing the above, he deduced the entropy of radiation[14] and compared it to the entropies of ideal gases and dilute solutions. The result of the comparison showed that "the entropy of monochromatic radiation of sufficiently low density varies with the volume according to the same law as the entropy of an ideal gas or a dilute solution"[16].

He combined this fact with Boltzmann's principle, which deduced as the probability of finding radiation energy E in the portion v of volume v_0:

$$W = \left(\frac{v}{v_0}\right)^{E/h\nu}. \tag{10}$$

From this he concluded that "monochromatic radiation of low density (within the range of Wien's radiation formula) behaves thermodynamically as if it consisted of mutually independent energy quanta of magnitude $h\nu$" [17].

The paper ends with the indication of three supporting pieces of evidence of this view of 'light-quantum': Stokes' rule in photoluminescence; the generation of cathode rays in the photoelectric phenomena; and the ionization of gases due to the irradiation of ultraviolet light. These phenomena correspond to the following statement in his introduction to this paper [15]:

> [I]t is quite conceivable, despite the complete confirmation of the theory of diffraction, reflection, refraction, dispersion, etc., by experiment, that the theory of light, operating with continuous spatial functions, leads to contradictions when applied to the phenomena of emission and transformation of light.

The decrease in frequency ν of emitted light compared with that of absorbed light in Stokes' rule could be interpreted as a loss of absorbed light energy during the process occurring inside the material. Also, the emission of electrons in the photoelectric phenomena, and in the ionization of gases by the ultraviolet light irradiation, would not take place without a local concentration of light-energy exhibited by the 'light-quantum'.

---

[14]The application of thermodynamics to black-body radiation was first carried out by Boltzmann in his derivation of Stefan–Boltzmann's law. Following Boltzmann's method, Wien defined the temperature and the entropy of radiation in his paper of 1894.

In the photoelectric phenomena, the light-quantum energy, which penetrates a body's surface layer, can be converted into the kinetic energy of electrons, which subsequently brings about their emission from the body. In leaving the body, each electron has to do some work P. Therefore, the maximum kinetic energy of the emitted electron should be $h\nu - P$.

From this, Einstein proposed his famous photoelectric equation:

$$eV = h\nu - P, \tag{11}$$

where e is the electron charge, V is the positive electric potential of the body raised just enough to prevent the emission of electrons from it. Making use of the approximation $P = 0$, Einstein estimated the value of V from existing data, and obtained "a result that agrees in order of magnitude with the results" of Philipp Lenard's experiment of 1902. Although Lenard's results provided qualitative evidence for Eq. (11), it was not until 1916 that the crucial evidence was provided by Robert Millikan from his experiments.

5. Einstein's First Quantum Theories of Matter and Radiation

It is in his 1906 paper [28] that the full derivation of Planck's radiation formula from Einstein's statistical thermodynamics was presented for the first time. This time, he turns to the collection of resonators, just as Planck did, instead of radiation itself. Einstein showed that, if the energy of resonators is permitted to take a continuous value, the resulting radiation formula is necessarily that of Rayleigh–Jeans' Eq. (8), but if we restrict the energy of resonators to integral multiples of $h\nu$, then Planck's formula (Eq. (9)) results.

The outline of his logic is as follows. As for the relation between the mean energy $\overline{E}$ of resonators with proper frequency ν and energy density ρ of component ν of the radiation, he assumes that Planck's result from the Maxwell's theory holds:

$$\overline{E} = \frac{c^3}{8\pi\nu^2}\rho. \tag{12}$$

In accordance with Planck's method, the expression for $\overline{E}$, and therefore ρ, can be obtained from the thermodynamic relationship $1/T = \partial S/\partial\overline{E}$, if the entropy S of resonators is expressed as a function of $\overline{E}$. In order to accomplish the latter, one can utilize Boltzmann's principle:

$$S = k \log W, \tag{13}$$

where W is the number of divisions of the total energy to each resonator.

Einstein's point of departure from Planck's method is in the expression for W. According to Einstein's statistical thermodynamics discussed in Section 3, W is obtainable from Eqs (3) and (4), by using the relationship

$$\int_{E_\alpha}^{E_\alpha + dE_\alpha} dx_\alpha \, d\xi_\alpha = const \cdot dE_\alpha \tag{14}$$

in terms of the variables describing the resonator α. The result is:

$$W = \int_{H}^{H+\delta H} dE_1 \cdots dE_n, \tag{15}$$

where H is the total energy of n resonators.

If one integrates directly Eq. (15), one gets the equipartition law $\overline{E} = kT$ for the mean energy value per resonator [45], therefore Eq. (12) yields Rayleigh–Jeans' formula Eq. (8) for ρ. On the other hand, it is only when one restricts each value of $E_1, \ldots, E_n$ to integral multiples of $h\nu$, thus reducing the multiple integral to a multiple summation, that one can obtain Planck's result:

$$W = \frac{(n + p - 1)!}{(n - 1)!p!}, \quad \text{where } p = \frac{H}{h\nu}. \tag{16}$$

That is to say, one obtains Planck's Eq. (9), only when one introduces into the right-hand side of Eq. (14), instead of *const*, the state-density $\omega(E_\alpha)$ of resonator α as a discrete entity (i.e. in the form of a sum of delta-functions, in modern mathematical language).

With regard to this part of Einstein's 1906 paper, Kuhn states [46]:

> That passage is the first public statement that Planck's derivation demands a restriction on the classical continuum of resonator states. In a sense, it announces the birth of the quantum theory.

Thus, in accordance with Kuhn, quantum discontinuity was first introduced in the form of physical theory by Einstein in 1906. At the same time, Einstein's statistical thermodynamics became the first quantum theory, which may have been what Einstein had in mind ever since he began the statistical trio.

Einstein's first quantum theory was readily applied in the following year to the problem of specific heats, which resulted in the 1907 paper [30], where his statistical thermodynamics is applied to the collection of thermal vibrations of atoms in solids. Here, in accordance with what he wrote to Marič in his letter dated 23 March 1901, and cited in Section 1, the vibrations of atoms are treated in the same way as the resonators in the thermal radiation problem.

With the classical continuum of state-density, one obtains the equipartition law for the vibration energy, which yields the Dulong–Petit law[15] for specific heat. According to Einstein's view, however, this law corresponds to Rayleigh–Jeans formula Eq. (8) for radiation. Therefore, the correct specific heat value corresponding to Planck's law, Eq. (9), could only be obtained by introducing discreteness into the state-density of the vibrating atoms. The temperature dependence of the value of specific heat, thus obtained by Einstein, shows that, irrespective of the species of matter, the specific heat value should come nearer to zero as the temperature goes to zero.

Walther Nernst, professor of physical chemistry at the University of Berlin, was astonished at the agreement of Einstein's theory with his experimental results, especially with the 'Nernst's theorem' concerning the temperature dependence of thermodynamic quantities in the low temperature region. Therefore, he convened the first Solvay conference on 'Radiation Theory and Quanta' in 1911. It was after this conference that contemporary physicists came to consider seriously the theory of quanta. The proceedings

---

[15]Dulong–Petit's law states that the product of the specific heat [cal/g K] and the atomic weight [g/mol] is constant for different simple solids. This law was discovered by P.L. Dulong and A.T. Petit in 1819, and was utilized to determine the atomic weight of various elements.

of this conference [50] left a lasting impression on physicists, two fruits of which were Niels Bohr's theory of atomic structure in 1913 [5] and Louis de Broglie's theory of matter waves in 1923 [10].

At the conference, after his lecture on "The Present State of the Problem of Specific Heat", Einstein stated [34]:

> [I]t also turned out that classical mechanics [. . .] can no longer be viewed as a useful schema for theoretical representation of all of physical phenomena. This raises the question of which general laws of physics we can still expect to be valid in the domain with which we are concerned. To begin with, we will all agree that the energy principle is to be retained. In my opinion, another principle whose validity we must maintain unconditionally is Boltzmann's definition of entropy through probability.

As the two laws Einstein referred to above correspond to the first and the second laws of thermodynamics, this statement confirms my conjecture that the core ingredient of Einstein's research programme was thermodynamics. It may also clearly explain the reason why he constructed the statistical thermodynamics of 1903–1904, aiming at the first quantum theory.

Einstein discussed 'Nernst's theorem' again in 1914 paper [12], where he treated a collection of molecules carrying one resonator each. What is interesting about this paper is that the chemical distinction between molecules is reduced to the difference in discrete energy values that the resonators carried by them can take. Moreover, as will be discussed in Section 7, his concern with Nernst's theorem (later known as 'the third law of thermodynamics') led him to the quantum theory of ideal gas in 1924–1925 [31].

Having finished the construction of the general theory of relativity, in 1916 Einstein returned to the problem of thermal radiation. The resultant paper [13] consists of three parts, the first of which is entitled "Planck's Resonator in a Field of Radiation". In this part, Einstein distinguished two kinds of energy change taking place in resonators over a short time interval, the first of which is due to the emission of radiation, and the second is due to the work done by the electric field of the incident radiation on the resonator.

The second part of the paper is entitled "Quantum Theory and Radiation". This section treats a collection of identical molecules in static equilibrium with thermal radiation. In accordance with Bohr's theory of 1913, each molecule is assumed to be in discrete energy states. It should be remarked here that, whereas Bohr and his colleagues, e.g. Arnold Sommerfeld, treated the electronic motion within a single atom, Einstein treated a collection of molecules in thermal equilibrium. Here we can see a clear difference in their approaches to quantum theory, which brought forth their subsequent disagreement about the character of quantum theory. We will return to this problem in the final section of this chapter.

In a somewhat similar manner to the 1914 paper, Einstein assumed that the molecules interact with radiation in just the same way as resonators do. Corresponding to the first kind of energy change of the resonator, he put the number of transitions per unit time from higher energy state m to lower n as $A_m^n N_m$, where N_m is the number of molecules in the state m. Corresponding to the second kind, he put the number of transitions per unit time from lower state n to higher m, and from higher m to lower n, as $B_n^m N_n \rho$ and $B_m^n N_m \rho$ respectively, with ρ being the radiation energy density. The values of A_m^n and $B_m^n \rho$ are identified, in the next paper to be discussed below, as the probability of transition from state m to n per unit time per atom. Therefore, we can see that the concept of 'transition probability' was first introduced into physics by Einstein in these papers.

Assuming the statistical equilibrium between the transitions from m to n and from n to m, i.e.

$$A_m^n N_m + B_m^n N_m \rho = B_n^m N_n \rho, \tag{17}$$

and assigning to the relative values of N_m and N_n their probabilities at thermal equilibrium with constant temperature T,

$$\frac{N_n}{N_m} = \frac{p_n}{p_m} e^{(\varepsilon_m - \varepsilon_n)/(kT)}, \tag{18}$$

he could derive the same type of relationship between ρ and T as in Planck's Eq. (9). Applying Wien's displacement law to this relationship, he could also obtain Bohr's frequency condition $\varepsilon_m - \varepsilon_n = h\nu$.

At this point, Einstein remarked [13, pp. 215–216]:

> [T]he natural connection to Planck's linear oscillator (as a limiting case of classical electrodynamics and mechanics) seem to make it highly probable that these are basic traits of a future theoretical representation. [Italics added]

This remark is believed to be the root of Bohr's 'correspondence principle' mentioned in Bohr's 1918 paper on the atomic spectra, where this paper was cited [6].

Einstein also pointed out that the first kind of energy change is simply Rutherford's law of radioactive decay. Therefore, the first kind of change (later called 'spontaneous emission') is regarded as corresponding to the particle-like behavior of radiation. On the other hand, the second kind of change (later called 'induced emission and absorption'), which derives from the resonator's interaction with the electric wave of radiation, is regarded as corresponding to the wave-like property of radiation. Thus, the deduction of Planck's formula in this paper seems to be based on reversed inference from the fact that Planck's formula leads both to particle-like and wave-like behaviour in the fluctuation of thermal radiation, with which the molecules are kept at equilibrium.

The last part of this paper is entitled "Remark on the Law of Photochemical Equivalence". The latter law, first expounded in his 1912 paper [33], means that the decomposition of one molecule in a photochemical process correponds to the absorption of one light quantum $h\nu$.

In another 1916 paper, entitled "On the Quantum Theory of Radiation" (the same paper was published again in 1917), Einstein posed the question [27]:

> Does the molecule receive an impulse when it absorbs or emits energy ε? For example, let us look at emission from the point of view of classical electrodynamics. When a body emits energy ε it suffers a recoil (momentum) ε/c if the entire amount of radiation energy ε is emitted in the same direction.

He answered this question himself as follows [27, pp. 221–222]:

> It turns out that we arrive at a theory that is free of contradictions, only if we interpret those elementary processes as completely directed processes. Here lies the main result of the following considerations. [Einstein's italics]

Here, Einstein treated the motion of molecules immersed in thermal radiation. In the same type of reasoning as that used in the Brownian movement of a suspended mirror in his 1909 paper, he derived the same relationship as that obtainable by combining Eqs (6) and (7). He calculated both sides of this equation independently, utilizing the quantum theory of radiation of the preceding paper. Calculations of the left-hand side (i.e. $\overline{\Delta^2}$) was done in the rest coordinate system, while the right-hand side (i.e. P) was done in the moving coordinate system by utilizing special relativity.

The result of the calculations showed that it is only when one allows the association of momentum transfer $(\varepsilon_m - \varepsilon_n)/c$ that both sides of the equation agree. In this way, Einstein showed that his 'light quantum' carries, not only energy $h\nu$, but also momentum $\frac{h\nu}{c}$, thus reducing the 'light quantum' to the modern concept of a photon. However, the acceptance of this concept by the wider physical community required the discovery of the Compton effect in 1923.

6. Interlude: Two Research Traditions in Physics at the Turn of the Century

Usually, modern science, especially physics, is regarded as constructed during the 'Scientific Revolution' that took place during the sixteenth and seventeenth centuries. However, it is also well known that, during the nineteenth century, modern science underwent another transformation, due to the Industrial Revolution of the eighteenth and nineteenth centuries and the accompanying institutionalization of science, i.e. the establishment of various scientific communities and the emergence of professional scientists. Stephen Brush, famous historian of science, termed this process of the transformation of science during the nineteenth century and the first half of the twentieth century "the second scientific revolution" [9].

Among the transformations that took place in science, the most conspicuous was the emergence of new scientific disciplines such as chemistry, thermodynamics, electromagnetism, and physiology. Although the pioneers in these new fields were British technologists/scientists during the Industrial Revolution, and French during the French revolution, systematic scientific research in these new fields was begun by German scientists in the nineteenth century. The term 'physics' (Physik) itself, as the name of a subject, was introduced during the German university reformation in the early nineteenth century [51]. During the latter half of that century, as a result of university reformation, research universities, which carried out research in the new scientific fields, were built all over Germany.

In this process, the unification of experimental and theoretical research was also pursued. Another famous historian of science, Russell McCormmach, once stated [52]:

> Early in the nineteenth century, German mathematical physics, rational mechanics, and applied mathematics tended to be practiced by people trained in mathematics; and experimental physics tended to be practiced by those trained in chemistry. The two groups published in different places and worked in distinct, essentially noncommunicating sciences. Many early organizers of German physics discipline such as Johann Christian Poggendorff and Heinrich Gustav Magnus strongly condemned the infusion of mathematical theory into physics. [...] In the 1830's and 1840's the division between mathematicians and experimentalists became less sharp, due in part to the development of a method of mathematical physics that drew attention to the common purposes of the two groups.

I have called this largely German, mathematical-experimental tradition of the later nineteenth century 'the chemico-thermal tradition' [1]. The reason I used the term "chemico-" here is that the members of this tradition were often interested, as their roots might suggest, in the chemical phenomena, especially of physical chemistry.

The main body or scientific community of this tradition can be identified with the Helmholtzian School, which consisted of colleagues, students, and followers of Hermann von Helmholtz at the Berlin Physical Society, the universities of Heidelberg and Berlin, and the Physical-Technological Imperial Institute. Among its celebrated members were Clausius, Kirchhoff, Helmholtz, Hertz, Wien, Boltzmann, Planck, and Ernst Mach, most of whom studied hydrodynamics, the kinetic as well as thermodynamical gas theory, chemical thermodynamics, the theory of thermal radiation, and the electromagnetic-field theory. Some of them, especially Kirchhoff, Mach, and Hertz, engaged in the 'criticism of mechanics' and tried to reformulate mechanics into an empirical theory based primarily on the picture of a continuous medium. From my point of view, the famous debate between atomistics and energetics, which concerned the foundations of thermodynamics and its relationship with other branches of physics, was an intra-school controversy.

Nothing quite like the German chemico-thermal tradition existed elsewhere. The separation of physical sciences into mathematical and experimental continued in Britain, France, and Holland. Men like John Rayleigh, James Jeans, Henri Poincaré, and Hendrik Lorentz worked mainly in the classical sciences [48], i.e. in astronomy, harmonics, mathematics, optics and dynamics. Nevertheless, after the discovery of cathode rays, mathematical and experimental physicists gradually began to interact in these countries.

The style of interaction, however, was very different from that in Germany. Unlike the Germans, they worked together by infusing experimental methods into the classical sciences. These attempts met with great success: in 1892, Lorentz offered his electron theory, in 1896 Pieter Zeeman discovered the effect named after him, in 1897 Joseph J. Thomson identified the electron, and in 1911 Ernest Rutherford identified the atomic nucleus and constructed his atomic model. I shall call this line of research 'the particle-dynamical tradition' and identify its main body or scientific community as the Cavendish School.

It would be wrong to suppose that the particle-dynamical tradition had no adherents in Germany. In Germany there were groups of physicists and mathematicians who concentrated solely on the electron theory. Emil Wiechert, Max Abraham, Sommerfeld, and the mathematical physicists of the Göttingen electron-theory seminar may be counted as members of the particle-dynamical tradition.

Generally speaking, the chemico-thermal tradition constituted the core of the second scientific revolution, and scientists in this tradition placed great importance on the concept of entropy. While the particle-dynamical tradition inherited much of the legacy of the (first) Scientific Revolution from the time of Newton, the scientists in this tradition treasured the concept of the ether. In this division, Einstein stood within the chemico-thermal tradition. In contrast, Bohr, who studied under Rutherford, can be regarded as a member of the particle-dynamical tradition. This point may be relevant to the later controversy between Einstein and Bohr over the evaluation of quantum mechanics.

7. Einstein's Contributions to Early Quantum Theory

Very often, the history of early quantum theory is described as a refining process of Bohr's theory of atomic structure. But, as Martin J. Klein put it [44], that picture of the development "lacks just that variety and complexity peculiar to the principal achievement of twentieth century physics". As we have seen, the wave–particle duality, the most peculiar to the quantum theory, was first introduced by Einstein. Moreover, he was the first to introduce the energy quantization of matter, expressed by the introduction of the discreteness in the state-density of his statistical thermodynamics, thus constructing the first quantum theory in 1906.

His construction of the quantum theory of radiation introduced the germ of the so-called correspondence principle. Also, in that theory, the concept of transition probability was introduced in connection with the concept of resonators being carried by molecules. The former concept (transition probability) inspired Werner Heisenberg to construct his 'matrix mechanics' in 1925, while the latter concept (resonators carried by molecules) may be the origin of Hendrik A. Kramers' virtual oscillator model (he was working with Heisenberg at that time).

On the other hand, Einstein's first quantum theory was connected, through its application to the problem of specific heats, with Nernst's theorem, i.e. the third law of thermodynamics. The requirement of the latter theorem, as well as the avoidance of Gibbs' paradox (i.e. the production of 'entropy of mixture'[16] by mixing the same type of gas), led Einstein to construct the quantum theory of the ideal gas in 1924–1925, when he applied to the ideal gas the same statistics that Satyendra N. Bose applied to thermal radiation. This theory stimulated Erwin Schrödinger to look seriously into de Broglie's theory of matter waves, which subsequently led Schrödinger to construct his wave mechanics in 1926.

The above overview suggests that the most important steps in the construction of quantum mechanics were made, not by Bohr, but by Einstein. As I noted in the previous section, Bohr's attention was concentrated on internal electronic processes in atoms and molecules, supplemented by some philosophical considerations thereupon. This peculiarity must be closely related to his early studies in Britain at Thomson's and Rutherford's laboratories.

On the other hand, Einstein's concern spreads so wide from the physical chemistry to the space–time structure that his aim was to construct a consistent theory that was sustainable from every point of view. He also frequently discussed chemical and thermal effects in the material sciences. He inherited these concerns from reading books and articles by Kirchhoff, Helmholtz, Hertz, Wien, Boltzmann and Planck, in what I called the 'chemico-thermal tradition' in the last section.

As is pointed out in Chapter 4, even his attempt at the special theory of relativity seems not unrelated to his aim of constructing quantum theory. Moreover, he had in mind up to the end of his life the hope of constructing his own theory of the microscopic world from his unified field theory. Therefore, even his construction of the general theory of relativity itself might have been a step towards constructing his own microscopic theory, which was after all his lifelong aim throughout his scientific career [41].

8. Postlude: Einstein's Attitude toward Quantum Mechanics

Heisenberg, one of the pioneers of quantum mechanics, once commented on Einstein's attitude to it [38]:

---

[16]The 'entropy of mixture' is defined as the difference between the sum of the entropies of two separated substances and the entropy of their mixture.

In the course of scientific progress it can happen that a new range of empirical data can be completely understood only when enormous effort is made to change one's philosophical framework and to change the very structure of the thought process. In the case of quantum mechanics, Einstein was no longer willing to take this step, or perhaps no longer able to do so.

This kind of view was shared among the scientists of the Copenhagen school, i.e. the colleagues, students and followers of Bohr. Contrary to that, Arthur Fine, famous philosopher of science, expressed the view [37]:

It is Bohr who emerges the conservative, unwilling (or unable?) to contemplate the overthrow of the system of classical concepts and defending it by recourse to those very conceptual necessities and *a priori* arguments. [...] Whereas, with regard to the use of classical concepts, Einstein's analytical method kept him ever open-minded.

In the fifth (1927) and the sixth (1930) Solvay conferences, Einstein and Bohr exchanged discussions on their views of quantum mechanics. This discussion is described in Bohr's memoirs expressing the latter's triumph [4].

Seen from Einstein's side, however, the process was very different. According to the notes enclosed in a letter to Lorentz after the fifth conference, Einstein argued there [65]:

[I]f the state function were interpreted as expressing probabilities for finding properties of an individual system, then the phenomenon of the collapse of the wave packet would represent a peculiar action-at-a distance. [...] It represents a peculiar nonlocalized mechanism, which violates relativity. [...] These problems are not of the theory itself but of the interpretation according to which the theory gives a complete statistical description of individual systems. The alternative is to interpret the state function as providing information only about the distribution of an ensemble of systems and not about features of the individual system themselves.

Also, in a letter to Sommerfeld just after the fifth conference, he wrote [39]:

On 'Quantum Mechanics' I think that, with respect to ponderable matter, it contains roughly as much truth as the theory of light without quanta. It may be a correct theory of statistical laws, but an inadequate conception of individual elementary processes.

During these two conferences, Bohr rebutted Einstein's argument using the doctrine of disturbance, stating that certain simultaneous determinations were not possible because any one of them would inevitably disturb the physical situation so as to preclude the other. Therefore, the purpose of the EPR paper [11], i.e. Einstein's paper of 1935 co-authored with Boris Podolsky and Nathan Rosen, was to neutralize this doctrine of disturbance. On the essence of this paper, Einstein wrote to Schrödinger [64]:

Consider a ball located in one of two boxes. An incomplete description of this 'reality' might be 'The probability is one-half that the ball is in the first box'. A complete description would be 'The ball is in the first box'. Assuming a principle of separation, i.e. 'the contents of the second box are independent of what happens to the first box', and an obvious conservation law for the number of ball, one can find out by looking in the first box whether or not the ball is in the second box. If a theory only allows, in these circumstances, probabilistic assertions, that theory is incomplete.

Fine commented on the EPR paper, "I think it is fair to conclude that the EPR paper did succeed in neutralizing Bohr's doctrine of disturbance" [35].

The standard statistical interpretation using ensemble and hidden parameters has been confronted by difficulties. In 1964, John Bell showed that statistical interpretation is actually in numerical inconsistency with quantum theory when applied to a coupled system treated in the EPR paper [3].

According to Fine, however, the interpretation Einstein entertained was not the standard one but one described by the prism model, in Fine's terminology. He drew attention to a footnote in Einstein's 1936 article [29]:

> A measurement on *A*, for example, thus involves a transition to a narrower ensemble of systems. The latter (hence also its ψ function) depends upon the point of view according to which the reduction of the ensemble of systems is carried out.

Starting from this footnote, Fine constructed his prism model theory representing Einstein's idea. According to Fine, this model produces the same results as does the quantum mechanics, and thus does not conflict with Bell's theorem [36].

References

The following abbreviations are used: *AEPS*, P.A. Schilpp ed., *Albert Einstein, Philosopher–Scientist*, 1st ed. (Evanston, 1949), 2nd ed. (New York, 1951), 3rd ed. (La Salle, 1969); *Blackbody*, T.S. Kuhn: *Black-Body Theory and the Quantum Discontinuity, 1894–1912* (Chicago, 1978); *CPEE, The Collected Papers of Albert Einstein, English Translation*, vol. 1- (Princeton, 1987-); *HSPS, Historical Studies in the Physical and Biological Sciences*; *Letters*, J. Renn & R. Schulmann eds., *Albert Einstein–Mileva Marič, Love Letters* (Princeton, 1992); *Miraculous*, J. Stachel ed., *Einstein's Miraculous Year* (Princeton, 1998); *Shaky*, A. Fine: *The Shaky Game*, 2nd ed. (Chicago, 1996).

[1]   S. Abiko: *op. cit.* note 21, on 5-7.
[2]   S. Abiko: "On the Chemico-Thermal Origins of Special Relativity", *HSPS*, 22 (1991), 1–24 on 10.
[3]   J. Bell: "On the Einstein Podolsky Rosen Paradox", *Physics*, 1 (1964), 195–200.
[4]   N. Bohr: "Discussion with Einstein on Epistemological Problems in Atomic Physics", *AEPS*, pp. 199–242.
[5]   N. Bohr: "On the Constitution of Atoms and Molecules (Part I)", *Phil. Mag.* 26 (1913), 1–25, on p. 2 of which this *Proceedings* is cited.
[6]   N. Bohr: "On the Quantum Theory of Line Spectra, Part I: On the General Theory", *Kgl. Danske Vidensk. Selsk. Skr., Nturvid. og mathem. afd.*, 8, 4-1 (1918), 1–36.
[7]   M. Born: "Einstein's Statistical Theories", *AEPS*, pp. 161–177 on p. 166.
[8]   S.G. Brush: *The Kind of Motion We Call Heat*, 2, Amsterdam: North Holland, 1976, pp. 666–672.
[9]   S.G. Brush: *The Kind of Motion We Call Heat, Book 1*, Amsterdam, 1976, pp. 35–51.
[10] L. de Broglie; "Waves and Quanta", *Nature* 112 (1923), 540; this *Proceedings* is referred to in Louis de Broglie; *Physicien et Penseur*, Paris (1953), 458.
[11] A. Einstein, B. Podolsky & N. Rosen: "Can Quantum-Mechanical Description of Physical Reality Be Considered Complete?" *Physical Review*, 47 (1935), 777–780.
[12] A. Einstein: "Contributions to Quantum Theory", *CPEE*, 6, Doc. 5, pp. 20–27.
[13] A. Einstein: "Emission and Absorption of Radiation in Quantum Theory", *CPEE*, 6, Doc. 34, pp. 212–217.
[14] A. Einstein: "Kinetic theory of thermal equilibrium and of the second law of thermodynamics"; "A theory of the foundation of thermodynamics"; "On the general molecular theory of heat", *CPEE*, 2, Docs. 3–5, pp. 30–77.
[15] A. Einstein: *Miraculous*, p. 178.
[16] A. Einstein: *Miraculous*, p. 187.
[17] A. Einstein: *Miraculous*, p. 191. In the quotation, Einstein's original notation $(R/N)\beta\nu$ is replaced by the modern $h\nu$.
[18] A. Einstein: "Notes", pp. 32–33.
[19] A. Einstein: "Notes", pp. 44–45.
[20] A. Einstein: "Notes", pp. 46–47.
[21] A. Einstein: "Notes" (trans. P. A. Schilpp), *AEPS*, pp. 1–95, on which Einstein's original German text and Schilpp's English translation are printed on opposite pages to each other.

[22] A. Einstein: "Notes", *AEPS*, 3rd ed., pp. 50–53.

[23] A. Einstein: "Notes", pp. 48–51. In the quotation, supplementations in parentheses are omitted, and the modern notation k of Boltzmann's constant is used instead of R/N in Einstein's original text, where R is the gas constant and N the Avogadro's number.

[24] A. Einstein: "On a Heuristic Point of View Concerning the Production and Transformation of Light", *CPEE*, 2, Doc. 14, pp. 86–103; *Miraculous*, pp. 177–198.

[25] A. Einstein: "On the Motion of Small Particles Suspended in Liquids at Rest Required by the Molecular-Kinetic Theory of Heat", *CPEE*, 3, Doc. 16, pp. 123–134; *Miraculous*, pp. 85–98.

[26] A. Einstein: "On the Present Status of the Radiation Problem", *CPEE*, 2, Doc. 57, pp. 357–375.

[27] A. Einstein: "On the Quantum Theory of Radiation", *CPEE*, 6, Doc. 38, pp. 220–233 on p. 221.

[28] A. Einstein: "On the Theory of Light Production and Light Absorption", *CPEE*, 2, Doc. 34, pp. 192–199.

[29] A. Einstein: "Physics and Reality", *Journal of Franklin Institute*, 221 (1936), 313–347; *Ideas and Opinions*, New York, 1954, pp. 290–323 on p. 317.

[30] A. Einstein: "Planck's Theory of Radiation and the Theory of Specific Heat", *CPEE*, 2, Doc. 38, pp. 214–224.

[31] A. Einstein: "Quantentheorie des einatomigen idealen Gases", *Sitzb. Preuss. Akad. Wiss. Phys.-math. Klasse* (1924), 261–267; "Quantentheorie des einatomigen idealen Gases. 2. Abhandlung", *ibid.* (1925), 3–14.

[32] A. Einstein: "Review", *CPEE*, Doc. 37, pp. 211–213 on p. 212.

[33] A. Einstein; "Thermodynamic Proof of the Law of Photochemical Equivalence", *CPEE*, 4, Doc. 2, pp. 89–94.

[34] A. Einstein et al.: "Discussion", *CPEE*, 3, Doc. 27, p. 426.

[35] A. Fine, *Shaky*, p. 35.

[36] A. Fine: *Shaky*, p. 52.

[37] A. Fine: *Shaky*, p. 19; see also S.G. Brush: "The Chimerical Cat", *Social Studies of Science*, 10 (1980), 393–447.

[38] W. Heisenberg: "Preface", *The Born-Einstein Letters*, New York, 1971, p. x; see also G. Holton: "Werner Heisenberg and Albert Einstein", *Physics Today*, 53 (2000), 38–42.

[39] A. Hermann ed.: *Albert Einstein / Arnold Sommerfeld Briefwechsel*, Stuttgart, 1968, letter 53 of Nov. 9, 1927, the English trans. from *Shaky*, p. 29.

[40] A. Einstein & L. Hopf: "Statistical Investigation of a Resonator's Motion in a Radiation Field", *CPEE*, 3, Doc. 8, pp. 220–230.

[41] M. Jammer: "Einstein and Quantum Physics", in *Albert Einstein: Historical and Cultural Perspectives, The Centennial Symposium in Jerusalem*, ed. G. Holton & Y. Elkana, Princeton, 1982, pp. 39–58.

[42] H. Kangro: "Planck, Max Ernst Ludwig", *The Dictionary of Scientific Biography*.

[43] M.J. Klein: "Einstein and Wave–Particle Duality", *Natural Philosopher*, 3 (1964), 3–49.

[44] M.J. Klein: *op. cit.* note 35, on 3.

[45] T. Kuhn: *Blackbody*, pp. 182–183.

[46] T.S. Kuhn: *Blackbody*, p. 170.

[47] T.S. Kuhn: *Blackbody*, p. 172.

[48] T.S. Kuhn: "Mathematical versus Experimental Traditions in the Development of Physical Sciences", *Essential Tension*, Chicago, 1977, pp. 31–65.

[49] T.S. Kuhn: "Preface", *Blackbody*, p. viii.

[50] P. Langevin & M. de Broglie eds.: *La théorie du rayonnement et les quanta. Rapports et discussions de la réunion tenue à Bruxelles, du 30 octobre au 3 novembre 1911, sous les auspices de M.E. Solvay*, Paris: Gauthier-Villars (1912).

[51] C. Jungnickel & R. McCormmach: *Intellectual Mastery of Nature*, vol. 1, Chicago, 1986, Chap. 1.

[52] R. McCormmach: "Editor's Forword", *HSPS*, 3 (1971), p. iv.

[53] M. Planck: "Über irreversible Strahlungsvorgänge. 5. Mitteilung", *Sitzb. Preuss. Akad. Wiss.* (1899), 440–480.

[54] M. Planck: "On the Theory of the Energy Distribution law of the Normal Spectrum", H. Kangro ed., trans. D. ter Haar & S.G. Brush, *Planck's Original Papers in Quantum Physics*, German and English edition, London: Taylor & Francis, 1972, p. 40, also Note 32 on pp. 55–56.

[55] M. Planck: *Vorlesungen über die Theorie der Wärmestrahlung*, Leipztig, 1906, §149.

[56] P.A. Schilpp: "Vorwort zur Deutschen Ausgabe", *Albert Einstein als Philosoph und Naturforscher*, Stuttgart: Kohlhammer, 1955, pp. ix–xi on p. x.

[57] *Letters*, pp. 41, 43.
[58] *Letters*: p. 37.
[59] *Letters*: p. 47.
[60] *Letters*: p. 54.
[61] *Letters*, p. 32.
[62] *Letters*, p. 45.
[63] *Letters*, pp. 35–36.
[64] Letter of June 19, 1935, from Einstein to Schrödinger, quoted in *Shaky*, p. 36.
[65] *Shaky*, pp. 28–29.

Physics Before and After Einstein
M. Mamone Capria (Ed.)
IOS Press, 2005
© 2005 The authors

Chapter 9

The Quantum Debate: From Einstein to Bell and Beyond

Jenner Barretto Bastos Filho

Department of Physics, Federal University of Alagoas, Maceió-Alagoas, Brazil

1. Introduction

Einstein's thought processes are very complex ones. Not even a century of diversified research on him and his work has been enough to explore all his epistemological and methodological insights. One might argue that this is the situation for all the great thinkers of humankind: in other words, the thoughts of any great thinker necessarily involve a high degree of complexity. This is true, but in Einstein's case we can assert that the most representative statements attributed to him have been interpreted by the scholars in an inadequate way. So, our feeling is that in spite of the vast literature about Einstein's work, it is necessary to improve the present Einsteinian hermeneutic to overcome the serious problems appearing in the Tower of Babel that has been built up round Einstein's memory.

Consider, for example, Einstein's famous statement according to which "God does not play dice". This is frequently interpreted as the 'conservative' attitude of a great thinker who was unable to overcome the grip of the classical concept of physical reality, despite overwhelming experimental evidence to the contrary. Although it is possible to criticize Einstein, for example, because of the 'super-determinism' of his special theory of relativity, we believe that to qualify him as a 'conservative' constitutes a misunderstanding, which is, at the same time, an epistemological obstacle hindering the comprehension of the depth of his thought. In the present chapter, we wish to document and explain this misunderstanding.

Einstein was not only an extraordinary scientist who marked indelibly the course of the development of physics with his seminal work on relativity and quantum theory. Einstein's intellectual activity was very broad and highly comprehensive, including themes as diversified as the epistemological implications of twentieth-century physics, education, pacifism, freedom, citizenship, the intellectual and political autonomy of the individual, and so on. Einstein is also famous for his ideal of a 'world government', firmly based on justice, peace and prosperity for all people. In a world with apparently insurmountable social, economical, cultural and regional inequalities, and in which harmful, dogmatic and even fundamentalist attitudes play a dominant role, Einstein's dream seems to be somewhat naïve and utopian. But Einstein was aware of the immense difficulties that had to be overcome in order to implement his ideas. He argued that, regrettably, the struggle against one 'organized power' necessarily requires another 'organized power' (see [30, Chapter 25, Section 25b, p. 539]). However, everyone must play his – or her – role.

It is crucial to emphasize that Einstein's pacifist attitude was an important aspect of his lucid *rationalism*. In simple and direct terms, Einstein considered war to be barbaric and at the same time the worst way to solve humankind's complex problems. Peace, critical discussion, freedom of thought and of expression, tolerance with respect to the diversity of opinions (though not tolerance of intolerance), and the autonomy of the individual are no mere marginal part of his rationalism. On the contrary, these aims and values are a central part of it.

In order to better situate Einstein's *rationalism*, we must connect it strongly to his *realism*. This approach is best adapted to providing the means to confront his scientific research programme with rival ones, principally Bohr's.

We shall organize this chapter as the study of the confrontation among scientific research programmes. This concept was proposed by Imre Lakatos, in order to account for some important difficulties arising in Popper's falsificationism. Although Lakatos' philosophy is chronologically posterior to Einstein's death in 1955, we think that it is very useful to interpret several points of Einstein's thinking.

2. The Idea of a Scientific Research Programme

The methodology of scientific research programmes constitutes an attempt due to Lakatos to correct some exaggerations and even inadequacies of the naïve version of Popper's falsificationism. According to Popper, the truth of a given scientific theory cannot be definitively proved, no matter how much the theory has been empirically corroborated. In other words, even in cases of extensively and broadly corroborated theories, the truth of these theories cannot be taken for granted. In fact, the existence of an empirical refutation invalidating these theories is always conceivable. On the other hand, one can provide proof that a theory is false: just one counterexample is enough. According to Popper, the method of science consists of bold *conjectures* followed by severe attempts at *refutation*, and in this process the critical discussion and intellectual honesty of the scientists involved play a central role.

Several authors[1] have criticized naïve falsificationism by arguing that the eventual falsehood of a given theory also cannot be conclusively proven, because the observational statements which constitute the basis of the refutation may be revealed to be false in the light of future developments.

Lakatos also criticized Popper's views. According to Lakatos, naive falsificationism does not correspond to the real development of science. Lakatos partially adopts the rationalistic commitment of Popper according to which we must not put up with contradictions. Thus Popper and Lakatos agree in their struggle against Hegel, who raised contradiction to the category of a supreme virtue. Although supporting Popper in his criticism of Hegelian irrationalism, Lakatos disagrees with him as regards refutation on the basis of only one counterexample: scientists do not abandon their theories when a counterexample arises. Lakatos asserts that it is also possible to progress in science by working

---

[1] See [14] for an outline of the debate.

on the basis of inconsistent provisional foundations. For example, Bohr's atomic theory contradicts classical electrodynamics, but in spite of this fact it allowed an important advance in science, in the context of a rational scientific programme in which the *correspondence principle* plays a central role.

Lakatos' methodology of scientific research programmes can be outlined as follows. A given scientific theory should be seen as the union of a *hard core* and a *protective belt*. The *hard core* is the unchangeable part of the theory, so considered by methodological decision of adherents of the theory. The *protective belt*, on the other hand, is the changeable part, which may be variously modified and adjusted during the development of the theory. When this accommodation process leads to new possibilities and results, it is said that the adherents of the theory are working in the context of a positive heuristic. Otherwise, when the accommodation process hinders the development of the theory, by making regressive steps and stumbling on insurmountable difficulties, then it is said that the adherents of the theory are working in the context of a regressive or negative heuristic.

This flexibility allows for greater freedom of investigation because one empirical test is normally unable to invalidate a great idea. There is a very complex dialogue between theory and experiment, necessarily accompanied by an equally complex confrontation among webs of theories. In short, all experimental tests involve a confrontation of webs of theories. The concept of an *objective reality* is preserved, but one can decide whether an experimental test is "crucial" only with reference to an accepted theoretical framework. Of course, this limitation characterizes any theory, and cannot be interpreted as an argument in favour of the thesis of the "dissolution" of reality, or other instrumentalist and positivist claims. It is essential to emphasize that the existence of reality does not constitute a result, but a starting point, and as a consequence it is impossible to prove the existence of reality. Realism is a postulate, so it cannot be refuted by any experiment. Of course, the same holds for the opposite philosophical view.

3. Einstein's Scientific Research Programme

Physics can be considered a web of diversified scientific research programmes. This statement has nothing to do with cultural relativism. On the contrary, our starting points are: (1) the existence of an objective reality independent of ourselves and of our desires (realistic adoption), and (2) the philosophical assumption that the world is comprehensible, or in other words, that human reason plays a central role in the understanding of the world (rationalist adoption). We emphasize here that the term *rationalism* must be interpreted in a very broad sense. It differs from seventeenth-century rationalism in that it comprehends both rationalism and empiricism, as entangled intellectual approaches to knowledge. In this broad sense, the epistemologies due to Bachelard, Popper and Lakatos, for example, are all rationalist.

The statement that physics consists of an entangled web of diversified scientific research programmes might lead to the conclusion that instrumentalist and positivist approaches would be equally acceptable. But we wish to emphasize that this is not the case. Instrumentalist and positivist approaches to reality contain an explicit cognitive renunciation that hinders the search for knowledge in depth.

The web of science is embedded inside a rationalistic and realistic attitude, admitting both the complexity of human thought and the complexity of reality. But it is important to say that in spite of scientific theories being human constructions, this does not undermine the autonomy of reality itself.

What is the Einsteinian Scientific Research Programme?

The answer to this question is not easy, given the great complexity of Einstein's thought, but fortunately we can outline Einstein's research objectives here. Einstein's ontology is clear: physical theories are about what things are and not about what we can say about what things are. Einstein does not confuse ontology with epistemology.

Another important point of his research programme can be expressed by one of his famous statements: "The world of our sense experiences is comprehensible. The fact that it is comprehensible is a miracle".[2] Sometimes this statement appears in an alternative form: "The most incomprehensible thing about the world is that it is comprehensible".[3]

Let us give one provisional statement of the *hard core* of the realistic and rationalistic Einsteinian research programme:

E_1: *Physical theories are attempts at saying how things are. The world is comprehensible.*

Clearly the above statement is a very general one. We need to say something else about the term *comprehensibility*. Of course, the statement E_1 seems to be not enough to characterize uniquely Einstein's programme. In fact, E_1 is also perfectly adaptable to the Galilean, Cartesian, Newtonian, Leibnizian, Maxwellian and several other scientific programmes. However, according to Einstein, quantum objects are concrete entities existing in a space–time where causality holds. Thus, in order to express Einstein's thought we ought to enrich E_1. Our second tentative statement is:

E_2: *Physical theories (including quantum theory) are attempts at saying how things are (including quantum objects). The objective world (including the quantum world) is comprehensible. By the simultaneous help of space–time and causal conceptual categories we can study this comprehensible world.*

E_2 is more precise than E_1, and exhibits the peculiarity of Einstein's approach. To better characterize this peculiarity we will compare Einstein's and Niels Bohr's scientific research programmes.

To make explicit Einstein's claims in favour of objectivity and independence of reality we end this section by quoting his reality criterion stated in his famous EPR paper:

If, without in any way disturbing a system, we can predict with certainty (i.e., with probability equal to unity) the value of a physical quantity, then there exists an element of physical reality corresponding to this physical quantity [16].

4. The Debate on Quantum Mechanics: A Chronological Outline

A precise and complete chronological outline of the debate on quantum mechanics is beyond the scope of this chapter. The literature on the quantum mechanics debate is enormous. This debate has been studied from various point of views by philosophers, scientists, scientists with philosophical inclinations, intellectual people of several inclinations

---

[2]Einstein [18, p. 292] *apud* Lindley [27, p. 4].
[3]Einstein *apud* Gal-Or [21, p. 166].

and so on. However, it is possible to offer a rough outline of some important events that marked the history of quantum theory.

First, we can say that quantum theory constitutes a collective construction of one hundred scientists (perhaps two or three hundred). In a relatively recent essay [36,38], a well-known living scholar made one difficult choice by proposing a list of 12 principal names. However, this choice is so restrictive, and excludes several important contributors to the theory. The list covers three generations of physicists divided respectively, into an old generation, an intermediate generation and a young generation. All of these 12 physicists are dead. Those belonging to (1) the old generation are Max Planck (1858–1947) and Arnold Sommerfeld (1868–1951); (2) the intermediate generation are Albert Einstein (1879–1955), Paul Ehrenfest (1880–1933), Max Born (1882–1970), Niels Bohr (1885–1962) and Erwin Schrödinger (1887–1961); and (3) the young generation are Louis de Broglie (1892–1987), Wolfgang Pauli (1900–1958), Werner Heisenberg (1901–1976), Pascual Jordan (1902–1980) and Paul Dirac (1902–1984).

It is widely accepted that the construction of quantum mechanics can be considered as taking place in two principal periods: (i) 'old' quantum mechanics (1900–1924) and, (ii) new and orthodox quantum mechanics (1924–1927).

Roughly speaking, in the first period a provisional quantum theory was constructed in which the classical tradition of physics was combined with the postulates of quantization of energy and angular momentum of Pythagorean inspiration. Very important results were achieved, such as the derivation of the black-body radiation formula by Planck (1900), a new and original explanation for the photoelectric effect by Einstein (1905), the formulation of Bohr's theory of the atom (1913) and several others. In spite of this fact and of a clear awareness of the beginning of a revolutionary period, with the creation of the quantum and relativistic theories, no claim was made for a radical reappraisal of microscopic reality. The scientific community conceived that physics could be understood on the basis of the same general principles consolidated by the classical tradition since the days of Galileo and Newton. According to this tradition, roughly speaking, phenomena were considered as taking place in an objective space–time and as ruled by causal laws, principally the conservation laws.

The second period, however, constituted a radical change. Over a very short time (1924–1927), a debate was launched on the problems of the wave–particle duality, leading to the interpretation of Heisenberg's theorem, to the physical interpretation of the Ψ-function (the solution of Schrödinger's equation), and to Born's statistical interpretation of the wave function. As a result, a new conception of the micro-world was adopted under the leadership of Niels Bohr.

The physicists of the Copenhagen School of thought (Bohr, Heisenberg and Pauli)[4] claimed that a radical change of view on microscopic reality was necessary: nature could not be properly understood by adopting the conception of reality consolidated by classical physics. The new conception grew to dominate physics. But a highly qualified minority, including Einstein, de Broglie, Ehrenfest, Schrödinger, Planck and von Laue, disagreed with it.

The central debate, broadly speaking, was a confrontation between Bohr and Einstein, who considered the Copenhagen interpretation to be an extravagant idea with harm-

---

[4]Or rather the Copenhagen-Göttingen School (Bohr, Heisenberg, Pauli and Max Born).

ful consequences. In 1927, during an important conference held in Como, Italy, in honour of Alessandro Volta, Bohr explained his *Complementarity Principle*.

In 1935 Einstein, Podolsky and Rosen published a famous paper arguing the incompleteness of quantum theory. The paper received, after a few months, a reply from Bohr himself.

In 1964, John Stuart Bell made very important progress with his famous inequality which offered a criterion, at least in principle, to decide experimentally whether Einstein or Bohr was right.

In recent years, the debate has been revisited by physicists and several experiments have been carried out, among them the famous Aspect's experiments in 1982. In spite of widespread opinion that these experiments decided in favour of Bohr's conception, including a very strange non-locality property of quantum reality, there has been much qualified criticism of this hasty conclusion.

Over the last fifteen years there have been several international conferences on the foundations of quantum theory; I wish to mention here at least the International Conference on Bell's Theorem and Foundations of Modern Physics (Cesena, Italy, 1991) [44] and the International Conference on the Frontiers of Fundamental Physics (Olympia, Greece, 1993) [1,2].

5. Niels Bohr's Scientific Research Programme

Niels Bohr's scientific research programme can be summarized in the following way:

B_1: *Classical theories are attempts at saying how things are. The objective classical world is comprehensible. By using both space–time and causal categories we can study the classical world, but* not *the quantum world.*

Let us give a quotation from Bohr himself:

B_2: *"I advocated a point of view conveniently termed "complementarity", suited to embrace the characteristic features of individuality of quantum phenomena, and at the same time to clarify the peculiar aspects of the observational problem in this field of experience. For this purpose, it is decisive to recognize that, however far the phenomena transcend the scope of classical physical explanation, the account of all evidence must be expressed in classical terms"[10, p. 39].*

According to Bohr, quantum phenomena are not comprehensible in the same sense as classical phenomena. In classical physics, objects are spatial and temporal entities that are ruled by causal laws. In this way, classical phenomena are comprehensible according to causal laws (conservation laws) in space–time. Thus in a classical context, the conceptual categories of *space–time* and *cause* can be used together to study physical phenomena.

However, according to Bohr the situation changes drastically when we are dealing with quantum phenomena. In microphysics, the categories of space–time and cause can be used only in a mutually exclusive manner, according to Bohr's *complementarity principle*. Bohr argued in favour of the "indispensable use of classical concepts [...] even though classical physical theories do not suffice" [9, p. 701]. This is how Henry Stapp described Bohr's programme:

According to Niels Bohr, quantum theory must be interpreted, not as a description of nature itself, but merely as a tool for making predictions about observations appearing under conditions described by classical physics [41, p. 255].

In other words, although quantum phenomena cannot be described by simultaneous use of the space–time and causal concepts, the use of these and other classical physics concepts is unavoidable. If in a given experiment the space–time aspect is implemented, then the causal aspect disappears, and vice versa. In quantum mechanics the causal and space–time categories are mutually exclusive but complementary.

The same holds for particle and wave categories of classical physics: in experiments where corpuscular aspects are implemented, causal aspects are excluded; analogously, in the context of experiments in which wave aspects are implemented, corpuscular aspects are excluded. Once more, they are mutually exclusive categories, but they are complementary aspects in the context of mathematical formalism.

6. Einstein and Objective Reality

Let us come back to Einstein. In order to better outline his beliefs, it is necessary to take into account some aspects of his struggle against the adherents of the so-called "dissolution of reality" thesis, which he considered one of the most harmful and regrettable things for the development of science, for education in the broad sense of the term, and for the intellectual development of individuals. In a famous letter written in April 1938 to his friend Maurice Solovine Einstein wrote about his disappointment:

> In Mach's time a dogmatic materialistic point of view exerted a harmful influence over everything; in the same way today, the subjective and positivistic point of view exerts too strong an influence. The necessity of conceiving of nature as an objective reality is said to be superannuated prejudice while the quanta theoreticians are vaunted. Men are even more susceptible to suggestion than horses, and each period is dominated by a mood, with the result that most men fail to see the tyrant who rules over them [20, p. 85].

The attitude of considering objective reality as a "superannuated prejudice" was completely unacceptable to Einstein. He considered this attitude to be the effect of brainwashing and thus incompatible with genuine realist and rationalist commitments.

Einstein's reaction to the "dissolution of reality" thesis gives rise to one of the most dangerous and perverse misunderstandings in the history of the interpretations of his thought. In spite of his genius and the revolutionary character of his contributions, Einstein is frequently remembered for his supposedly "conservative" attitude firmly based on classical ideas and his "old" concept of reality. Einstein was discontented with this situation. In 1938 he wrote in confidence to Solovine, in a letter where he revealed his plan to defeat what he considered harmful tendencies in physics:

> I am working with my young people on an extremely interesting theory with which I hope to defeat modern proponents of mysticism and probability and their aversion to the notion of reality in the domain of physics. But say nothing about it, for I still do not know whether the end is in sight [20, p. 91].

To accuse Einstein of conservatism because of his concept of physical reality is a mistake. In order to clarify this important point let us consider his theory of the photoelectric effect.

Einstein explained why Maxwell's electromagnetic theory was unable to explain the photoelectric phenomenon: in this phenomenon the central role is played by the *frequency*, not the *intensity*, of light. For this reason we must go beyond Maxwell's electromagnetism. In order to solve this important problem he proposed a new and completely revolutionary idea according to which a spatially concentrated entity carrying the energy $E = h\nu$, where h denotes Planck's constant and ν the frequency of the associated wave, is entirely destroyed in the process, but its energy is entirely conserved, being utilized for: (i) transferring a given energy to an electron of the metallic cathode in order to overcome the potential barrier of the metal, and (ii) for transferring a given kinetic energy to the released electron. Einstein's simple formula is:

$$h\nu = \Phi + E_{\mathrm{kin}},$$

where Φ denotes the energy required to overcome the potential barrier of the metal that constitutes the cathode and E_{kin} denotes the kinetic energy acquired by the released electron.

In 1905, this idea was highly revolutionary. Nobody knew about the existence of a microscopic entity that during the effect is totally destroyed while its energy is conserved. Physicists did not know of anything like this. It is important to emphasize that classical objects such as particles and waves were radically different from this completely new object introduced by Einstein in this highly revolutionary conjecture. Thus, Einstein introduced in physics *the first formulation of particle–wave duality*. Classical physics did not know anything about duality. Particles were dimensionless entities like the points of Euclidean geometry, and waves were spatial and temporal entities characterized by their corresponding frequencies and wavelengths.

Einstein can be considered a precursor of the extraordinary phenomena featuring the annihilation and creation of particles in the high-energy range. A microscopic entity carrying the energy $E = h\nu$ being completely destroyed while its energy is conserved anticipates by a few decades these important phenomena in the high-energy range. Today these phenomena are well known, but in 1905 they were unknown and hard to understand.

It is strange to call 'conservative' the physicist who anticipated the annihilation and creation phenomena, gave the first formulation of the wave–particle duality, and introduced an entirely new object in physics. It is still more surprising, and ironic, that his "old" conception of reality is blamed for his "conservatism"!

But this is not the end of the history. Einstein's papers on relativity, his doctoral thesis and his seminal contribution to the theory of Brownian motion, all of them published during Einstein's *annus mirabilis*, showed, without ambiguity, the strong and broad revolutionary character of his work. In particular, he was the one who definitively established the atomic paradigm in physics, a highly revolutionary event in itself. Thus the epithet of 'conservative' frequently attributed to Einstein cannot be justified.

Einstein knew that a world of intense *becoming*, characterized by the annihilations and creations of the new physics, cannot be understood by extravagant philosophical views such as the "dissolution of reality" thesis. The reality is extremely complex, but this complexity does not mean absurdity. For him the *becoming* shown in high-energy physics experiments co-exists with the *being* guaranteed by conservation laws. Thus the two seminal, although antithetical, philosophical programmes, the Heraclitean and the Parmenidean ones, are both fertile and co-exist in physics. Einstein's strong belief in the objective character of reality can explain his irritation with the adherents of the idealistic mood. The comparison made by him of these adherents with "horses" seems rude, but it arose from a legitimate and constructive irritation. Einstein was seriously worried by the low intellectual standard of the majority of members of the scientific community, who seemed to him highly vulnerable to manipulation and brainwashing.

Let us consider something related to Einstein's famous aphorism according to which "God does not play dice". In a letter dated 30 November 1926 Max Born wrote to Einstein:

> My idea to consider Schrödinger's wavefield as a "Gespensterfeld" in your sense of the word proves to be more useful all the time [. . .]. The probability field propagates, of course, not in ordinary space but in phase space (or configuration space).[5]

Einstein's answer shows his disappointment with the constantly increasing level of abstraction of the quantum formalism, with no intuitive counterpart. Einstein wrote to Born on 4 December 1926:

> Quantum mechanics is very impressive. But an inner voice tells me that is not yet the real thing. The theory produces a good deal but hardly brings as closer to the secret of the Old One. I am at all events convinced that He does not play dice. Waves in 3-n dimensional space whose velocity is regulated by potential energy (e.g. rubber bands) [. . .][6]

The ideas inspired by Einstein's *Führungfeld*, by the inherent duality of all quantum objects due to de Broglie and by the concept of a probability field propagating in a mathematical configuration space, led to a great degree of mathematical abstraction without any corresponding intuitive counterpart. This situation has constituted a serious obstacle which hindered a clear comprehension of how these quantum objects interact in space–time and how they are governed by causal laws.

We must stress that mathematical abstraction in itself is not against causal and space–time explanation. General relativity, for example, involves a high degree of mathematical abstraction and, no doubt, constitutes a causal and space–time theory. With respect to quantum mechanics, mathematical abstraction is accompanied by almost an absence of a physical intuitive counterpart and also by serious ambiguities of interpretation.

Einstein's philosophical position cannot be considered conservative: he had strong motives to deny the new interpretation due to Born.

7. An Interlude: Physics as a Complex Web of Entangled Scientific Research Programmes

Physics is a science that can be considered an entangled and complex web of diversified scientific research programmes. For example, the Parmenidean and Heraclitean programmes are both present in physics today; both historically played a very important role in the development of this science. However, the complexity of physical science goes beyond these two important programmes. We can also assert that the atomistic programme as well as the programme based on continuous conceptual categories such as

---

[5]Born cited *apud* [30, Chap. 25, p. 526].
[6]Einstein *apud* [30, Chap. 25, p. 527].

plenum, ether and *field* have both been enormously fertile for the progress of scientific development, especially physics.

Several other programmes are equally important: the Cartesian and Leibnizian programmes based on conservation laws; the Newtonian programme based on point masses acting at a distance through absolute time and space; the Faraday–Maxwell programme based on continuous electromagnetic waves propagating through the ether at the velocity of light; Einstein's programme, Bohr's programme, the Pythagorean programme, the geometric programme and several others.

The *hard core* of the Pythagorean research programme is the idea that the essence of reality is mirrored in whole numbers and their ratios. In short, according to Pythagoreans, numbers rule all things. However, the Pythagoreans stumbled on a very great difficulty which was, at that time, an insurmountable contradiction and constituted a traumatic discovery, with serious consequences for their programme. They discovered the incommensurability of the diagonal of the square and its side, i.e. that the ratio (d/a), where d denotes the diagonal of the square and a denotes its side, cannot be written as a ratio of whole numbers. In fact, if we allow that the ratio (d/a) is a rational number, then this can be shown to lead to a contradiction according to which a certain integer is both even and odd. *The unavoidable conclusion is that rational numbers do not cover all of conceptual reality.*

It is essential to emphasize that the incommensurability problem appearing in the context of arithmetic is not itself a geometric problem. This circumstance played a central role in the context of the history of the competition between the arithmetic and the geometric programmes; the reason for this can be easily understood. For example, when Socrates (the main character of Plato's dialogue *Meno*) asks the slave boy how long is the side of a square having twice the area of another given square, the answer is that the side of the square having a double area is equal to the diagonal of the other square. The exact solution can be drawn on the earth (or on the sand) without ambiguity and without contradiction. Therefore, the incommensurability problem does not arise geometrically.

Plato also noted this extremely important fact and therefore rated geometry as superior to arithmetic as a world-view. This was, according to Popper, the Platonic and Euclidean Programme (see [33, Chap. 2]). Classical mechanics and the general theory of relativity are frequently considered as emblematic examples of the extraordinary success of the geometric programme in the history of physics.

In spite of the this traumatic affair, the Pythagorean programme resurrected from its ashes – like the phoenix – and played an important end even decisive role in several developments of physics. The branches in which the Pythagorean central idea had most success were electrolysis, the physics of oscillations (normal modes of vibration, as in a vibrating string) and quantum mechanics.

Einstein himself, in his *Autobiographical Notes*, commented in Pythagorean terms about the extraordinary Pythagorean realization of Bohr's theory of the atom. Einstein said that Bohr's theory was "the highest form of musicality in the sphere of thought". The emphasis on music comes, no doubt, as a reference to the Pythagorean Programme. The quantum energy levels enumerated by $n = 1, 2, 3, \ldots$, corresponding to the stationary states of the atom whose transitions occur by 'quantum jumps', i.e. in discrete quantities (emission of photons of well-defined frequencies) are pure Pythagorean music. In the forthcoming section I shall argue that Bohr, Einstein and de Broglie were partially committed to aspects of the arithmetic Pythagorean Programme.

8. Einstein, Bohr and de Broglie

In this section we can see how the Pythagorean programme worked in the context of old quantum mechanics. It is often emphasized that Bohr's theory (1913) contradicts classical electrodynamics. Bohr supposed that electrons perform stationary circular orbits around their nuclei, each with a given constant value of energy. On the other hand, classical electrodynamics holds that an electric charge in circular motion around the nucleus undergoes the action of the centripetal force and, as a consequence, loses its energy and falls into the nucleus. This means that the supposed stability of the atom in Bohr's theory is very strange from the viewpoint of classical electrodynamics.

In order to surmount this paradox, one may attempt to bring into Bohr theory the de Broglie relation

$$p = \frac{h}{\lambda} \tag{1}$$

where p denotes the linear momentum of the particle (an electron in this case), h is the Planck constant and λ is the wavelength of the associated de Broglie wave. As we know, if we consider the nucleus as being at rest (an approximation in which the motion of the nucleus can be neglected due to the fact that the mass of the nucleus is much bigger than the mass of the electron), the total energy of the hydrogen atom (kinetic plus potential energy) is given by

$$E = \frac{p^2}{2m} - \frac{e^2}{r}, \tag{2}$$

where p and m are, respectively, the linear momentum and mass of the electron charge and r is the radius of the supposed circular motion.

If we assume the periodicity condition

$$\lambda = 2\pi r, \tag{3}$$

then the combination of Eqs (1), (2) and (3) gives rise to

$$E = \frac{h^2}{8\pi^2 mr^2} - \frac{e^2}{r}. \tag{4}$$

In order to find the value of r compatible with the minimum value of energy of the hydrogen atom we must derive Eq. (4) with respect to r according to the requirement of minimum energy

$$\left(\frac{dE}{dr}\right)_{r=a} = 0. \tag{5}$$

By performing the calculations we arrive at the result

$$a = a_{\text{Bohr}} = \frac{h^2}{4\pi^2 me^2}. \tag{6}$$

Replacing the value of r in Eq. (4) by the value of a_{Bohr} given by Eq. (6) we obtain the value of the ground state of the hydrogen atom

$$E = E_{Bohr} = -\frac{2\pi^2 m e^4}{h^2}. \tag{7}$$

This procedure can be easily generalized to all stationary states by assuming the relation

$$n\lambda = 2\pi r, \tag{8}$$

where $n = 1, 2, 3, \ldots$. In an analogous way, by combining Eqs (1), (2) and (8) we obtain:

$$E_n = \frac{n^2 h^2}{4\pi^2 2m r^2} - \frac{e^2}{r}. \tag{9}$$

The condition

$$\left(\frac{dE}{dr}\right)_{r=a(n)} = 0 \tag{10}$$

leads to a set of values $\{a(n)\}_n$ with $n = 1, 2, 3, \ldots$, i.e.

$$\{a(n)\}_n = \frac{n^2 h^2}{4\pi^2 m e^2}. \tag{11}$$

Replacing the value of r in Eq. (9) by the values of $\{a_{min}\}_n$ given by Eq. (11) and E_n in (9) by $\{E_{min}\}_n$ we obtain

$$\{E_{min}\}_n = -\frac{|E_{Bohr}|}{n^2}. \tag{12}$$

It is important to note that Eqs (11) and (12) give rise, respectively, to Eqs (6) and (7) in the case where $n = 1$, i.e. where a_{Bohr} is the radius *minimum minimorum* and E_{Bohr} is the energy *minimum minimorum* among all the stationary states of the hydrogen atom.

The point to be emphasized here is that even though in the context of a simple theory like this one, Bohr's theory of 1913, the de Broglie relation (Eq. (1)) of 1923–4 constitutes an important explanatory principle providing a better understanding of the mysteries of existence of these stationary states, as well as a possible overcoming of the contradiction between Bohr's theory and the theoretical framework of classical electrodynamics. Surely the de Broglie relation is a general quantum law expressing the inherent objective duality of the quantum objects. When (1) is associated with the periodicity condition expressed by (8) and the minimum condition expressed by (10), it leads, in a natural way, to the stationary states of the hydrogen atom. This means that the understanding of the existence of stationary states of an atom goes beyond the theoretical framework of classical electrodynamics, like the way in which the Planck–Einstein dualistic relation

$$E = h\nu, \tag{13}$$

which is absolutely necessary to explain the photoelectric effect, goes beyond Maxwellian electromagnetism. It is interesting to note that Eqs (1) and (13) are 'twin brothers' expressing the inherent duality of the quantum objects. These relations are touchstones of the quantum theory independent of any particular interpretation.

9. Aspects of the Quantum Debate

Ontology has to do with what kind of things there are, epistemology with our knowledge of things. Of course, our knowledge of the things is different from what the things are. Any objective theory of reality ought to presuppose something about what kind of things there are, independent of our knowing them.

Bohr raised an objection. According to him, it is impossible to separate the things (i.e. the objects of quantum theory) and their interaction with the measuring instruments acting on these things (objects). He argued in favour of

> [...] *the impossibility of any sharp separation between the behavior of atomic objects and the interaction with the measuring instruments which serve to define the conditions under which the phenomena appear* ([10, pp. 39–40]. The italics are Bohr's).

Bohr's principal reason for defending this "impossibility" depended on the supposed "uncontrollability of the interactions between objects and measurement instruments". He wrote:

> Notwithstanding all novelty of approach, causal description is upheld in relativity theory within any given frame of reference, but in quantum theory the uncontrollable interaction between the objects and the measuring instruments forces us to a renunciation even in such respect [10, p. 41].

According to him, the reason for the renunciation of a causal description in quantum theory concerns the indivisibility of the universal quantum of action expressed by Planck's constant h. This influential idea of Bohr's was combined with the Heisenberg's argument that "in the range of microphysics the ontology of classical physics does not work". Concerning this point, the meaning of the term *phenomenon* according to Bohr is radically different from any other meaning that can be given to this word when the quantum entities are objectively considered as existing, even independently of any measurements. On this important point, Bohr wrote:

> As a more appropriate way of expression I advocated the application of the word *phenomenon* exclusively to refer to the observations obtained under specified circumstances, including an account of the whole experimental arrangement [10, p. 64].

Bohr also argued that this attitude does not imply an arbitrary renunciation, but a clear recognition of the impossibility of a more detailed analysis of atomic phenomena:

> [...] in quantum mechanics, we are not dealing with an arbitrary renunciation of a more detailed analysis of atomic phenomena, but with a recognition that such an analysis is *in principle* excluded. ([10, p. 62]. Cf. Stapp's statement quoted in Section 5.)

Because of Bohr's renunciation of a quantum ontology, the orthodox interpretation of quantum theory he proposed introduced a new methodological choice, which required that (1) quantum reality has to be construed using the classical categories; and (2) in the quantum world, classical categories are mutually exclusive.

His famous reply to Einstein was summarized in the following final passage:

> In fact, it is only the mutual exclusion of two experimental procedures, permitting the unambiguous definition of complementary physical quantities, which provides room for new physical laws, the coexistence of which might at first sight appear irreconcilable with the basic principles of science. It is just this entirely new situation as regards the description of physical phenomena that the notion of *complementarity* aims at characterizing [9] (in [10, p. 61]).

This methodological choice constituted an instrumentalist approach to physical science, the principal implication being a programmatic renunciation to presupposing a given quantum reality independent of any observers. Only relevant are "the observations obtained under specified circumstances, including an account of the whole experimental arrangement". According to Bohr the only possible quantum reality is that which is explicitly referred to a macroscopic description of the experimental conditions. In other words, in the domain of quantum reality any statement about these objects, without specifying the conditions under which the observations are made, is meaningless.

Einstein's opposition to this view was not based on pure logic. Einstein argued that the quantum-mechanical description accounts for the behaviour of a large number of atomic systems (an *ensemble*). But his intuition (his inner voice) led him in another direction: he thought that an exhaustive theoretical description of individual phenomena had to be developed. Although Einstein considered that from a strictly logical point of view Bohr's arguments were acceptable, they were not acceptable to his "scientific instinct". In a paper published in 1936, he wrote:

> To believe this is logically possible without contradiction; but it is so very contrary to my scientific instinct that I cannot forego the search for a more complete conception [17] (in [10, p. 61]).

Several "impossibility theorems" were proposed in order to affirm Bohr's point of view in physics. The most famous of these was due to von Neumann, which worked as an epistemological obstacle hindering the search for a causal completion of quantum mechanics. But a simple concrete model of spin-1/2 particles and spin measurements [5,36–38] was enough to destroy this presumed impossibility. As Selleri wrote:

> This model reproduces all the quantum mechanical prediction for spin measurements (experimental results and probabilities) and therefore does exactly what von Neumann's theorem attempts to forbid, i.e., provide a causal completion of quantum mechanics [40, p. 48].

David Bohm's famous papers of 1952 played an extremely important role in overcoming the epistemological barrier established by von Neumann's authoritative impossibility theorem. By commenting on Bohm's papers in a humorous and ironical way, Bell declared in 1981: "In 1952 I saw the impossible done" [6].

In spite of the well-known explanation of the existence of stationary states in the context of orthodox quantum mechanics (1927–1930), Bohm's theory can be considered as another possibility of solution of the contradiction between classical electrodynamics and Bohr's atom theory (Appendix A). We remember that according to electrodynamics electrons moving in a circular motion around a nucleus undergo centripetal force and as a consequence of this must lose their energy and fall into the nucleus. For this reason Bohr's atom theory cannot explain the existence of stationary states; consequently, it is also unable to explain, in a completely satisfactory way, the stability of the atom: Bohr's atom theory just *postulates* – as a starting point – the existence of such states. In spite of this, Bohr's theory can calculate stationary states of energy (see Section 8, Eq. (12)).

In 1952, much criticism was levelled at Bohm's ideas. His last statement on the subject is in the book written with Basil Hiley and published in 1993 (after Bohm's death in 1992). Bohm and Hiley wrote:

> Moreover if one is not aesthetically satisfied with this picture of a static electron in a stationary state, one can go to the stochastic model given in Chapter 9. In this model the particle will

have a random motion round an average $p = \nabla S$, and the net probability density in this random motion comes out as $P = |\Psi(x)|^2$ [13, p. 43].

David Bohm's ideas were very important. One of the most relevant results of these ideas was to establish that it is possible to overcome the "impossibility proofs", such as von Neumann's theorem. An article clearly showing this fact is by Franco Selleri [37], where a simple model of a statistical ensemble of spinning spheres is enough to reproduce all the results (eigenfunctions, eigenvalues, probabilities) of the quantum theory of spin one-half. This result, which at the same time constitutes a criticism of both Bohr's complementarity and von Neumann's impossibility theorem, means that, at least from a methodological point of view, the idea of hidden variables is immensely valuable.

Karl Popper also protested against Bohr's view. He wrote: "But this doctrine is simply false: quantum theory is as objective as any theory can be" [32, p. 120], and also, "Quantum theory is exactly as objective as any other physical theory" [32, p. 121].

Gell-Mann also argued that:

Niels Bohr brainwashed a whole generation of physicists into believing that the problem had been solved fifty years ago [22, p. 29].

Penrose, too, manifested his preference and arguments for a realistic interpretation of quantum mechanics and an objective description of reality. He wrote:

Yet it must also be emphasized that, in my view, the standard theory is indeed quite unsatisfactory philosophically. Like Einstein and his hidden-variable followers, I believe strongly that it is the purpose of physics to provide an *objective* description of reality [31, p. 106].

Penrose's ideas, which he considered as going "into dangerously speculative territory", seem to be very close, broadly speaking, to those of the Einsteinian programme, except with respect to *indeterminacy*. Penrose argued that "[...] I do not regard *indeterminacy*, in the ordinary sense of that word, as being necessarily objectionable". I agree with this speculation because I believe that we live in a world of propensities, in which there are objective probabilities (concrete and real tendencies), not something existing in the mind of somebody (see [34,3]).

Penrose considers the co-existence of the deterministic evolution of the solutions of Schrödinger's equation and the non-deterministic jump that characterizes the quantum collapse "an absurd concoction". He argues in favour of attempts to write an improved theory and also in favour of a possible explanation of the collapse in gravitational terms. His research programme can be summarized as follows:

I would not dispute that some changes in classical general relativity must necessarily result if a successful union with quantum physics is to be achieved, but I would argue strongly that these must be accompanied by equally profound changes in the structure of quantum mechanics itself. The elegance and profundity of general relativity is no less than that of quantum theory. The successful bringing of the two together will never be achieved, in my view, if one insists on sacrificing the elegance and profundity of either one in order to preserve intact that of the other. What must be sought instead is a grand union of the two – some theory with a depth, beauty and character of its own (and which will be no doubt recognized by the strength of these qualities when it is found) and which includes both general relativity and standard quantum theory as two particular limiting cases [31, p. 112].

I think that the above lucid quotation helps us to better analyse the state of the art of the discussion on the enormous existing difficulties. In the present stages of the development of general relativity and quantum theory, it can be said that they are extraordinarily good theories but that they are also incommensurable. If we believe that an eventual conciliation between them is possible, then radical improvements are necessary. It is unlikely that just one will have to be modified. The resulting theory (if possible, but I share Penrose's optimism) should contain both general relativity and standard quantum theory as particular limiting cases. This hope once more is in agreement with Einstein's statement:

> There could be no fairer destiny for any [...] theory than that it should point the way to a more comprehensive theory in which it lives on, as a limiting case.[7]

John Stuart Bell also expressed his dissatisfaction several times with respect to orthodox quantum mechanics as based on ambiguous concepts such as 'measurement'. Instead of an orthodox theory based on *observable* elements, we need a theory centred on *'beable'* ones. His programme is clearly outlined in this passage:

> In particular we will exclude the notion of 'observable' in favour of that of '*be*able'. The *be*ables of the theory are those elements which might correspond to elements of reality, to things which exist. Their existence does not depend on 'observation'. Indeed observation and observers must be made out of beables [7, p. 174].

He follows this statement with:

> I use the term '*be*able' rather than some more committed term like 'being' to recall the essentially tentative nature of any physical theory. Such a theory is at best a *candidate* for the description of nature. Terms like 'being', 'beer', 'existent', etc., would seem to me lacking in humility. In fact 'beable' is short for 'may-beable' [7, p. 174].

The last book written by Bohm, in collaboration with Hiley, explicitly affirms the need for an ontological interpretation of a quantum theory. Bohm died suddenly in 1992 and the book was published in January 1993. In the preface Hiley wrote:

> It was quite clear from the outset that it was not going to be possible to return to the concepts of classical physics and we found it necessary to make some radical new proposals concerning the nature of reality in order to provide a coherent ontology.[8]

In this book Bohm's thought is considerably more wide-ranging than in his 1952 papers. His idea of *implicate order* is extensively explained in the last chapter, which is entitled "Quantum Theory and Implicate Order". But Bohm and Hiley affirm that the ontological approach presented in the book does not mean any return to classical concepts. They emphatically hold that the quantum potential concept, for example, essentially belongs to a quantum rather than to a classical context.

Concluding this section, I would like to emphasize once again that the 'dissolution of reality' brainwashing, which still dominates the scene of the scientific community, has found important opponents. Einstein himself was the most outstanding example, but he died in 1955. Concerning physical reality, Popper was an important realist and rationalist philosopher, even though his political views should be – and have been – severely criticized. Popper died in 1994. Bohm, who died in 1992, emphasized the necessity of an objective ontological theory without having to assume an outside observer. Bell, who died

---

[7] Einstein as cited in [33] (inscription to Chapter 1, p. 32).

[8] Hiley, preface to [13].

in 1990, had his 'beable' programme which is a realistic and rationalist one in the sense of the Einsteinian and Popperian conceptions. There are several research groups committed to a rationalist and realistic approach to physical inquiry. Selleri, who has published several books and articles and organized several conferences in which researchers belonging to all the tendencies (for or against realism) are invited, is an important example of a living physicist committed to realism and rationality. Penrose, with his vision of a future theory which must encompass both general relativity and quantum mechanics as limiting cases, in the spirit of a generalized correspondence principle, offers a concrete possibility. In the next section we will consider other aspects of the generalized 'beable' programme which can be summarized by the following claim: all physical theories are attempts at an objective description of nature. Quantum theory is no exception. In short, the 'beable' programme is the same as a search for a *quantum theory without observers*.

10. A Quantum Theory Without Observers

In the last section, I documented the views of some of the physicists who criticized the orthodox interpretation of quantum physics ascribing observers a central role in the theory. Bohm, Hiley, Selleri, Gell-Mann and Penrose, for example, argued in favour of different programmes that are committed to excluding 'the observer' from quantum physics. In other words, all these programmes require an ontological interpretation in which the observer does not play any essential role.

Following Bell's idea – abandoning the notion of 'observable' in favour of that of 'beable' – over the last two decades some authors have made an effort to understand the emergence *from* the quantum world (characterized by a peculiar entanglement of states) *to* the classical world with its separated and disentangled objects. This emergence can also be understood as a passage *from* the "coherent" world (ruled by the superposition principle and by interference of quantum amplitudes of probability) *to* the "decoherent" classical world which moulds our ordinary physical intuition. Among the authors working on theories of this kind are Griffiths, Omnès, Gell-Mann, Hartle and Zurek (see [23]).

The modern decoherence theories have the important objective of overcoming the Copenhagen wave-function collapse. In my opinion, these authors broadly accept the 'beable' programme, as shown by the following questions asked by them: "What is it in the laboratory that corresponds to a wave function, or to an angular momentum operator?" [23, p. 26], and later: "What are the 'beables' [...] of quantum theory – that is to say, the physical referents of the mathematical terms?" [23, p. 26]. They recognize that the connection of the mathematical structures of quantum theory with physical reality through the concept of measurement leads to immense difficulties and "ludicrous" consequences: "When quantum mechanics is applied to astrophysics and cosmology, the whole idea of using measurements to interpret its predictions seems ludicrous" [23, p. 26].

With respect to wave-function collapse, Haroche writes:

> [...] the proponents of the modern decoherence theories prefer to view this wavefunction collapse [Copenhagen interpretation] as a real physical process caused by the coupling of the measuring apparatus to its environment. For all practical purpose, of course, the orthodox and decoherence points of view are equivalent, because the decoherence time is infinitesimal for any measurement that ultimately involves a macroscopic apparatus [24, p. 41].

I think there exists a misunderstanding in this kind of evaluation. The question is not centred on a possible "practical equivalence" of the decoherence point of view and the wave-function collapse. If one adopts the decoherence point of view, it is obvious that an eventual and very special quantum relaxation must necessarily take place during a finite time as a physical process. The central question is that a consistent decoherence theory must be formulated independently of measurements. Only in this case the "ludicrous" aspects censored by Griffith and Omnès (more than 60 years after Einstein and Schrödinger) are removed.

The same author continues:

> Decoherence becomes more and more efficient as the size of a system increases. It protects with a vengeance the classical character of our macroscopic world [24, p. 42].

Griffiths and Omnès write:

> In the consistent-histories approach, the concept of measurement is not the basis for interpreting quantum theory. Instead, measurements can be analyzed, together with other quantum phenomena, in terms of physical processes. And there is no need to invoke mysterious long-range influences and similar ghostly effects that are sometimes claimed to be present in the quantum world [23, p. 26].

Griffiths and Omnès consider that the decoherence point of view constitutes a good way to progress. They argue that:

> [...] calculations carried out by one of us and by Gell-Mann and Hartle, indicate that, given a suitable consistent family, classical physics does indeed emerge from quantum theory [23, p. 31].

In spite of the rational aim of overcoming instantaneous action at a distance, ghostly effects and subjective interpretations like the attribution of physical effects to conscience, decoherence theories do not constitute – in my opinion – a completely satisfactory point of view. Decoherence is simply a reinterpretation of the Copenhagen orthodoxy, trying to overcome some of its absurdities. From this viewpoint, decoherence theories seem to offer no solid basis to unify and approximate quantum theory and classical general relativity in the sense proposed by Penrose.

We conclude this section by quoting two very interesting passages of Gerard 't Hooft [43].

> To this day, many researchers agree with Bohr's pragmatic attitude. The history books say that Bohr has proved Einstein wrong. But others, including myself, suspect that, in the long run, the Einsteinian view might return: that there is something missing in the Copenhagen interpretation. Einstein's original objections could be overturned, but problems still arise if one tries to formulate the quantum mechanics of the entire universe (where measurements can never be repeated), *and* if one tries to reconcile the laws of quantum mechanics with those of gravitation [43, p. 13].

> You may have already suspected that I still believe in the hidden variables hypothesis. Surely our world must be constructed in such an ingenuous way that some of the assumptions that Einstein, Bell and others found quite natural will turn out to be wrong. But how this will come about, I do not know. Anyway, for me the hidden variables hypothesis is still the best way to ease my conscience about quantum mechanics [43, p. 15].

All these circumstances show that the Einstein Programme lives, and that the mythology of a "loser" Einstein versus the "winner" Bohr must be replaced by a new and deeper understanding of Einstein's thought.

11. Do 'Crucial' Experiments Exist?

The obsession of the overwhelming majority of scientists with the search for crucial experiments is well known. These scientists search for experiments that can show, decisively, who is right and who is wrong, or which is the correct theory and which is (are) the false(s) one(s). Experimental control is a consolidated practice of science and the dialogue between theory and experiment plays a central role in the development of science. This dialogue gives rise to new possibilities, but very rarely leads to the last word on a subject (see [15,8], [28, p. 323], [29, pp. 44–46], [4]). A set of 'crucial experiments' is a confrontation among several webs of theories and this situation is completely different from the kind of confrontation imagined by some naïve realists.

The history of science shows several characteristic examples. The Fizeau and Foucault experiments in 1850 'decided' in favour of the wave character of light, but in the twentieth century Einstein 'invented' duality for light and de Broglie 'invented' duality for all 'particles'. Also, Michelson and Morley's experiments cannot be considered as crucial in discarding the existence of an ether.

The situation is even more confusing in the context of the confrontation *locality* versus *non-locality*. As we know, Einstein argued that "physics should represent a reality in space–time, free from any spooky action at a distance" [19, p. 158]. With a great acuteness and awareness of the EPR paradox (see Appendix B), Schrödinger wrote on the strange quantum entanglement:

> It is rather discomforting that the theory should allow a system to be steered or piloted into one or the other type of state at the experimenter's mercy in spite of his having no access to it [35].

Bell was able to find a formula – the famous Bell inequality (see Appendix C) – which gives rise to the possibility of a possible "crucial experiment" to settle the nature of this entanglement. In other words, Bell's inequality offered the possibility of an experimental confrontation of quantum mechanics (theory implying entanglement) with local theories or local models. Some extremely difficult experiments were carried out, but whether any of them can be considered as "crucial" cannot be considered as decided among scholars in the field. Hiley and Peat, for example, put the important questions:

> 1. Whether Bell's notion of locality is too restrictive; and 2. Whether in fact the experiments actually measure what they intend to measure [25, p. 14].

Although the overwhelming majority of physicists believe that the final word has already been said, we will finish this section by quoting one of the dissenting voices:

> In EPR experiments, the experimenters do what they are asked to do: find conditions in which Bell's inequalities were infringed! Nobody, it seems, puts any restraints on the methods they used, or asks them to publish full data, including the runs that do not quite work. The magicians know how to *produce* their illusions (albeit not quite perfectly – witness those "anomalies"), but why do they still not *understand* them? [42, p. 358].

12. Concluding Remarks

The debate on the foundations of quantum theory is inconclusive. My feeling is that the basic problems of modern physics are enormously complex, and are still largely unsolved. Regrettably, a great number of scientists thinks that the final word on the subject has been given through some supposedly 'crucial' experiments. The great difficulty of the problems involved suggests that this view is a naive one. The Einsteinian programme of an eventual unification has still several adherents.

We have arrived at the following general conclusions:

(1) A large proportion of the misunderstanding, myths, brainwashing and conceptual confusion on the quantum theory comes from a limited understanding of the science involved, which is unable to conceive the enormous complexity of this kind of activity.

(2) Physics (and, specifically, quantum mechanics) is a very complex science made up of several scientific research programmes. In physics there are rival (competitive) programmes as well as cooperative ones. They form a web of theories and methods, but this does not imply inconsistency or any accommodation to contradictions. In incorporating rival programmes, physicists always aim at consistency, i.e. eliminating any contradictions.

(3) In this way, the co-existence and fertility of programmes like "Everything is Field" with Programmes like "Everything is Particle" does not mean a contradiction. This co-existence shows only the enormous complexity of physical reality.

(4) This complexity cannot be interpreted as any kind of "cultural relativism".

(5) The myth of the "dissolution of reality" constitutes dangerous brainwashing which threatens the objectivity of science and hinders the elimination of contradictions.

We believe that Einstein's thought was headed firmly in this rational and realistic direction. When Einstein said that, "It is theory that decides what is observable", he wanted to emphasize this important aspect of the complexity of the dialogue theory/experience which was a central part of his thought.

Appendix A. Bohm's Theory

As we know, in general a complex function $\Psi = \mathrm{Re}(\Psi) + i\,\mathrm{Im}(\Psi)$, where $i = (-1)^{1/2}$ and $\mathrm{Re}(\Psi)$ and $\mathrm{Im}(\Psi)$ are real quantities, can be written in polar form as

$$\Psi = R \exp\left(\frac{2\pi i S}{h}\right) \tag{A.1}$$

Inserting (A.1) into Schrödinger's equation

$$\frac{ih}{2\pi}\frac{\partial \Psi}{\partial t} = -\frac{h^2}{8\pi^2 m}\nabla^2\Psi + V\Psi \tag{A.2}$$

we obtain, after separating purely real terms and purely imaginary terms, the following two equations

$$\frac{\partial S}{\partial t} + \frac{(\nabla S)^2}{2m} + V - \frac{h^2}{8\pi^2 m}\frac{\nabla^2 R}{R} = 0, \tag{A.3}$$

$$\frac{\partial R^2}{\partial t} + \nabla.\left(\frac{R^2 \nabla S}{m}\right) = 0, \tag{A.4}$$

where S is the classical action of the system, V is the classical potential acting on the particle of mass m, and ∇ is the gradient operator which in Cartesian coordinates is given by

$$\nabla = \frac{\partial}{\partial x}\hat{e}_x + \frac{\partial}{\partial y}\hat{e}_y + \frac{\partial}{\partial z}\hat{e}_z,$$

where $\hat{e}_x$, $\hat{e}_y$, $\hat{e}_z$ are the unit vectors respectively in the directions x, y and z, and ∇^2 is the Laplacian operator which is a scalar operator resulting from the scalar product $(\nabla.\nabla)$, which in Cartesian coordinates is given by

$$\nabla^2 = \frac{\partial^2}{\partial x^2} + \frac{\partial^2}{\partial y^2} + \frac{\partial^2}{\partial z^2}.$$

If we consider the Hamilton–Jacobi theory according to which

$$\frac{\partial S}{\partial t} = -E \tag{A.5}$$

and

$$\nabla S = p, \tag{A.6}$$

Equation (A.3) can be written in the form

$$E = \frac{p^2}{2m} + V - \frac{h^2}{8\pi^2 m}\frac{\nabla^2 R}{R}. \tag{A.7}$$

Without the last term, the above equation becomes the well-known classical energy of the system which is the sum of kinetic plus potential energies. In the context of the WKB approximation the term $-(h^2/8\pi^2 m)(\nabla^2 R/R)$ is neglected (see [13, p. 28]). Alternatively, we can say that the classical Hamilton–Jacobi theory is reproduced from quantum mechanics in the first order of $(h/2\pi)$ and not in the zero-th order (see [26, Chap. III, § 17]).

Bohm argues that the last term plays an essential role. This term is the quantum potential

$$Q = -\frac{h^2}{8\pi^2 m}\frac{\nabla^2 R}{R}. \tag{A.8}$$

The quantum Hamilton–Jacobi Eq. (A.3) then becomes

$$\frac{\partial S}{\partial t} + \frac{(\nabla S)^2}{2m} + V + Q = 0. \tag{A.9}$$

On the other hand, Eq. (A.4) is usually interpreted in analogy with the equation of continuity

$$\frac{\partial \rho}{\partial t} + \nabla . j = 0$$

of hydrodynamics and of electromagnetism as one expressing the conservation of probability. The quantity R^2 (density of probability) is analogous to the density of charge (or mass) ρ and the quantity $R^2 \nabla S / m$ (density of current of probability) is analogous to the classical quantity j (superficial density of current). Let us go back to Eq. (A.9). If we accept the above description, we can allow that the equation of motion of the particle of mass m is given by

$$\frac{m \, dv}{dt} = -\nabla V - \nabla Q.$$

The above equation has the following meaning: the force acting on the particle of mass m has two components, each of a different nature. The first one is classical force $(-\nabla V)$ and the second is quantum force $(-\nabla Q)$. It is important to emphasize that according to Bohm and Hiley this description does not represent any return to the concepts of classical physics (see the quotation by Hiley in Section 9). The quantum potential Q constitutes a new concept which cannot be reduced to classical concepts.

In order to apply Bohm's ideas in quantum mechanics, we consider, for example, the s-states in a hydrogen atom. As we know, s-states have an orbital quantum number equal to zero and their corresponding wave-functions are real quantities. From Eq. (A.1) we can see that when this takes place then $S = 0$. Consequently, this circumstance leads to $p = 0$. In this case we obtain $E = V + Q$. According to this view, this result means that in the stationary states s electrons are at rest. This also means that the effects produced by the classical potential V on the electron are exactly balanced by the effects produced by the quantum potential Q on the same electron. Consequently the electron does not fall into the nucleus.

Appendix B. Orthodox Quantum Theory and the Problem of Measurement

Let us consider a pair of particles $\{\alpha, \beta\}$ described by the singlet state

$$|\Psi_S\rangle = (2)^{-1/2}\{|u^+\rangle|v^-\rangle - |u^-\rangle|v^+\rangle\}, \tag{B.1}$$

where $|u^\pm\rangle$ and $|v^\pm\rangle$ are eigenstates of the z-component of the spin operators $S_{\alpha z}$ and $S_{\beta z}$, respectively, where the subscript z denotes the z-component of the spin operator, the subscript α refers to the particle α and the subscript β refers to the particle β. The eigenvectors $|u^\pm\rangle$ and $|v^\pm\rangle$ are represented by their corresponding spinors.

We have

$$S_{\alpha z} = \frac{h}{4\pi}\sigma_\alpha; \qquad S_{\beta z} = \frac{h}{4\pi}\sigma_\beta, \tag{B.2}$$

where σ_α and σ_β are the 2×2 Pauli matrices respectively corresponding to the particles α and β.

Let us consider the square of the total spin of the system $\{\alpha, \beta\}$ whose corresponding operator is given by

$$S^2 = (S_\alpha + S_\beta)^2 \tag{B.3}$$

and the z-component operator of the total spin of the system $\{\alpha, \beta\}$

$$S_z = S_{\alpha z} + S_{\beta, z}. \tag{B.4}$$

Straightforward calculations lead to

$$S^2|\Psi_S\rangle = 0|\Psi_S\rangle, \tag{B.5}$$

$$S_z|\Psi S\rangle = 0|\Psi_S\rangle, \tag{B.6}$$

$$S_{\alpha z}|u^\pm\rangle = \pm\frac{h}{4\pi}|u^\pm\rangle, \tag{B.7}$$

$$S_{\beta z}|v^\pm\rangle = \pm\frac{h}{4\pi}|v^\pm\rangle. \tag{B.8}$$

From (B.5) and (B.6) it follows that:

$$\langle\Psi_S|S^2|\Psi_S\rangle = 0, \tag{B.9}$$

$$\langle\Psi_S|S_z|\Psi_S\rangle = 0, \tag{B.10}$$

i.e. in the singlet state the average values of the operators S^2 and S_z are both equal to zero.

The system made up of particles $\{\alpha, \beta\}$, which were initially together, decays and as a consequence particle α goes to a direction diametrically opposite to the direction in which particle β goes. In this new situation the particles are separated in space, but the system $\{\alpha, \beta\}$ continues to be described by the singlet state $|\Psi_S\rangle$.

We now suppose that a classical apparatus A performs a measurement on particle α *before* the classical apparatus B performs a measurement on particle β.

According to the usual orthodox quantum mechanics (associated with Bohr's theory of measurements) this measurement leads the initially entangled singlet state $|\Psi_S\rangle$ to one of two possible disentangled states, respectively, $|u^+\rangle|v^-\rangle$ or $|u^-\rangle|v^+\rangle$.

In other words, after the above measurement one of the two possibilities happens, whereas the other 'collapses'. Following this description, we have two possibilities for the average value of the operator S^2 in the new state. They are:

First possibility:

$$\langle u^+v^-|S^2|u^+v^-\rangle = \frac{h^2}{4\pi^2}, \tag{B.11}$$

Second possibility:

$$\langle u^-v^+|S^2|u^-v^+\rangle = \frac{h^2}{4\pi^2}. \tag{B.12}$$

If we compare (B.9) with (B.11) or, alternatively, (B.9) with (B.12), we conclude immediately that a measurement made by classical apparatus A on particle α is enough to cause a change in physical reality in the system $\{\alpha, \beta\}$. This measurement makes the value of the operator S^2 of system $\{\alpha, \beta\}$ jump *from zero to $h^2/4\pi^2$*.

Now we perform a second measurement by classical apparatus B on particle β *after* the measurement performed by classical apparatus A on particle α.

This also leads to two possibilities. They are:

First possibility:

$$S_{\beta z}|u^+v^-\rangle = -\frac{h}{4\pi}|u^+v^-\rangle.$$ (B.13)

Second possibility:

$$S_{\beta z}|u^-v^+\rangle = +\frac{h}{4\pi}|u^-v^+\rangle.$$ (B.14)

We note that this second measurement is irrelevant because in the first case above, $-h/4\pi$ is the eigenvalue of the operator $S_{\beta z}$ belonging to the eigenvector $|v^-\rangle$ and this eigenvalue was implemented by the first measurement made by the classical apparatus A on particle α, which led to Eq. (B.11). The same can be said with respect to the second possibility. In this case, $+h/4\pi$ is the eigenvalue of the operator $S_{\beta z}$ belonging to the eigenvector $|v^+\rangle$ and this eigenvalue was implemented by the first measurement made by the classical apparatus A on particle α, leading to Eq. (B.12). In short, in both cases the second measurement is irrelevant and does not change the physical reality of the system $\{\alpha, \beta\}$. This change of physical reality is effected by the *first* measurement.

Let us suppose that the first measurement is made at time t_A and that the second measurement is made at time t_B, where $t_B > t_A$. The quantity $\delta t = (t_B - t_A)$ may be considered infinitesimal; thus, no matter how small time δt is after the first measurement, any second measurement cannot change the physical reality. We remember that, according to the Copenhagen School's interpretation, the physical reality is contained in the state vector. Following this description (originally by Bohr) the collapse of the wave function is conceived as a form of instantaneous action at a distance.

In 1935 Schrödinger analysed a similar situation and wrote (ironically):

> Attention has recently (A. Einstein, B. Podolsky, and N. Rosen, *Phys. Rev.* **47** (1935), 777) been called to the obvious but very disconcerting fact that even though we restrict the disentangling measurement to *one* system, the representative obtained for the *other* system is by no means independent of the particular choice of observations which we select for that purpose and which by the way are *entirely* arbitrary. It is rather discomforting that the theory should allow a system to be steered or piloted into one or the other type of state at the experimenter's mercy in spite of his having no access to it.[9]

Finally, we stress that the famous EPR paper of 1935 was centred on the concept of completeness: Einstein and his collaborators tried to prove that quantum mechanics, as formulated in 1927, was an incomplete theory. Bell's seminal work of 1964, on the other hand, concentrated more on the concept of locality. He was able to offer a criterion for deciding whether quantum mechanics works when correlated particles happen to be separated by arbitrarily large distances. In this Appendix we have shown that if quantum mechanics and Bohr's theory of measurement are both correct, then instantaneous action at a distance follows as an unavoidable consequence.

---

[9]Schrödinger [35] *apud*, epigraph to Selleri [39].

Appendix C. Bell's Inequality

There are very clear demonstrations of Bell's inequality. We restrict ourselves to some comments. We recommend here to the reader the review article by Selleri [39] which includes several examples of this important theorem.

Here, we consider the following situation: we study a large number N of decays of type $\varepsilon \rightarrow \alpha + \beta$ forming a given ensemble. Observer $\hat{O}_\alpha$ performs a measurement on particle α and obtains a given value of the dichotomic observable $A(a)$; in an analogous way, observer $\hat{O}_\beta$ performs a measurement on particle β and obtains a given value of the dichotomic observable $B(b)$. The distance between particles α and β is considered to be arbitrarily large.

For the first decay we have the result (A_1, B_1), for the second decay the result (A_2, B_2), for the third the result (A_3, B_3), etc. We can define the correlation function,

$$P(a, b) = (N)^{-1} \sum_i A_i B_i, \tag{C.1}$$

where the sum is considered from $i = 1$ to $i = N$. The dichotomic observables assume the values $A_i = \pm 1$ and $B_i = \pm 1$. Consequently, $A_i B_i = \pm 1$. This implies, of course, that

$$-1 \le P(a, b) \le +1. \tag{C.2}$$

With respect to observable A we can take the arguments a and a' which are assumed to be experimental parameters, fixed in the structure of the apparatus in any given experiment, and the same holds for observable B and its arguments b and b'.

We can form correlation functions involving the arguments

$$P(a, b); \quad P(a, b'); \quad P(a', b); \quad P(a', b').$$

One specific combination of the above four correlation functions is enough to express the independence of the physical reality of sub-systems separated in space by arbitrarily large distances. This occurs for quantity Δ given by

$$\Delta = \left| P(a, b) - P(a, b') \right| + \left| P(a', b) + P(a', b') \right| \le 2 \tag{C.3}$$

which is Bell's inequality. In short, we can say that if quantity Δ has a value < 2, then the locality is preserved and systems that are separated in space are physically independent. Otherwise, if $\Delta > 2$, Bell's inequality is violated and the systems separated in space are not independent.

We apply now this ideas to the singlet state $|\Psi_S\rangle$. In this case the observable A is $(\sigma_\alpha . a)$ where σ_α are the 2×2 Pauli matrices and the point denotes a scalar product. We have:

$$\sigma_\alpha = \sigma_{\alpha x} \hat{e}_x + \sigma_{\alpha y} \hat{e}_y + \sigma_{\alpha z} \hat{e}_z,$$
$$a = a_x 1 \hat{e}_x + a_y 1 \hat{e}_y + a_z 1 \hat{e}_z,$$

where 1 is the 2×2 unit matrix. In the same way, the observable B is $(\sigma_\beta . b)$ and,

$$\sigma_\beta = \sigma_{\beta x} \hat{e}_x + \sigma_{\beta y} \hat{e}_y + \sigma_{\beta z} \hat{e}_z,$$
$$a = b_x 1 \hat{e}_x + b_y 1 \hat{e}_y + b_z 1 \hat{e}_z.$$

Our correlation function here is the average value of the operator

$$\{(\sigma_\alpha . a) \otimes (\sigma_\beta . b)\}$$

calculated in the singlet state $|\Psi_S\rangle = (2)^{-1/2}\{|u^+\rangle|v_-\rangle - |u^-\rangle|v^+\rangle\}$. We remember that the observable $(\sigma_\alpha . a)$ acts only on the spinors $|u^\pm\rangle$ while the observable $(\sigma_\beta . b)$ acts only on the spinors $|v^\pm\rangle$. So we have:

$$P(a,b) = \langle\Psi_S|(\sigma_\alpha . a) \otimes (\sigma_\beta . b)\}|\Psi_S\rangle = -(a_x b_x + a_y b_y + a_z b_z)$$

$$\therefore \langle\Psi_S|(\sigma_\alpha . a) \otimes (\sigma_\beta . b)\}|\Psi_S\rangle = -a.b = -\cos(a,b).$$

In general this result violates Bell's inequality. The maximum violation takes place for angles $(a,b) = \pi/4$; $(a,b') = 3\pi/4$; $(a',b) = \pi/4$; $(a',b') = \pi/4$. For these angles we have $\cos\pi/4 = \sqrt{2}/2$; $\cos 3\pi/4 = -\sqrt{2}/2$.

The corresponding quantity Δ will be

$$\Delta = \left|P(a,b) - P(a,b')\right| + \left|P(a',b) + P(a',b')\right|$$

$$= \left|\frac{\sqrt{2}}{2} - \frac{-\sqrt{2}}{2}\right| + \left|\frac{\sqrt{2}}{2} + \frac{\sqrt{2}}{2}\right| = \sqrt{2} + \sqrt{2} = 2\sqrt{2}.$$

The singlet state is entangled, i.e. the physical reality of particles α, β which are separated in space at arbitrarily large distances, are not independent. In other words, the singlet state leads to a violation of Bell's inequality.

References

[1] Barone M. & Selleri F. (Eds) 1994: *Frontiers of Fundamental Physics*, Plenum Press, New York, London.

[2] Barone M. & Selleri F. (Eds) 1995: *Advances in Fundamental Physics*, Hadronic Press, Palm Harbor, Florida, USA.

[3] Bastos Filho J.B. & Selleri F. 1995: "Propensities, Probability, and Quantum Theory", *Foundations of Physics* **25** (5), pp. 701–716.

[4] Bastos Filho J.B. 2003: "Os Problemas Epistemológicos da Realidade, da Compreensibilidade e da Causalidade na Teoria Quântica", *Rev. Bras. Ens. Fis.*, **25** (2), pp. 125–147 [Available in the site: http://www.sbf.if.usp.br].

[5] Bell J.S. 1966: "On the problem of hidden variables in quantum mechanics", *Rev. Modern Physics*, **38**, pp. 447–452.

[6] Bell J.S. 1982: "On the impossible pilot wave", *Foundations of Physics*, **12**, pp. 989–999.

[7] Bell J.S. 1987: *Speakable and Unspeakable in Quantum Mechanics*, Cambridge University Press.

[8] Ben-Dov Y. 1994: 'Local Realism and the Crucial Experiment', In: *Frontiers of Fundamental Physics*, Ed. by M. Barone & F. Selleri, Plenum Press, New York and London, pp. 571–574.

[9] Bohr N. 1935: "Can Quantum-Mechanical Description of Physical Reality Be Considered Complete?" *Phys. Rev.*, **48**, p. 696.

[10] Bohr N. 1961: *Atomic Physics and Human Knowledge*, Science Editions, Inc., New York.

[11] Bohm D. 1952: "A Suggested Interpretation of the Quantum Theory in Terms of 'Hidden' Variables, I", *Phys. Rev.*, **85**, pp. 166–179.

[12] Bohm D. 1952: "A Suggested Interpretation of the Quantum Theory in Terms of 'Hidden' Variables, II", *Phys. Rev.*, **85**, pp. 180–193.

[13] Bohm D. & Hiley B.J. 1993: *The Undivided Universe (An Ontological Interpretation of Quantum Theory)*, Routledge, New York, London.

[14] Chalmers A.F. 1999: *O que é a Ciência Afinal?*, Editora Brasiliense, São Paulo. [Brazilian translation of the original *What Is This Thing Called Science?*, 1982.]

[15] Duhem P. 1914: *La Théorie Physique. Son Object, sa Structure* [originally published in 1906], 2nd Edition, Vrin, 1989, Paris.

[16] Einstein A. 1935: 'Can Quantum-Mechanical Description of Physical Reality Be Considered Complete?' *Phys. Rev.*, **47**, pp. 777–780.

[17] Einstein A. 1936: 'Physics and Reality', *Journal of the Franklin Institute*, **221**, p. 349.

[18] Einstein A. 1954: *Ideas and Opinions*, Bonanza Books, New York.

[19] Einstein A. 1971: In: M. Born (Ed), *The Born-Einstein Letters*, Macmillan, London.

[20] Einstein A. 1993: *Letters to Solovine 1906–1955*, Citadel Press Book, New York.

[21] Gal-Or B. 1981: *Cosmology, Physics and Philosophy*, Springer Verlag Inc, New York.

[22] Gell-Mann M. 1979: In: *The Nature of the Physical Universe, the 1976 Nobel Conference*, Wiley, New York.

[23] Griffiths R.B. & Omnès R. 1999: 'Consistent Histories and Quantum Measurements', *Physics Today*, August 1999, pp. 26–31.

[24] Haroche S. 1998: 'Entanglement, Decoherence and the Quantum/Classical Boundary', *Physics Today*, July 1998, pp. 36–42.

[25] Hiley B.J. & Peat F.D. 1987: 'General Introduction: The Development of David Bohm's Ideas from the Plasma to the Implicate Order', In: *Quantum Implications (Essays in Honour of David Bohm)*, Ed. by B.J. Hiley & D. Peat, Routledge, London and New York, Ch. 1, pp. 1–32.

[26] Landau L. & Lifchitz E. 1966: *Mécanique Quantique*, Éditions Mir, Moscou.

[27] Lindley D. 1993: *The End of Physics (The Myth of a Unified Theory)*, Basic Books (A Division of Harper Collins Publishers), New York.

[28] Mamone Capria M. 1999: 'La Crisi delle Concezioni Ordinarie di Spazio e di Tempo: La Teoria della Relatività', In: *La Costruzione dell'Immagine Scientifica del Mondo (Mutamenti nella Concezione dell'uomo e del Cosmo dalla Scoperta dell'America alla Meccanica Quantistica)*, Ed. M. Mamone Capria, La Città del Sole, Napoli, pp. 265–416.

[29] Mamone Capria M. 1999: 'Introduzione. La Scienza e L'Immagine del Mondo', In: *La Costruzione dell'Immagine Scientifica del Mondo (Mutamenti nella Concezione dell'uomo e del Cosmo dalla Scoperta dell'America alla Meccanica Quantistica)*, Ed. M. Mamone Capria, La Città del Sole, Napoli, pp. 11–54.

[30] Pais A. 1995: *Sutil é o Senhor... (A Ciência e a Vida de Albert Einstein)*, Editora Nova Fronteira S.A., Rio de Janeiro. [Brazilian edition translated from the original *Subtle Is the Lord: The Science and the Life of Albert Einstein*, Oxford University Press, 1982.]

[31] Penrose R. 1987: 'Quantum Physics and Conscious Though', In: *Quantum Implications (Essays in Honour of David Bohm)*, Ed. by B.J. Hiley & D. Peat, Routledge, London and New York, Ch. 6, pp. 105–120.

[32] Popper K.R. 1982: *Quantum Theory and the Schism in Physics*, From the Postscript to 'The Logic of Scientific Discovery', Ed. by W.W. Bartley (III), Hutchinson, London, Melbourne, Sidney, Auckland, Johannesburg.

[33] Popper K.R. 1989: *Conjectures and Refutations (The Growth of Scientific Knowledge)*, Routledge, London and New York [Revised and corrected edition].

[34] Popper K.R. 1990: *A World of Propensities*, Thoemmes, Bristol.

[35] Schrödinger E. 1935: "Discussion of probability relations between separated systems", *Proc. Camb. Phil. Soc.*, **31**, pp. 555–563.

[36] Selleri F. 1983: *Die Debatte um die Quantentheorie*, Vieweg, Braunschweig.

[37] Selleri F. 1985: 'Bohr's Complementarity in the Light of Modern Researches in the Foundations of Quantum Theory', *Storia d. Scienza*, **2**, pp. 545–556.

[38] Selleri F. 1986: *Le Grand Débat de la Théorie Quantique*, Flammarion, Paris.

[39] Selleri F. 1988: 'History of the Einstein-Podolsky-Rosen Paradox', In: *Quantum Mechanics versus Local Realism*, Ed. F. Selleri, Plenum Publishing Corporation, New York, pp. 1–61.

[40] Selleri F. 1990: *Quantum Paradoxes and Physical Reality*, Kluwer Academic Publisher, Dordrecht, Boston, London.

[41] Stapp H.P. 1987: 'Light as Foundation of Being', In: *Quantum Implications (Essays in Honour of David Bohm)*, Ed. B.J. Hiley & D. Peat, Routledge, London and New York, Ch. 15, pp. 255–266.

[42] Thompson C.H. 1998: 'Behind the Scenes at the EPR Magic Show', In: *Open Question in Relativistic Physics*, Ed. F. Selleri, Apeiron, Montreal, pp. 351–359.

[43] 't Hooft G. 1997: *In Search of Ultimate Building Blocks*, Cambridge University Press.
[44] van der Merwe A. Selleri F. and Tarozzi G. 1992: *Bell's Theorem and the Foundations of Modern Physics*, World Scientific, Singapore, New Jersey, London, Honk Kong.

Physics Before and After Einstein
M. Mamone Capria (Ed.)
IOS Press, 2005
© *2005 The authors*

Chapter 10

Special Relativity and the Development of High-Energy Particle Physics

Yogendra Srivastava

Physics Department and INFN, University of Perugia, Perugia, Italy
&
Physics Department, Northeastern University Boston, MA, USA

1. Introduction

Lenard: *Relativity violates common sense.*
Einstein: *What is seen as common sense changes with time.*
Lenard: *At best relativity has limited applicability.*
Einstein: *On the contrary, an essential aspect of relativity is its universality.*[1]

We shall try to describe here the crucial role which Einstein's seminal 1905 paper on special relativity played in our understanding of the basic forces of nature. The space–time continuum embodied in special relativity was essential in reconciling the conflict, then present, between the classical laws of Newtonian–Galilean mechanics and Maxwell's electromagnetic theory. But this was just a beginning. The universal applicability of the new kinematics and the mechanics implied by special relativity went far beyond this, by opening the floodgates to other interactions. Special relativity, coupled with the "old" and the "new" quantum mechanics developed during the first quarter of the twentieth century, led to the realization that there are four fundamental interactions in nature: electromagnetic (EM), weak, nuclear and gravitational. Gravitational interactions require going beyond special relativity and form the subject matter of Einstein's general relativity (discussed in Chapter 5). Our attention will be focused in this chapter around the development of what is often called the *standard model* of *particle physics* or *high-energy physics*, which deals with the other three basic interactions. Much of the material presented here is, by its very nature not new. However, there are a few observations that are new and we hope an attentive reader will find them and find them stimulating.

---

[1] This dialogue took place between the German Nobelist Philipp Lenard and Albert Einstein with Max Planck chairing the session at the Bad Nauheim conference in September 1920. More can be found in [6]. Lenard, a rabid Einstein hater and anti-Semite, responsible for Einstein not getting the Nobel prize for several years, had gone to this conference to demolish this "arrogant and insulting" opponent.

This chapter is organized as follows. In Section 2, a simple example is presented which exposes the basic clash between the laws of Galilean–Newtonian mechanics and the laws of Maxwell's electromagnetic theory and shows the need for a new law for the addition of velocities. In Section 3, some details of the new relativistic mechanics and continuous space–time Poincaré–Lorentz transformations along with the concepts of the light cone and space-like, light-like and time-like events are presented. In Section 4 the notion of discrete space–time symmetries is described. Section 5 deals with Dirac's brilliant synthesis of relativity with quantum mechanics, which revolutionized the whole field of particle physics and modern cosmology through the concept of an anti-particle (or anti-matter) and the realization that the number of particles need not be conserved. In Section 6, we present the notion of real and virtual processes present in relativistic quantum mechanics and quantum field theory. In Section 7, a brief description of the EM, weak and strong (nuclear) interactions is presented. In Section 8, we describe the bare essentials of the standard model of fundamental interactions, some of its successes and the one (fundamental) missing link, the Higgs particle. In Section 9, we explore why one is impelled to go beyond the standard model (SM). One popular scheme beyond the SM, called *supersymmetry*, is presented in Section 10, as well as its relevance to the issue of the energy of the vacuum. In Section 11 we describe studies of higher dimensional space–time theories and their implications for the basic forces. A short description of non-commutative geometry is presented in Section 12 along with a realistic example of how such non-commutativity may arise dynamically. In Section 13, the old, outstanding problem of the definition of the position operator for a massless particle (such as the photon) is tackled, which leads to inherent correlations and non-commutativity. In Section 14, we return to a discussion of the breakdown of discrete symmetries and their implication for the space–time continuum. We also discuss here fundamental consequences that a breakdown of *CPT* (presently being vigorously pursued by experimentalists) would have for non-locality and the loss of Lorentz invariance itself. We close in Section 15 with present problems and future prospects.

2. Special Relativity and Maxwell's EM Theory

Consider a particle with electric charge q and (rest) mass m in a constant electric field E, say along the x-direction. Let the particle start from rest at time zero. Then, Newton's equation (for the x-component of the force) equates the time rate of change of the momentum ($p(t)$) of the particle at time t to the electric force

$$F = \frac{dp(t)}{dt} = m\frac{dv(t)}{dt} = qE \quad \text{(Newton)}. \tag{1}$$

We can solve for the velocity $v(t)$ as a function of time, and we find that the velocity increases linearly with time t:

$$v(t) = \left(\frac{qE}{m}\right)t, \tag{2}$$

and would equal the speed of light c at a time T such that

$$T = \frac{mc}{qE}, \tag{3}$$

after which the velocity of the charged particle will exceed the speed of light c. Thus, if there is a *maximum* speed (viz., the speed of light c), there is a clear inconsistency between Newtonian mechanics (the left-hand side of Eq. (1)) and the EM force (the right-hand side) of the same equation.

A related problem is with the speed of light as seen by two different observers. The Galilean law of addition of two velocities (in one space dimension) says that the resultant velocity u when we add two velocities v_1 and v_2 is their sum

$$u = v_1 + v_2 \quad \text{(Galilean)}. \tag{4}$$

Equation (4) implies that the speed of light would differ for two observers by the relative speed between the two observers. If we insist that the speed of light c be the same for all observers, we need to modify the above equation. Special relativity modifies Eq. (4) to

$$u = \frac{v_1 + v_2}{1 + v_1 v_2 / c^2} \quad \text{(Einstein)}. \tag{5}$$

For speeds $v_1, v_2 \ll c$, Eq. (5) reduces to the Galilean law Eq. (4). The reader is encouraged to verify that when either v_1 or v_2 is c, u is always c. Hence, the remarkable Eq. (5) leads to the speed of light being the same for all observers, as indeed is found experimentally (e.g. by Michelson and Morley).

Special relativity changes the relationship between the momentum p and the velocity v by the Lorentz factor γ

$$p(t) = m\gamma(t)v(t); \qquad \gamma(t) = \frac{1}{\sqrt{1 - v(t)^2/c^2}}, \tag{6}$$

so that Eq. (1) is changed to

$$F = \frac{dp(t)}{dt} = m\frac{d(\gamma(t)v(t))}{dt} = qE \quad \text{(Einstein)}, \tag{7}$$

and Eq. (2) is replaced by

$$v(t) = c\frac{t/T}{\sqrt{1 + (t/T)^2}}. \tag{8}$$

For times $t \ll T$, the speed v remains small and classical non-relativistic mechanics holds. On the other hand, for large times, $t \gg T$, v asymptotically approaches c, but cannot exceed c. This result is amply verified in high-energy machines where charged particles are accelerated under an electric field. At present, values for v have been reached which fall short of c by 1 part in 10^{10}, providing a remarkable confirmation of special relativity.

We may conclude this section by noting that as far as applications to particle physics are concerned, we have excellent experimental evidence that Einstein's special theory of relativity is correct, confirming that

   (i) The speed of light c (in the vacuum) is the same in all (inertial) reference systems. (Unification of EM, weak and strong interactions via a gauge theory, is based on the assumption that not only the photon but *all* massless gauge particles, e.g. the gluon, also propagate with the speed of light c.)

   (ii) Any material particle (i.e. one possessing a non-zero mass) may be accelerated to approach c but never really reach it.

3. Relativistic Kinematics and the Nature of Space–Time Events

Special relativity deals with "events" in the four-dimensional space–time continuum (cf. Chap. 5, §4). Consider two events A and B with their 4-coordinates $x_A^\mu = (ct_A; \vec{x}_A)$ and $x_B^\mu = (ct_B; \vec{x}_B)$, where the first coordinate is the time and the second set is the position (in space) of the events, in some reference frame. The invariant separation between the events is given by the scalar

$$I_{AB} = (x_A - x_B)^2 = c^2(t_A - t_B)^2 - (\vec{x}_A - \vec{x}_B)^2, \tag{9}$$

which would be the same in all inertial reference frames. Depending upon the sign of I, the two events fall into three disjoint classes:

(i) If $I > 0$, we say that A and B are "time-like" separated. The events are within each other's "light-cone" so transmission of messages is possible.

(ii) If $I = 0$, A and B are "light-like" situated.

(iii) If $I < 0$, A and B are said to be "space-like" situated. No messages can be received or sent between them.

Consider now the trajectory of a particle of mass m being described by its 4-coordinate $x^\mu = (ct, \vec{x})$, where $\vec{x}$ gives its 3-position coordinates at a given time t, and its 4-momentum is $P^\mu = (E(p)/c, \vec{p})$. Here, E and $\vec{p}$ are the total energy and momentum of the particle in some reference frame. The relativistic (generalization of Newton's) equation of motion in terms of the 4-force F^μ reads

$$F^\mu = \frac{dP^\mu}{d\tau}, \tag{10}$$

where τ is the proper time. If the particle is "free", i.e. is subject to no force ($F^\mu = 0$), then Eq. (10) tells us that P^μ is independent of τ. In fact, using the definition of the 4-velocity U^μ

$$\frac{dx^\mu(\tau)}{d\tau} = U^\mu = \frac{P^\mu}{m}, \tag{11}$$

we may solve for the coordinates of a free particle as

$$x^\mu(\tau) - x^\mu(0) = \frac{P^\mu}{m}\tau. \tag{12}$$

Thus, the proper time τ describes in an invariant manner how the trajectory evolves

$$c^2\tau^2 = \eta_{\mu\nu}\big(x^\mu(\tau) - x^\mu(0)\big)\big(x^\nu(\tau) - x^\nu(0)\big), \tag{13}$$

where the Minkowski metric is represented by $\eta_{\mu\nu}$ (with $\eta_{00} = 1$ and $\eta_{ij} = -\delta_{i,j}$ and $\eta_{0i} = 0$), and Einstein's convention on repeated indices is adopted. A free particle is then described by a time-like 4-vector P^μ lying on its "mass shell"

$$P^2c^2 = \big(P_\mu P^\mu\big)c^2 = \big(mc^2\big)^2 = E^2 - |\vec{p}|^2c^2. \tag{14}$$

From the above equation, we find that the energy of a free particle $E(p)$ (in a given frame) is given in terms of its momentum $\vec{p}$ in that frame via

$$E(p) = \pm\sqrt{(mc^2)^2 + |\vec{p}|^2 c^2}. \tag{15}$$

In the rest frame of the particle, where $\vec{p} = 0$, Eq. (15) gives the famous equation $E = mc^2$ (upon choosing the positive sign for the square-root).

However, what about the negative sign of the energy in Eq. (15)? In classical physics, one simply imposes the physical condition that free particles must have only positive energies and one throws away the negative energy solutions as not being physical. In quantum mechanics, one needs to include both positive and negative energy solutions in order to have a complete set of states. This introduces a new and essential complication in formulations of relativistic quantum mechanics.

A related problem is with the trajectory equation Eq. (12). In principle, both types of motions, i.e. moving forward in time ($\tau > 0$), and moving backward in time ($\tau < 0$), are possible. However, causality forces us to consider only motions forward in time ($\tau > 0$).

Both problems haunted Dirac when he formulated a relativistic quantum mechanical equation for the electron. As we shall see in Section 5, Dirac was able to interpret both these difficulties as an essential further positive element lacking in classical physics. That is: a (positive energy) positron (the anti-particle of an electron) *à la* Dirac is simply an electron of negative energy moving backwards in space and time. This was a truly revolutionary move, which changed forever the way we envisage our universe.

4. Discrete Transformations

Along with continuous rotations and Lorentz transformations, we have a set of discrete transformations possible in a relativistic theory, called (i) *parity* (P), (ii) *time-reversal* (T) and (iii) *charge conjugation* (C):

(i) P: Parity transformation means we change all the spatial coordinates $\vec{r}$ to $-\vec{r}$, so that a right-handed coordinate system becomes its mirror image, a left-handed coordinate system.

(ii) T: Time reversal means that we change the sign of the time, i.e. t is changed to $-t$. Physically, this sends past to the future and vice versa; it is as if we are viewing a film backwards.

(iii) C: Here the electric charge q of a particle is changed to its opposite, i.e. $-q$.

The relevance of these transformations may be explicitly appreciated in classical EM through the Lorentz force equation

$$\vec{F} = \frac{d\vec{p}}{dt} = q\left[\vec{E} + \frac{1}{c}\vec{v} \times \vec{B}\right]. \tag{16}$$

The above equation is invariant under P, since $\vec{F}$, $\vec{p}$, $\vec{E}$, $\vec{v}$ change sign as a "proper" vector should under $\vec{r}$ to $-\vec{r}$, whereas $\vec{B}$, being an "axial" (or pseudo-) vector does not. Physically, it means that the laws of EM do not change if we view electrodynamics from a right-handed or a left-handed coordinate system. Nature does not distinguish its left from its right through its electromagnetic interactions. Thus, P is a "good" symmetry for EM.

Idem for T. That is, when t goes to $-t$, $\vec{F}$, $\vec{E}$ remain unchanged whereas $\vec{v}$ and $\vec{B}$ do change their sign, so that the system remains invariant under time-reversal. Nature does not distinguish its past from its future through its EM interactions. Also, T is a "good" symmetry for EM.

C is also a good symmetry for EM since as q is sent to $-q$, the Lorentz force remains the same provided both $\vec{E}$ and $\vec{B}$ change sign.

Thus, for EM, individually the three symmetries P, T and C are good symmetries. Without proof here, we simply state that for nuclear interactions the same remains true.

It was widely believed therefore that nature respected these symmetries individually. It was a great contribution of T.D. Lee and C.N. Yang therefore, in 1956, to have boldly suggested that the separate conservation of these symmetries was in the nature of the interactions (EM and nuclear to be precise), whereas we needed to check experimentally whether other interactions might exist in nature which violate these symmetries. Were that so, nature would be able to distinguish its right from its left, the future from its past, and the sign of the electric charge. Lee and Yang were absolutely correct, of course, since the "weak" interactions were soon shown to violate P and C separately. A few years later, violations of T were also experimentally found. In a later section, we shall discuss these symmetry breakdowns and the important role they played in the development of the standard model of fundamental interactions. Since in the present chapter we are highlighting the role of Lorentz invariance (part and parcel of special relativity) in this development, here we only state that in any local, Lorentz-invariant theory, the *PCT* theorem is rigorously valid, i.e. nature must respect the *combined* operations of right to left, particles to anti-particles, and past to future. So far this theorem seems to hold. But intensive experimental research effort has been and continues to be dedicated to finding even a small but observable violation of the above theorem. Were it to be violated, we would have to discard our cherished picture of a local Lorentz-invariant Lagrangian field theory.

5. Dirac's Synthesis of Relativity and Quantum Mechanics

If you are not confused by quantum physics, you have not understood it.
(Attributed to Niels Bohr)

Circa 1928, Dirac wrote down the following relativistic equation to describe a free electron

$$i\hbar \frac{\partial \Psi}{\partial t} = H_D \Psi, \tag{17}$$

where the Hamiltonian H_D is written in terms of four 4×4 hermitean matrices $\vec{\alpha} = (\alpha_1, \alpha_2, \alpha_3)$ and β:

$$H_D = -i\hbar c \vec{\alpha} \cdot \nabla + mc^2 \beta. \tag{18}$$

The requirement that the relativistic energy-momentum condition Eq. (15) be satisfied for the eigenvalues leads to these four matrices being unimodular and mutually anti-commuting. However, as alluded to in Section 3, Dirac was stuck with solutions for both signs of energy. When an EM interaction was added to Eq. (18), Dirac found that for every solution for a particle of charge q, there was also a solution of charge $-q$, i.e. the property of charge conjugation. This led him to advance the hypothesis of the existence of an anti-electron (the positron). Generally, special relativity requires that for every particle there exists an anti-particle with the opposite sign of charge but with the same mass. We have literally hundreds of examples of particles verifying this generic result.

Lest there be confusion, let us see what charge conjugation actually says and prove that it is neither trivial nor obvious. If a particle has positive charge, under an external electric field it would accelerate in the direction of the field. Under the same field, its anti-particle would accelerate in the opposite direction. Thus, matter and anti-matter (with opposite charges) have opposite trajectories in electrodynamics. On the other hand, in a gravitational field, say on the earth, both particles and anti-particles would fall towards the earth. There is no anti-gravity (yet) and the gravitational "charge" (the inertial mass) is the same for matter and anti-matter. Hence, under gravity both matter and anti-matter trajectories are the same. As stated in the Introduction, we shall ignore gravity effects in this chapter even though they are fascinating in their own right.

As can be found in good textbooks (see, for example, [5,10]), detailed computations show that under an EM field a positron of positive energy is an electron of negative energy going backwards in space and time. For an observable particle and its anti-particle we now assume that both have only positive energies.

Since the above is true for all fields, one may ask why the photon field which is the carrier of electromagnetism and has zero mass, and thus is ultra-relativistic, does not exhibit its anti-particle. The simple reason is that the photon field carries no charge, i.e. is neutral, and thus its anti-particle is the same as its particle.

Dirac was able to deduce from the above two novel phenomena: matter–anti-matter *annihilation* and matter–anti-matter *pair production*. Since a pair particle–anti-particle has zero (total) charge, quantum mechanically there would be a finite probability that such a pair can annihilate and reappear as radiation carrying the same energy-momentum as the initial pair, or vice versa. Note that these processes are absent in non-relativistic quantum mechanics where the anti-matter degrees of freedom are ignored. We stress that special relativity coupled with quantum mechanics allows for the possibility of conversion of matter (through the intervention of anti-matter degrees of freedom) into radiation or whatever else is allowed. The bottom line is that, generally, the number of material particles need not be conserved. In a typical high-energy collision, say of a proton and an anti-proton, the average number of final particles produced may be as large as 200–400.

There is another aspect of relativistic quantum theories, the concept of real and virtual processes, that played an important role in the development of particle physics. We shall discuss this in the next section.

6. Real and Virtual Processes

It is a basic tenet of quantum mechanics that if we "observe" an object it behaves classically. Since interaction times are extremely short, most of our knowledge about particle physics comes through scattering processes. A typical scattering experiment is conceived as follows. Initially (at time $t \to -\infty$), two (classical) free particles $a(p_a)$ and $b(p_b)$, infinitely far apart, with 4-momenta p_a and p_b are sent towards each other. When sufficiently near, the two particles interact. After the reaction, we wait (theoretically until time $t \to +\infty$) with our "detectors" sufficiently far away from the interaction region and detect the produced (classical, free) set of n particles with definite 4-momenta: $c_1(p_1), c_2(p_2), \ldots, c_n(p_n)$. Through a measurement of its rate of production, we determine the probability of such a transition.

The initial and final sets of particles are all classical with simultaneously well defined positions and momenta and all particles remain on their mass shells, i.e. $p_i^2 = (m_i c)^2$, where m_i denotes the mass of the ith particle. It is during the interaction, which is *not* observed, that all quantum effects are supposed to occur. Thus, the initial and final particles are "real", whereas the (unobserved) particles in the "intermediate state" are "virtual".

What the latter implies is that these particles need not be on their mass shells and need not follow classical trajectories. To paraphrase Feynman, an electron in the intermediate state can do whatever it wants: it can go into the past, into the future, inside or outside its light cone, it can move with any speed – even speeds greater than the speed of light. And in order to describe its motion completely, we must sum over "all such possible paths". Relativistic quantum theory requires that only the observed objects obey causality and make movements within their light-cones; unobserved (virtual) particles need not. For example, virtual particles can go backwards in time. This leads to the concept of a "closed particle loop", where through an initial interaction, say at time $t = T_I$ and position $\vec{x} = \vec{X}_I$, a particle is produced. As a virtual (intermediate state object), it can propagate along until some future time $t = T_F$ and position $\vec{x} = \vec{X}_F$ where it has a final interaction, after which it decides to go back in time and space to its initial value, where it annihilates itself. Calculations of such (and even more complicated) processes are routinely performed using *Feynman diagrams* to compute transition probabilites and are successfully compared with experimental data. To the uninitiated it may appear weird, but this procedure really works. One's faith is so strong that when it does not work, we suspect that the interaction chosen needs improvement, not the underlying picture. For example, the Fermi theory describes weak interactions quite well at low energies. At high energies, when it fails to agree with experiments, we change the Fermi interaction, but keep the relativistic rules.

To recapitulate: only in special relativity applied to classical (not quantum) physics do we have the requirement that material particles stay on their mass shells with a well-defined trajectory; not so for the virtual particles. It would appear that virtual processes are hidden for the practical reason that the interaction times are too short for them to be observable. We may entertain the theoretical possibility that, were interaction times large enough for an experimental probe, hitherto unseen novel phenomena might arise.

We illustrate one such curious prediction from quantum field theory for the Feynman–Stückelberg (F-S) propagator of the photon. For a free photon, the probability amplitude that we should find a photon at space–time point x if we knew that there were a photon present at the space–time point y, is given by the (complex) F-S propagator

$$D(x - y) = -\left(\frac{1}{4\pi^2}\right) P \frac{1}{(x - y)^2} - \frac{i}{4\pi} \delta\big[(x - y)^2\big], \tag{19}$$

where P stands for the principal value. The second Dirac δ function term in the above equation corresponds to the classical (relativistic) result that the proper time

$\tau = \sqrt{(x-y)^2}/c = 0$, i.e. the photon lies on its light cone. The first (principal value) term on the right side of Eq. (19) is the "quantum correction", allowing the photon to stray off the light cone.

On the other hand, if we Fourier-transform Eq. (19), we find the same probability amplitude in momentum ($p = \hbar k$) space to be

$$D(k) = \pi \delta(k^2) + i P \frac{1}{k^2}, \tag{20}$$

where the first (second) term on the right side of Eq. (20) is the Fourier transform of the first (second) respectively of the right side of Eq. (19). Now the δ function term in Eq. (20) is the classical (relativistic) result, the photon is on its mass shell $k^2 = 0$, whereas the second term in Eq. (20) (the principal value term) is the "quantum correction" which allows the photon to stray off its mass shell. What we learn is that quantum mechanically, if a photon is on its mass shell it cannot be on its light cone and, vice versa, if it is on its light cone it cannot be on its mass shell. On the other hand, classical special relativity requires both to be true simultaneously. Surely a curious result. In Section 13, we shall return to this problem through a quantum mechanical definition of the photon position operator and shall uncover a built-in non-locality which leads to some experimentally measurable predictions.

7. Basic Interactions: EM, Weak, Strong and Gravitational

As mentioned earlier, we have evidence for four types of basic interactions: gravitational, weak, EM and strong, in order of increasing strength of force. If we deal with fundamental particles of limited masses, the gravitational force is typically 10^{20} smaller than the other forces, which is why the standard model neglects it, and attention is concentrated on the other three forces. There is one caveat to bear in mind. Were one to probe the system to extremely small distances (less than, or equal to, the Planck length, 10^{-32} cm), the gravitational force would become significant and the space–time picture based on a flat Minkowski space would need modification. It is this fact which has led to various schemes of unification of all four forces, but we shall not discuss them here since we are experimentally far from being able to probe such distances. A few comments shall be made regarding this aspect in Section 9.

Regarding the other three forces, a flat four-dimensional space–time continuum implied by special relativity appears valid for all, but the underlying discrete symmetries for each are different. As yet we have no deep understanding of the reason for such disparities, but the phenomenology of the interactions is well delineated [15]:

(i) *Weak interactions*. Here none of the discrete symmetries P, C or T holds separately. All matter known so far is made of spin one-half particles divided into two classes, leptons and quarks, and both leptons and quarks participate in weak interactions.

(ii) *EM interactions*. All three discrete symmetries hold separately. Both leptons and quarks participate in EM interactions.

(iii) *Strong interactions*. Also, for these (nuclear) interactions, all three discrete symmetries hold separately. Leptons do not participate in nuclear interactions, but quarks do. Moreover, different types of quarks are not distinguished by the strong interactions ("nuclear democracy") but their electro-weak interactions are different.

Thus, any scheme such as the standard model which unifies these interactions must be very tightly constrained in a precise mosaic framework to respect their various discrete symmetries and other conserved quantum numbers such as the "lepton" and "baryon" numbers.

8. Unification of the Three Fundamental Forces: The Standard Model [26]

After intense work lasting over 40 years, with many false starts both on the experimental and the theoretical side, a fairly coherent unification of EM, weak and strong interactions emerged which is written in terms of a (special relativistic) scalar Lagrange density describing a local gauge theory (for the group $SU(2)_L \times U(1)_Y \times SU(3)_c$) of spin-one gauge bosons and three families of spin-one-half fermionic matter (leptons and quarks). It would be tedious to describe the details of the standard model (SM) here. The interested reader is referred to some excellent texts for a complete explanation [11,3]. Here we shall just discuss some salient features of interest.

At the "zeroth" level, all spin-one bosons and all spin-one-half fermions are massless to respect the full gauge symmetry. (Masslessness of gauge particles ensures that they generate forces of infinite range.) Only left-handed leptons and quarks interact with the gauge vector bosons of the group $SU(2)_L$, which explains the suffix L. Hence for this sector there is a maximum breakdown of parity, i.e. it would appear that nature distinguishes its left from its right on a fundamental level. Since the left-handed matter transforms as a "doublet" (that is, it has two components) whereas right-handed matter transforms as a "singlet" (one component), masses for Dirac fermions are precluded, since a Dirac mass term (which must behave as a scalar) needs a coupling between its left and its right counterparts which behave differently under this group. A local gauge symmetry precludes masses for all spin-one gauge bosons.

To introduce masses, the present scheme in the standard model invokes a *spontaneous breakdown* of symmetry by the introduction of a spin-zero doublet of Higgs bosons, which couples to the "electro-weak" sector ($SU(2)_L \times U(1)_Y$). Here the abelian group $U(1)_Y$ refers to the quantum number "hypercharge", a generalization of the concept of electric charge. The Higgs doublet, which has four real spin-zero fields, does not couple to the spin-one "gluons", the gauge particles pertaining to the group $SU(3)_c$, which generates the nuclear (strong) force, whose symmetry is therefore left intact.

The general idea about the *Higgs mechanism* may be appreciated from the phenomenon of superconductivity, from which it is derived.

At sufficiently low temperatures all metals become superconductors, with surprising results. For example, within a superconductor a pair of two electrons (e^-e^-) can bind into a spin zero charge $q = -(2e)$ *Cooper pair*, something which does not happen for normal metals due to the mutual Coulomb repulsion between like charges such as two electrons. Another surprising result is that a magnetic field cannot penetrate into a superconductor (or if it is inside, it remains "trapped", that is, it cannot get out). Both these results are understood theoretically by constructing the energetics of an electrically charged spin-zero Cooper particle field in such a way that the energy minimum is obtained when the value of the field is nonzero. Technically, this means that it has a nonvanishing expectation value which leads to a spontaneous symmetry breakdown. This is enough to trigger the photon, the carrier of electromagnetism, to develop a mass within a superconductor. This mass barrier is what prevents a static magnetic field from getting in or out easily since it costs energy to surmount this energy barrier.

The SM Higgs mechanism runs along the same lines, by imposing on the Higgs field (the analogue of the spin-zero Cooper pair) potential to minimize its energy for a nonzero value of the field. It leads to the masses for three of the gauge bosons $W^\pm$ and Z^o, whereas the photon (and the gluons) remain massless. The Higgs field is also coupled to the fermions (leptons and quarks) in such a way that they also acquire masses through the nonzero value of the Higgs field. The three massive vector bosons "eat up" three of the Higgs fields and only one neutral Higgs field remains. Intensive search for this field, called the Higgs particle, continues.

The SM is extremely successful, especially in the electroweak sector. Due to the small value of the coupling constant, accurate perturbative calculations can be performed and compared to experiments. For example, the anomalous magnetic moment of the *muon* (the weightier sister of the electron) has been measured to the accuracy of a few parts in a million, and it agrees perfectly with the SM calculations.

The strong ($SU(3)_c$) sector pertaining to the interaction between quarks and gluons describes hadronic (i.e. strong nuclear and subnuclear interaction) physics. Analysis of the hadronic sector is more difficult since we do not observe free quarks and gluons; they are presumably confined. As yet, no convincing theoretical proof of confinement exists and it is an open problem. However, for this interaction a new phenomenon called *asymptotic freedom* has been discovered, which roughly states that under certain conditions, the effective coupling constant becomes so small that perturbative calculations can be performed. This has led to quite interesting results verified by high-energy experiments, such as "scaling" and particle production in the form of "jets". So all in all, most workers in this field consider the Quantum Chromodynamics (QCD) part of the SM to be correct and experimentally vindicated.

Accepting the above, then, the only missing link so far in the SM is the Higgs particle. If the Higgs particle is not found experimentally, the SM as formulated would obviously have to be revised. But even if a Higgs particle were to show up, fundamental issues about the SM need to be addressed, and this forms the topic of the next section.

9. Beyond, Below and Behind the Standard Model

Even if all the predictions of the SM (including a proper Higgs particle) were to be experimentally verified, one would feel compelled to look beyond it for several reasons.

The standard model has a large number of arbitrary parameters. There are three independent coupling constants for the three (cross-product) groups. Also, the masses of all the (twelve) fermions (leptons and quarks, which constitute all of known matter) are arbitrary, along with the mass of the symmetry-breaking agent Higgs. One feels uneasy to call such a theory "fundamental". Moreover, there is a technical problem in so far as this theory is not "stable". That is, in perturbation theory, which is our sole tool, these masses suffer major modifications due to higher-order corrections.

Guided by the above, the notion of *Grand Unified Theories* (GUT) has been studied extensively during the last three decades, the idea being that there existed a unifying local gauge group with one coupling constant, which would suffer a symmetry breakdown to give rise to the SM in its "low energy" limit. Popular examples of such attempts were the groups $SU(5)$ and $SO(10)$, whose group structures could allow for an effective low energy theory of the SM type.

Such GUT schemes predict novel features such as proton decay, since in GUT neither the lepton nor the baryon numbers are conserved. So far no convincing experimental evidence has been presented for the decay of the proton. Hence, we can deduce that the life-time of the proton is certainly above 10^{32} years. To put this number in perspective, it is a sobering thought that the age of the Universe is estimated to be about a few billion years, 20 orders of magnitude smaller than the lower limit on the life-time of the proton! If this limit could be improved, say by two orders of magnitude, I believe that mainstream physicists would agree that there is no such thing as the decay of a proton, and on this basis would consider GUT schemes to be fundamentally wrong.

Another direction, mainly motivated by string theory, which has found many followers [1], is to contemplate a higher-dimensional space–time where for some reason a contraction of space–time occurs, so that the spatial dimensions beyond the so-far observed three dimensions "curl up" and become unobservable with present-day "lower energy" probes. These theories have special predictions of their own. For example, in the static limit, the gravitational potential would no longer be of the familiar $1/r$ type but instead would change to $1/r^{1+\delta}$, where $\delta \neq 0$ would signal an underlying higher-dimensional space–time structure. So far, intensive searches have not revealed any such deviations.

In the next few sections we shall briefly comment only on those aspects of these extensions of the SM which have consequences transcending particle physics per se, where the possibility exists that non-high-energy experiments may shed light on these interesting issues. Examples are the energy of the vacuum, Lorentz invariance and non-commutativity of the coordinates.

10. Supersymmetry and the Structure of the Vacuum

Supersymmetry [13,28,2,9,27] is a boson-fermion symmetry that enlarges the Poincaré algebra to a super-Poincaré graded algebra. In practice, this means that the number of bosonic degrees of freedom is exactly equal to the fermionic degrees of freedom. Hence, for every fermion there is an associated boson and vice versa. Example: for an electron there is an associated scalar (and pseudo scalar) *selectron*; for the spin-1 photon there is an associated (Majorana) spin $1/2$ *photino*, for each quark there is a *squark* (spin zero); the gluons have their spin-one-half partners the *gluino*, etc.

This scheme has several pleasing features so long as supersymmetry (SS) is exact (i.e. it is unbroken). One example concerns the energy of the vacuum. A bosonic oscillator of frequency ω has the well-known zero point energy $(1/2)\hbar\omega$. In field theory, for each value of momentum $\mathbf{k}$, there is thus a zero point energy $(1/2)\hbar\omega(\mathbf{k})$. A sum over all momenta then leads to a three-fold infinity in the total energy of the vacuum. In the standard theory, this infinite vacuum energy is subtracted by the normal ordering of the operators. In exact supersymmetry, this (positive) infinite vacuum energy by bosons is exactly cancelled by a compensating negative infinite vacuum energy provided by the fermionic oscillators, and hence the total vacuum energy is zero so long as supersymmetry is unbroken. In general, the degree of divergence in SS theories is much less severe than in theories without SS. Also, other arguments exist in favour of SS such as a resolution of the so-called hierarchy problem, that is, why there is such disparity in the observed masses of elementary particles, which range over five orders of magnitude? More ambitious schemes such as supergravity models have been constructed which couple (spin 2) gravity to gravitinos (with spin 3/2). The supergravity model is finite as a field theory, in contrast with the Einstein gravity theory (coupled to matter) which is divergent.

For the above reasons, SS standard models have been constructed and experimental searches for the missing particles have been vigorously pursued. So far there is no experimental evidence for any one of the superpartners. For example, the mass of the selectron has been pushed to beyond several hundred GeV/c^2. (If superpartners of the known particles do exist then their enormous rest mass would make their production feasible only at very high energy accelerators not yet in existence.) Moreover, if the partner of the electron is about a million times heavier than the electron, SS is very badly broken. Phenomenological SS breaking schemes proposed so far, which are variants of the Higgs scheme in the standard model, provide no clue to the underlying dynamics.

Initially SS had provided hope for an understanding of why the vacuum energy density of the universe is as small as it is (at present estimated to be roughly about 10^{-12} eV^4). If SS is broken at the scale of say a tenth of a TeV ($= 10^{12}$ eV), we have a mismatch between the observed and the expected SS vacuum energy (10^{44} eV^4) by over 56 orders of magnitude. A sobering result indeed!

It is fair to say that while SS is still theoretically quite appealing, currently we are very far from understanding at a basic level the breaking of SS which generates such diverse masses for its fundamental constituents. Let us hope that ongoing and future experiments will soon provide us with some clues, just as in the past theory had to wait until experiments in the 1960s and the 1970s provided the basic ingredients to lead theorists away from looking for symmetries the blindfold way, towards the construction of the standard model.

11. Higher Dimensions, Kaluza–Klein Excitations

In recent years, due to the superstring theory, which is formulated in 10 space–time dimensions, much effort has been devoted to a study of possible manifestations, albeit tiny, from the unseen six dimensions. The original Kaluza–Klein (KK) formulation envisaged an extra fifth (spatial) dimension apart from the standard 3 space and 1 time dimensions, which had several attractive features such as generating, along with gravitation, the EM interaction and an extra field called the "dilaton". The extra dimension was supposed to be "compactified" into a small bundle of the size of the Planck length (about 10^{-32} cm).

Much speculation has been made to enlarge the KK scale from the gravitational Planck length to the Fermi length of weak interactions (about 10^{-16} cm), in order to connect this sector with the SM symmetry breakdown scale. Under such a hypothesis, from the curled-up extra dimension, a whole tower of KK excitations (particles) are expected with masses at the level of hundreds of GeV/c^2 or more. This opens the possibility of their detection, either directly or through unexplained missing 4-momentum, at Fermilab and the Large Hadron Collider (LHC) machine at Geneva. The vacuum energy discrepancy discussed in the previous section remains a problem here as well.

Another outcome of higher dimensional space–time schemes is a change in the Newtonian gravitational potential, i.e. deviations from the well-known $1/r$ potential (cf. Section 9). Concentrated experimental efforts have been made to uncover any such modifications, but so far the evidence is negative.

Once again, as with SS, we have to wait patiently for some help from possible anomalies in the experimental data, which might point to progress in these exciting endeavours.

12. Ab Initio and Dynamical Non-Commuting Coordinates

Hypotheses have been entertained that the space–time coordinates do not commute with each other, that is [8]:

$$\left[x^\mu, x^\nu\right] = \Lambda^2 \, C^{\mu\nu}, \tag{21}$$

where Λ is some (presumably very small) length scale. Models have been invented and limits have been obtained on how small Λ must be for such non-commutativity of coordinates to have escaped detection so far.

It provides one with a fundamental length scale Λ which one may hope might provide a natural cutoff in relativistic quantum field theory. Much more work would be needed for a practical implementation of this interesting idea.

On the other hand there are several physics problems where effective non-commuting coordinates automatically emerge by virtue of the underlying dynamics or other physical contraints. Notable examples are: (i) vortices in superfluid ^{4}He films [32], (ii) quantum Hall effect charged magnetic vortices in two-dimensional electron liquids [4], (iii) quantum interference phase between two alternative paths (as in the Aharanov–Bohm effect) [21], (iv) high-energy charged particle beams stored in cyclotrons [29], and (v) the position coordinates of massless particles such as the photon and the graviton. We shall discuss below the case of charged particles in cyclotrons and its application to quantum beams. Non-commuting photon coordinates are discussed in the next section.

In particle accelerators, charged particles are accelerated by electric fields and kept in a circular orbit by a magnetic field. The coordinate $\vec{r}$ of a particle of charge e, mass m and velocity $\vec{v}$ may be decomposed into $\vec{\rho}$, which is the radius of curvature of the orbit and $\vec{R}$, which points to the orbit centre

$$\vec{r} = \vec{\rho} + \vec{R}; \quad \vec{\rho} = \frac{cm\gamma \, \vec{B} \times \vec{v}}{eB^2}, \tag{22}$$

where $\gamma = 1/\sqrt{1 - (v/c)^2}$ is the Lorentz factor of the particle (cf. Eq. (6)).

The Lorentz force on the charge is given by

$$m\frac{d(\gamma\vec{v})}{dt} = e\left[\vec{E} + \vec{v} \times \vec{B}/c\right], \tag{23}$$

where $\vec{E}$ is the static electric field used to focus the beam.

The orbit centre drifts due to the focusing field

$$\frac{d\vec{R}}{dt} = c\left(\frac{\vec{E} \times \vec{B}}{B^2}\right). \tag{24}$$

While the components of particle coordinate $\vec{r}$ commute with each other, neither the components of $\vec{\rho}$ nor those of $\vec{R}$ commute. In the plane transverse to the magnetic field (presumed along the z-axis), these commutation relations read

$$[\rho_x, \rho_y] = -[R_x, R_y] = iL^2, \tag{25}$$

where the length scale $L = \sqrt{(\hbar c)/(eB)}$, called the *Landau length*, is inversely proportional to the magnetic field. For two non-commuting coordinates, such as above, we have the *quantum Pythagorean theorem*, where their hypotenuse (the cyclotron radius length ρ for example) is in quantized units:

$$\rho = L\sqrt{2n+1}, \quad n = 0, 1, 2, \ldots. \tag{26}$$

Thus, we have a built-in quantum uncertainty ($\Delta\rho$ and ΔR) in the coordinates of particles circulating in a cyclotron beam. Some applications of the quantum aspects of beams have been made and found to be crucial for certain precision measurements, such as the anomalous magnetic moment of the muon [24,25].

In connection with the above result, let us make a general remark. All charged particles must necessarily possess non-commutative coordinates in a plane perpendicular to the ever-present (and varying) magnetic fields near the earth or within the cosmos. This leads to an interesting and unavoidable fuzziness in the coordinates of all charged particles around us (unless they are enclosed in a superconducting environment).

13. Photon Coordinates and Polarized Photon Beams

In the physical examples mentioned in the last section, there exists the ordinary Euclidean geometry of position vectors ($\vec{r}_1, \vec{r}_2, \ldots$) of the particles which make up the system. Euclidean coordinates may also appear (1) in the quantum mechanical wave functions of the system; (2) in the classical rulers employed to construct the length scales of the apparatuses used in measuring the properties of the quantum mechanical system. However, the quantum coordinates (such as vortex positions or the position of cyclotron orbit centers) are described by strictly non-commuting operators. Thus, both classical Euclidean geometry and quantum non-commutative geometry live tranquilly together in the theoretical description of the above quantum systems. The underlying space is Euclidean but the system coordinates are not.

Such tranquillity also exists in the quantum mechanical description of the (non-commuting) position coordinates of zero mass spinning particle beams. Examples are the zero mass, spin-1 photon; the spin-2 graviton and (if massless) spin-one-half neutrinos. We discuss below the case of zero mass photons [31,19,18,30,14,12,20], where the experimental technology (of quantum optics) [17] is most fully developed. An amusing result has been found [22] about two-photon coincidence counts for polarized photon beams. The statistics of the counts should reflect the non-commutativity of the photon positions.

The root of the complication associated with a massless spinning particle is an obvious one. A particle of spin s should have ($2s + 1$) components, whereas for a particle of zero mass, only 2 components of the spin angular momentum $\pm s$ are observable. These are the components of the spin angular momentum parallel (or anti-parallel) to the particle momentum. The standard designation for these components is "helicity", denoted by $\Lambda = \pm s$ (see Eqs (28) and (30)). The other unobservable components of the spin, by virtue of their own noncommuting algebra, play an interesting dynamical role nonetheless. In particular, they produce a basic "Zitterbewegung" (that is, a jitter or uncertainty) in the position of the photon.

As usual, let us define the total angular momentum as the sum of the orbital plus the spin angular momentum

$$\vec{J} = \vec{r} \times \vec{p} + \hbar \vec{\Sigma}, \tag{27}$$

where $\vec{\Sigma}$ is the spin of the photon and we have chosen an arbitrary origin of reference. For a photon of momentum $\vec{p}$, the helicity is defined as

$$\Lambda = \frac{\vec{p} \cdot \vec{\Sigma}}{|\vec{p}|}. \tag{28}$$

The motion of the photon turns out to be that of a helix, with the axis of the helix being the direction of the momentum $\vec{p}$ of the photon. The photon position $\vec{r}$ may be decomposed into a vector $\vec{R}$ which points to the helix axis plus a vector $\vec{\rho}$ normal to the axis.

$$\vec{r} = \vec{R} + \vec{\rho}. \tag{29}$$

$\vec{\rho}$ precesses (moves in a circle) at an angular velocity $\vec{\Omega}$, whose magnitude is given by $cp/\hbar$.

As in the last section, the components of $\vec{R}$ do not commute, neither do the components of $\vec{\rho}$. For a photon moving along the z-axis with helicity $+1$, for example,

$$[\rho_x, \rho_y] = -[R_x, R_y] = i\left(\frac{\lambda}{2\pi}\right)^2, \tag{30}$$

where λ is the wavelength of the photon. For a photon of opposite helicity but still moving in the $+z$-direction, the sign on the right is reversed. (For an unpolarized beam, it is hence zero.)

Quantum mechanically, the radius of the helix is fixed at $\sqrt{\rho_x^2 + \rho_y^2} = \lambda/(2\pi)$, but one cannot tell the separate values of ρ_x and ρ_y (since they do not commute). Likewise for $\vec{R}$. For its coordinates X and Y in the plane transverse to the direction of motion, we have the quantum Pythagorean theorem for its length

$$\sqrt{X^2 + Y^2} = \frac{\lambda}{2\pi}\sqrt{2n + 1}, \quad n = 0, 1, 2, \ldots. \tag{31}$$

According to Wigner [31] and Schwinger [20], $\vec{R}$ is the photon coordinate, and we have the important physical result that the position of a perfectly polarized photon in the plane perpendicular to its momentum is only defined within an area $A = (\lambda/2\pi)^2 = \rho^2$.

While all of the above is true, this effect cannot be observed directly since the origin is arbitrary for a single photon. Not so for a state of two photons, moving parallel to each other with the same helicity. Now the relative coordinate $\vec{R}_{12} = \vec{R}_1 - \vec{R}_2$ is in principle observable and it would also show "fuzziness" of the above type.

Let us consider a monochromatic beam of photons of a given helicity incident normal to a plane of photon detectors. In a two-photon coincidence count experiment of events in the perpendicualr plane, the distance between two photon counts should be quantized into a spectrum

$$D_n = \frac{\lambda}{2\pi}\sqrt{2(2n+1)}, \quad n = 0, 1, 2, \ldots . \tag{32}$$

Experiments have recently been suggested [22,23] to verify these predictions about such an inherent "nonlocality" for (polarized) photons, which follows from simple (special) relativistic and quantum mechanical considerations.

14. CPT Theorem and Lorentz Invariance

Let us now return to our earlier discussion of discrete symmetries which may have an important consequence for Lorentz invariance itself.

As stated earlier, until 1956 parity or right-left symmetry was considered fundamental: for QED and QCD it still is. But, since Lee and Yang, we know that through weak interactions we can tell our right from our left. In the decay

$$\pi^+ \rightarrow l^+ + \nu_L, \tag{33}$$

the emitted neutrino from a π^+ is a perfect left-hand screw. Vice versa, a weakly produced π^- would be a perfect right-hand screw. So, if we can tell the sign of the charge (π^+ versus π^-), we have defined a left-handed versus a right-handed screw in an absolute way (not just a convention).

The above reaction violates both parity P and charge conjugation C. But the system behaves properly under the combined operation of CP. That is, we cannot tell the difference if we reverse the sign of the charge and simultaneously change right to left. The ratio of the partial decay widths (or, relative transition rates) satisfies

$$\frac{\Gamma(\pi^+ \rightarrow l^+ + \nu_L)}{\Gamma(\pi^- \rightarrow l^- + \bar{\nu}_R)} = 1. \tag{34}$$

So to fix the sign of the charge (C violation) and tell right from left (P violation) independently, we need a reaction which also violates CP. Certain decays of the neutral kaons do just that. For example: The ratio of the partial decay widths

$$\frac{\Gamma(K_L \rightarrow \pi^- + e^+ + \nu_e)}{\Gamma(K_L \rightarrow \pi^+ + e^- + \bar{\nu}_e)} = 1.00648 \pm 0.00035. \tag{35}$$

Therefore, by rate counting alone, we can distinguish between an electron and a positron: there exists an *absolute* difference between the opposite signs of the electric charge.

Now we come to time reversal (T). It is easily shown that Eq. (35) also violates T but the product operation $\vartheta = CPT$ (or any permutation) is still a good symmetry even though C, P and T are separately violated. To repeat, CP, CT and PT are separately not conserved. But $\vartheta = CPT$, not just for the above process but for all fundamental processes observed so far. Indeed, ϑ is a quintessential symmetry and we have the *TCP* Theorem (cf. Section 4).

Physically, if we simultaneously interchange

  (i) right to left (P),
  (ii) particle to anti-particle (C), and
  (iii) past to future (T),

the system is left invariant.

If CPT-invariance is so fundamental, we should be able to prove it theoretically with very few assumptions. More importantly, if it is violated, at least one of the assumptions must not hold and that is of paramount interest.

While it is a truism that nothing can be proved rigorously in quantum field theory, fortunately the celebrated CPT theorem provides a counter-example. Roughly speaking, all that is required is for the interactions to be 'local' and Lorentz-invariant [15]. "Local" means roughly that interactions occur at a given space–time point. Not much, you would say, and you would be right; which is why all Feynman diagrams (from a local Hermitian Lagrangian) automatically satisfy it.

If there were violations of CPT in nature, this would have most basic consequences for the way we look at particle interactions. Experimentalists are pursuing it vigorously. The best bet is in high-precision K-decays [16], which would allow us to look for (some extremely) tiny deviations of this fundamental symmetry.

A breakdown of CPT would mean that either locality or Lorentz invariance (or both) would have to be discarded. If Lorentz invariance is the culprit, we do not know what to replace it with. Some authors have suggested [7] that in the EM energy density $(1/2)(\vec{E}^2 + \vec{B}^2)$, where both the electric and magnetic contributions are equal for reasons of Lorentz invariance, one should look for a tiny imbalance between the two terms. Apart from the conceptual problems which that raises, there is the practical problem (as alluded to in the earlier sections) that all around us, near the earth or in the cosmos, there are background magnetic fields which would make observation of any (fundamental) tiny disparity between the electric and magnetic terms very hard to interpret. It is fair to say that from the most precise observations in superconductivity, there appears to be no such disparity.

At present, we have no understanding of what would be behind, below or beyond Lorentz symmetry were it to cause violation of CPT. It would open a completely new chapter in particle physics. (For one thing, it would mean that a particle and its anti-particle can have different masses and, if unstable, different total decay rates.)

15. Conclusions and Future Prospects

Since its inception, special relativity has not only transformed the underlying kinematics of all physical processes but has also directed and motivated our search of fundamental interactions. Its successful coupling with quantum mechanics has led to the SM which describes in great precision basic phenomena in nature (sans gravitation). Very active research is underway all around the globe to unite gravitation as well. But Lorentz invariance itself may be compromised if it is responsible for CPT violation. Since the underlying base space for Einstein's general relativity is that of Minkowski with its attendant Lorentz symmetry, there may indeed be major revisions ahead in the way we think today. Only time will tell.

Acknowledgements

It is a pleasure to thank Giulia Pancheri, Eugene Saletan and Allan Widom for many discussions and help with this chapter. I also thank my student Silvia Chiacchiera for help with references.

References

[1] Arkani-hamed N., Dimopoulos S., Dvali G. 1998: "The Hierarchy Problem and New Dimensions at a Millimeter", *Phys. Lett.* **B429**, pp. 263–72.

[2] Bagger J.A. 1999: *Supersymmetry, Supergravity & Supercolliders*, Singapore, World Scientific.

[3] Barger V.D., R. Phillips J.N. 1987: *Collider Physics*, Redwood City (California), Addison-Wesley.

[4] Belissard L. 1988: *Ordinary quantum Hall effect and noncommutative cohomology*, in *Proc. on Localization in Disordered systems*, ed. Bad Schandau, Leipzig, Teubner.

[5] Bjorken J.D., Drell S.D. 1964: *Relativistic Quantum Mechanics*, New York, McGraw-Hill.

[6] Brian D. 1995: *Einstein. A Life*, New York, Wiley.

[7] Coleman S., Glashow S.L. 1999: "High-Energy tests of Lorentz Invariance", *Phys. Rev.* **D59**, pp. 116008 (14pp).

[8] Connes A. 1994: *Noncommutative Geometry*, San Diego, Academic Press, Harcourt Brace & Co.

[9] Duplij S., Bagger J., Siegel W. 2003: *Concise Encyclopedia of Supersymmetry*, Kluwer Academic, Dordrecht.

[10] Greiner W. 1990: *Relativistic Quantum Mechanics*, Berlin, Springer Verlag.

[11] Halzen F., Martin A.D. 1984: *Quarks & Leptons*, New York, Wiley.

[12] Hawton M. 1998: "Photon Position Operator with Commuting Components", *Phys. Rev.* **A59**, pp. 954–9.

[13] Hollik W., Ruckl R., Wess J. 1993: *Phenomenological Aspects of Supersymmetry*, Berlin, Springer Verlag.

[14] Jauch J.M., Piron C. 1967: "Generalized Localizability", *Helv. Phys. Acta* **40**, pp. 559–70.

[15] Lee T.D. 1983: *Particle Physics and Introduction to Field Theory*, Hardwood Academic Publishers, Chur (Switzerland).

[16] Maiani L., Pancheri G., Paver N. (eds) 1995: *The Second DAΦNE Physics Handbook*, SIS Pubblicazioni dei Laboratori Nazionali di Frascati, Frascati (Italy).

[17] Mandel L., Wolf E. 1995: *Optical Coherence and Quantum Optics*, New York, Cambridge University Press.

[18] Newton T.D., Wigner E.P. 1949: "Localized States for Elementary Systems", *Rev. Mod. Phys.* **21**, pp. 400–6.

[19] Pryce M.H.L. 1948: "The Mass-Centre in the Restricted Theory of Relativity and Its Connection with the Quantum Theory of Elementary Particles", *Proc. Royal Soc.* **A195**, pp. 62–81.

[20] Schwinger J. 1998: *Particles, Sources and Fields*, Percous Books, Reading, Vol. I, sec. (1-3).

[21] Sivasubramanian S., Srivastava Y., Vitiello G., Widom A. 2003: "Quantum Dissipation Induced Noncommutative Geometry", *Phys. Lett.* **A311**, pp. 97–105.

[22] Sivasubramanian S., Castellani G., Fabiano N., Widom A., Swain J., Srivastava Y.N., Vitiello G. 2004: "Noncommutative Geometry and Measurements of Polarized Two-Photon Coincidence Counts", *Annals of Phys.* **311**, pp. 191–203.

[23] Sivasubramanian S., Castellani G., Fabiano N., Widom A., Swain J., Srivastava Y.N., Vitiello G., to appear in *International Journal of Modern Optics*.

[24] Srivastava Y., Widom A.: "Coherent Betatron Oscillations and Induced Errors in the Experimental Determination of the Muon (g - 2) Factor", arXiv:hep-ph/0109020.

[25] Srivastava Y., Widom A. 2004: "Muon g - 2 Measurements and Non-Commutative Geometry of Quantum Beams", *9th International Symposium on Particles, Strings and Cosmology (PASCOS 03)*, Mumbai (Bombay) India, Jan. 2003. Published in *PRAMANA* **62**, pp. 667–670.

[26] Weinberg S. 2000: *The Quantum Theory of Fields*, New York, Cambridge University Press.

[27] Wess J., Bagger J. 1992: *Supersymmetry & Supergravity*, Princeton University Press (New Jersey).

[28] West P.C. 1986: *Introduction to Supersymmetry and Supergravity*, Singapore, World Scientific.

[29] Widom A., Srivastava Y.: "Electric field effects and the experimental value of the muon (g - 2) anomaly", arXiv:hep-ph/0111350.

[30] Wightman A.S. 1962: "On the localizability of quantum mechanical systems", *Rev. Mod. Phys.* **34**, pp. 845–72.

[31] Wigner E.P. 1939: "On the unitary representations of the inhomogeneous Lorentz group", *Ann. Math.* **40**, pp. 149–204.

[32] Yourgrau W., Mandelstam S. 1968: *Variational Principles in Dynamics and Quantum Theory*, New York, Dover (Chap. 13).

Physics Before and After Einstein
M. Mamone Capria (Ed.)
IOS Press, 2005

Chapter 11

Quantum Theory and Gravitation

Allan Widom [a], David Drosdoff [a] and Yogendra N. Srivastava [a,b]

[a] *Physics Department, Northeastern University, Boston, MA, USA*
[b] *Physics Department & INFN, University of Perugia, Perugia, Italy*

1. Introduction

The fundamental element employed to discuss the classical geometry of space–time is the proper time interval $d\tau$ read by a moving clock suffering a displacement dx^μ in space–time. Formally

$$c^2 \, d\tau^2 = -g_{\mu\nu} \, dx^\mu \, dx^\nu. \tag{1}$$

From the metric, one may deduce the curvature of space–time by following the usual rules of Riemannian geometry (cf. Chap. 5). The source of space–time curvature is the tensile strength of ponderable matter. The energy-pressure tensor $T_{\mu\nu}$ strains space–time, bending it into a geometry with a Ricci tensor $R_{\mu\nu} = R^\lambda_{\ \mu\lambda\nu}$. The Einstein gravitational field equations [7] describing this bending process read

$$R_{\mu\nu} - \frac{1}{2} g_{\mu\nu} R = \left(\frac{8\pi G}{c^4} \right) T_{\mu\nu}. \tag{2}$$

From the very inception of quantum mechanical theory, there was a notion that quantum mechanics and gravitational field theory would be closely intertwined. However, after over half a century of extensive research, quantum gravitational theory has not yet reached its final form.

The discussions between Einstein and Bohr concerning the completeness of the quantum viewpoint are well known ([5]; cf. Chap. 9). The formal ending of the conversations was induced by considerations on the role of gravity in the interpretation of the energy–time uncertainty relation, as will be reviewed in Sect. 2. In Sect. 3, the conventional formulation of quantum gravity theory is discussed. The non-linear nature of the resulting gravitational constraint equations have thus far defied exact solutions. The perturbation theory solutions converge only in the lowest order. Higher-order corrections diverge badly. The cosmological interpretation of the constraint equations are also briefly considered.

Quantum gravity is a gauge field theory. In a qualitative manner, one understands a gauge theory in terms of internal degrees of freedom contained within a particle. As a particle with internal structure moves through space–time, the internal degrees of freedom can be described by a moving point on a manifold. The moving point is called a gauge. For example, if the particle has a moving internal arrow which ends on a point lying on a circle, then the gauge reading is in $U(1)$. If an internal arrow ends on a point within the unit 3-sphere (with opposite surface points identified) then the gauge reading takes values in $SU(2)$. A gauge field is a vector bundle sensitive to the direction in which the internal arrow of the moving particle lies. By changing coordinates on the internal symmetry manifold of the particle and changing the basis vectors of the field, one performs a gauge transformation. Gauge transformations do not, however, change the nature of the physical motions. The internal manifold on which the arrows end can be described as a symmetry group. If the group is commutative then it is termed Abelian, otherwise it is non-Abelian. Electrodynamics is Abelian while gravitational, weak and strong interaction gauge fields are non-Abelian.

In Sect. 5, we consider a recent method of quantization in close analogy with non-Abelian gauge field theories such as the electro-weak $SU(2) \times U(1)$ or the strong $SU(3)$ Yang Mills theory of glue. While such theories produce reliable predictions for weak and electromagnetic forces and reasonable predictions for strong forces, they fail for gravity in that Newton's law has not yet been derived as a reasonable approximation in this framework. Gauge theory quantization procedures are thus far merely of mathematical (not physical) interest.

In spite of the difficulty of formulating quantum gravitational theory in a rigorous manner, it is still possible to calculate reliably in low orders of perturbation theory. The situation is similar to Fermi contact theories of weak interactions [10]. In low-order perturbation theory, Fermi theories survive even today as a reliable tool for computing (say) beta decays for small energy. In higher-order perturbation theory, the Fermi contact theories break down, and the more recent and fundamental electro-weak gauge theory must then be invoked. Similarly, we can reliably calculate in lowest order perturbation theory quantum gravitational processes such as graviton emission and absorption.

Some perturbation theory results are discussed in Sect. 6. While we do not yet have a more fundamental viewpoint allowing us to go to higher orders, it is clear that the lower-order results will retain validity. The application of low-order quantum mechanical theory to mechanical gravitational wave antennae is discussed in Sect. 7. The physical principles of the discussions of quantum gravity which follow are summarized in Sect. 8.

2. The Bohr–Einstein Controversy and Gravity

It is well known that Einstein and Bohr agreed on what the quantum mechanical formalism entailed, but disagreed on whether or not the theory represented a complete description of physical reality. The Bohr–Einstein discussions on this matter involved the application of the uncertainty relations to so-called "thought experiments" in order to examine the consistency of the quantum mechanical viewpoint. Recall the quantum uncertainty relation for two physical non-commuting quantities, A and B. The uncertainty principle asserts that

$$\frac{i}{\hbar}[A, B] = C \quad \text{implies} \quad \Delta A \Delta B \geq \frac{\hbar}{2}|\langle C \rangle|, \tag{3}$$

wherein $[A, B] = AB - BA$ and $\langle C \rangle$ denotes the mean value of the operator C. The usual application of Eq. (3) involves the position x and the conjugate momentum p_x; i.e.

$$\frac{i}{\hbar}[p_x, x] = 1 \quad \text{implies} \quad \Delta p_x \Delta x \geq \frac{\hbar}{2}. \tag{4}$$

Similarly, let T be a coordinate associated with a clock. In detail, one estimates a laboratory time t by examining a coordinate *clock time* operator T [12].

For example, the position T of the "hand" on an electric clock can be employed to read the time t. The time rate of change of any clock coordinate T is determined by a clock Hamiltonian H via the equation of motion

$$\frac{dT}{dt} = \frac{i}{\hbar}[H, T]. \tag{5}$$

The measured clock time may be calibrated according to

$$\left\langle \frac{dT}{dt} \right\rangle = \eta \quad \text{so that} \quad t_T \equiv (T/\eta). \tag{6}$$

If the clock coordinate T is synchronized according to Eqs (5) and (6), then Eq. (3) implies

$$\Delta E \Delta t_T \geq \frac{\hbar}{2} \tag{7}$$

representing the energy–time uncertainty relation.

In the final discussions between Bohr and Einstein, general relativity entered into the considerations in an essential manner. If Einstein had *not* previously worked out some consequences of general relativity (such as the equivalence principle), then Bohr would have lacked an argument to use in order to save the foundations of quantum mechanics.

Bohr and Einstein considered a clock which reads proper time τ via the coordinate operator T obeying

$$\frac{dT}{d\tau} = \frac{i}{\hbar}[Mc^2, T],$$

$$\left\langle \frac{dT}{d\tau} \right\rangle = 1, \tag{8}$$

$$\Delta M \Delta \tau_T \geq \frac{\hbar}{2c^2} \quad \text{where} \quad \tau_T \equiv T.$$

The clock is schematically drawn in Fig. 1.

The puzzling feature about clocks was the nature of the uncertainty relation $\Delta M \Delta \tau > (\hbar/2c^2)$. Einstein considered the notion that the mass of the clock could be lowered if some separate device within the clock emitted (say) a photon. Einstein did not understand why such a separate device should have an effect on how the clock actually reads the proper time. Bohr countered that the clock's *inertial* mass could be determined by the *gravitational weight* Mg when the clock sits on a scale at height h. The

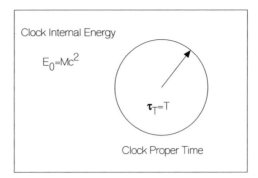

Figure 1. Shown is a macroscopic clock of mass M. The coordinate operator T (pictured schematically as a "pointer") is employed to measure the proper time τ.

Hamiltonian H of the clock at height h in a gravitational field may be written

$$H \approx Mc^2 + Mgh = Mc^2\left(1 + \frac{gh}{c^2}\right),$$

thus

$$\frac{dT}{dt} = \frac{i}{\hbar}[H, T] \approx \frac{i}{\hbar}[Mc^2, T]\left(1 + \frac{gh}{c^2}\right),$$

and therefore

$$\frac{dT}{dt} \approx \frac{dT}{d\tau}\left(1 + \frac{gh}{c^2}\right). \tag{9}$$

The *energy–time uncertainty relation* for an *arbitrary* clock coordinate T may be expressed equivalently in terms of the laboratory time (Eq. (7)) or in terms of proper time (Eq. (8)). The proper calibration factor for laboratory time in Eq. (6) is $\eta \approx 1 + (gh/c^2)$ which has been experimentally proved via the gravitational frequency (time measurement) shift [14].

It is ironic that Einstein's confusion concerning Eq. (8) was due to having left out the gravitational frequency shift, which he himself had theoretically invented. Bohr thanked Einstein for his wonderful theoretical effect, which in fact saved the quantum mechanical theory. That particular debate between Einstein and Bohr was then closed, with Einstein subdued. If the uncertainty relation can be overcome in a single experiment, then quantum mechanical theory is completely invalid. A single system which overcomes quantum uncertainties can be used as an apparatus to overcome quantum uncertainties in all quantum objects.

The lesson from Einstein's discussion, which only a few seem to realize, is that gravity played a central role in the foundations of quantum mechanics despite the small value of the coupling strength G. The gravitational field $g = (G M_{\text{earth}}/R^2)$ is not very small. Nevertheless, the final nature of quantum gravity is far from being completely understood.

3. Conventional Quantum Gravity

The most obvious form of quantum mechanical gravity employs simple canonical quantization [1] by fixing a gauge wherein space and time are appropriately separated. In the classical theory of gravity, it is well known that in a finite space–time region it is possible to express the metric in the synchronous $x = (r^1, r^2, r^3, ct) = (\mathbf{r}, ct)$ coordinate system [9]. It is

$$c^2 \, d\tau^2 = c^2 \, dt^2 - ds^2$$

$$ds^2 = \gamma_{ab}(\mathbf{r}, t) dr^a \, dr^b. \tag{10}$$

The spatial metric field γ_{ab} then refers to three-dimensional space at a fixed time. One may then compute from the curvature a Lagrangian depending on the spatial metric and the first time derivatives so that the conjugate momenta $\pi_{ab}(\mathbf{r})$ may be defined. The equal time commutation relations take the form

$$\frac{i}{\hbar} \left[\pi_{ab}(\mathbf{r}), \gamma_{cd}(\mathbf{r}') \right] = \delta_{abcd}(\mathbf{r}, \mathbf{r}'), \tag{11}$$

with the Dirac δ-function taken with respect to the volume element $\sqrt{\det |\gamma_{ab}|} d^3 \mathbf{r}$. While the conjugate momenta $\pi_{ab}(\mathbf{r})$ are linear combinations of the time derivatives of the gravitational fields $\dot{\gamma}_{cd}(\mathbf{r})$, the coefficients of the linear expansion depend non-linearly on the metric fields themselves. The operator constructions then have operator ordering problems. Also, there are six independent γ_{ab} metric fields and ten independent $g_{\mu\nu}$ metric fields. Out of the ten independent Einstein Eqs (2), four of them appear as "Gauss law" constraints on the wave function. With $N^\mu = (0, 0, 0, 1)$ in the synchronous gauge, the constraint equations are

$$\left\{ R_{\mu\nu} - \frac{1}{2} g_{\mu\nu} R \right\} N^\nu |\Psi\rangle = \left(\frac{8\pi G}{c^4} \right) T_{\mu\nu} N^\nu |\Psi\rangle. \tag{12}$$

The constraints are not simple but they are first class. (This means that they are self-contained and do not require new constraints for mathematical consistency.) The remaining six of Einstein's equations are dynamically derived from the Hamiltonian $H = \int \mathcal{H} \sqrt{\gamma} d^3 \mathbf{r}$. The constraint Eq. (12) implies a Schrödinger equation of the form $H|\Psi\rangle = 0$. The derivative with respect to time is removed from the Schrödinger equation so long as the synchronous gauge is employed. In fact, the energy *density* at each spatial point vanishes in the synchronous gauge

$$\mathcal{H}|\Psi\rangle = 0 \quad \text{for any } \mathbf{r}. \tag{13}$$

Thus, if one solves the constraint equations, then the quantum mechanical wave function can be regarded as known and all quantities of interest can in principle be computed. Newton's gravitational law is contained in the gravitational constraint equations at least in lowest-order perturbation theory. However, the constraints are non-linear (in fact non-polynomial) in the metric. The theory is thus subject to divergences in higher-order perturbation theory and little is known about exact solutions.

From the conventional quantum mechanical point of view, many workers have considered the constraint Eq. (13) to represent the Schrödinger equation for the "big" wave function Ψ of the *entire universe*. However, such a grand cosmological interpretation of the physical wave function is in conflict with the conventional Copenhagen view of quantum mechanical theory. (We once offered a $100 award to anyone who could find the wave function collapse in any of Bohr's writings. This concept was introduced later by von Neumann and hence we do not consider it to be part of the Copenhagen interpretation.) The conventional Copenhagen view appears to be the *only* view of quantum mechanics which is completely experimentally well founded. Not only is it consistent with general relativity, but in Bohr's opinion general relativity was an actual requirement for this consistency.

4. Quantum Mechanics and Cosmology

The conventional quantum rules imply some theoretical pitfalls in the quantum cosmology viewpoint which are briefly included in the following considerations:

(i) A measurement is an interaction between a quantum object and an apparatus which obeys classical laws to a sufficient degree of accuracy. All experimental data are classical. For example, suppose one's data are stored on a computer disk. If all the binary bits on the disk are in a superposition of reading 0 and 1, then the disk is *totally unreadable* and no data exists in reality. All actually measured data are classical. In the disk example, the bits must read as a given specific *fixed* string of symbols, e.g. $\ldots 011011100\ldots$, in order to represent real data. A quantum object is not *directly* observed. If one directly observes what a quantum object is doing, then it no longer behaves as a quantum object. In the two-slit electron beam interference experiment, if one does not directly observe through which slit an electron passes, then the electron obeys quantum mechanical laws. If one does observe through which slit the electron passes, then the electron obeys classical mechanical laws. In a general measurement process, a quantum mechanical object interacts with a classical measuring apparatus. Only the classical apparatus produces actual data whose statistical properties indirectly reflect the state of the quantum object. If Ψ is assumed to represent the wave function of the entire universe viewed as a quantum object, then there is little room left for the classical apparatus needed for analysing actual experiments. There are other theoretical versions of quantum mechanics which require many universes, but by design are rendered meaningless since there is absolutely no experimental method to verify the existence of these other worlds.

(ii) Much more generally, quantum mechanics makes probabilistic predictions which can only be verified by repeating an experimental situation often enough to obtain reliable statistics. The wave function, in and by itself, is not a property of a quantum mechanical system. The wave function is merely a representation of statistical predictions about an ensemble of experiments on many such quantum mechanical systems or an ensemble of repeated experiments on one quantum mechanical system. In any event, amplitudes are computed only to predict the statistical behaviour of an ensemble. One cannot possibly experimentally verify a statistical prediction about an ensemble of universal experiments presuming only one universe in the ensemble.

(iii) As regards *quantum* cosmology, it is important to distinguish between a set of complex amplitudes and an actual physical object. For example, dear reader, suppose that an expert quantum mechanic works for an automobile insurance company. The quantum mechanic has sets of complex numbers in his desk (wave functions) by which he can compute the differential cross-section for your automobile to collide with another automobile on the highway. Based on such probabilities (which the quantum mechanic makes sure will include diffraction effects) the insurance company sets your yearly insurance rates. Now (heaven forbid) suppose that your automobile does *indeed* collide with another automobile. Some workers in quantum mechanics would presume (*after* the accident is reported) that the wave function in the quantum mechanic's desk would collapse. We think it more likely that your insurance rates would *explode* because your automobile and the other *scattered* automobile *both* will have *collapsed*. The wave functions in the quantum mechanic's desk would (of course) remain *totally intact*. Why would the wave function collapse during an inelastic automobile *scattering* accident? A clear distinction should be made between the *wave function* Ψ and the *actual physical object* (a member of a statistical ensemble) under consideration. If Ψ represents the wave function of the universe and if the wave function should also be *distinguished* from the universe, then there is little worthwhile content left for quantum cosmology.

(iv) The metric in Eq. (1) becomes an operator in the conventional [1] quantized version of gravity. As a result of this operator representation, distances and time intervals are not always mutually commuting. There have been attempts at starting the quantum theory of space–time with non-commuting coordinates and time, i.e. $x = (\mathbf{r}, ct)$, and employing such non-commuting space–time coordinates directly into field functions [11]. This non-commutative geometric approach to gravity is interesting but is thus far experimentally irrelevant since the parameters of theory are designed to make the effects too small.

5. The Gauge Quantization Viewpoint

A comparison between quantum electrodynamics and quantum gravity is now in order. In quantum electrodynamics one may employ the temporal gauge

$$A^\mu = (\mathbf{A}, \Phi) \equiv (\mathbf{A}, 0),$$

$$\mathbf{E} = -\frac{1}{c}\left(\frac{\partial \mathbf{A}}{\partial t}\right) \quad \text{and} \quad \mathbf{B} = curl\mathbf{A}. \tag{14}$$

In the temporal gauge (in Gaussian units) the canonical conjugate commutation relations are

$$\left[E_i(\mathbf{r}), A_j(\mathbf{r}')\right] = 4\pi i \hbar c \delta_{ij}\delta(\mathbf{r} - \mathbf{r}'). \tag{15}$$

There are only three vector potential components, rather than four, in A^μ. This requires a Gauss law constraint on the wave function which may be written in Gaussian units as

$$div\mathbf{E}|\Psi\rangle = 4\pi\rho|\Psi\rangle. \tag{16}$$

Coulomb's law is an exact consequence of the linear Gauss law constraint in Eq. (16). Gauss law constraint in quantum electrodynamics (being linear) can be exactly implemented by choosing wave functions which depend only on the transverse parts of the vector potential $\langle\mathbf{A}|\Psi\rangle = \Psi[\mathbf{A}_{\text{trans}}]$.

On the other hand, Newton's law in quantized Einstein theory is merely a *lowest-order perturbation theory* consequence of the non-linear synchronous gauge constraint Eqs (12) or (13). No comparable useful exact statement can be made concerning Newton's gravitational law and the allowed wave functions $\langle \gamma | \mathcal{H} | \Psi \rangle = \hat{\mathcal{H}} \Psi [\gamma] = 0$.

Newtonian gravity follows as a perturbation theory result of approximately solving the Schrödinger equation, and eventually perturbation theory diverges. Most of the other theorist's weapons concerning Gauss law constraints were sharpened by computations in quantum chromodynamics (QCD). In this case there are eight colour electric and eight colour magnetic fields. In the temporal gauge we have

$$\mathbf{E}_n = -\frac{1}{c}\left(\frac{\partial \mathbf{A}_n}{\partial t}\right), \quad n = 1, 2, \ldots, 8.$$

$$\left[E_n(\mathbf{r}), A_m(\mathbf{r}')\right] = i\hbar c \delta_{nm} \delta(\mathbf{r} - \mathbf{r}').$$

(17)

In Eq. (17), Lorentzian units have been employed. The QCD Gauss law constraint's read

$$Div_{\mathbf{A}}\mathbf{E}_n|\Psi\rangle \equiv \left(div\mathbf{E}_n + \sum_k f_{nmk}\mathbf{E}_m \cdot \mathbf{A}_k\right)|\Psi\rangle = \rho_n|\Psi\rangle. \tag{18}$$

Does the non-linear colour Gauss law constraint Eq. (18) produce a colour Coulomb law? The QCD Gauss law is non-linear and cannot be solved exactly. Perturbation theory does produce a colour Coulomb law which is usually presumed to be valid at short distances. For large distances, perturbation theory is no longer considered valid since one finds experimental quark confinement. This is often described by an approximately linear confinement potential. In general, QCD confinement remains theoretically not proved.

From a physics point of view, the QCD Gauss law constraints demand that the total colour at every point in space vanish. For example, the colour of glue (if it exists) would have to be exactly cancelled by the colour of the quarks. From a physics point of view, the quantum gravitational Gauss law Eq. (12) requires that the total energy-momentum density at every point in space vanishes. For example, the energy-momentum density of ponderable matter has to be *exactly* cancelled by the energy-momentum of the gravitational field, and this must be true at every point in space for a given time. Nobody suggests that ponderable matter is confined. Only quarks seem to be confined. Not much is understood about solving non-linear Gauss law constraint problems, so the problem remains mathematically and physically open.

With regard to the general relativity problem, Einstein was well aware that assigning an energy-momentum density to the gravitational field was not a simple proposition. To assign a stress tensor to gravitational waves, Einstein introduced a *pseudo-tensor* admitting partial defeat in enforcing the usual true tensor character $T^{\mu\nu}$ to pressure energy and momentum densities. In the merely pseudo-tensor manner it is possible to derive the Newtonian force law from single graviton exchange without worrying unduly about whether or not quantum gravity confines matter. In QCD, both the Yang–Mills field and the quarks are described by colour vectors. To have gluons (which are supposed to be confined) carry colour one follows Einstein's path and assigns pseudo-colour-vectors to glue.

There is a "toy model" of gravity theory wherein the Gauss law constraint can be exactly solved [2,3]. Unfortunately the model is not sufficiently physical to compare with an experiment [8]. One treats the connection

$$\Gamma^{\mu}_{\nu\sigma} = \frac{1}{2}g^{\mu\lambda}\{\partial_{\nu}g_{\lambda\sigma} + \partial_{\sigma}g_{\lambda\nu} - \partial_{\lambda}g_{\nu\sigma}\} \tag{19}$$

as a gauge field in a manner closely analogous to a gauge field $\mathbf{A}$. Let us again employ the synchronous gauge with a spatial length scale at time t,

$$ds^2 = \gamma_{ab}(\mathbf{r}, t)\, dr^a\, dr^b. \tag{20}$$

It is then possible to write the spatial metric in terms of three-dimensional vectors $\mathbf{E}_a$ as

$$\gamma_{ab}(\mathbf{r}, t) = \delta_{AB}E_a^A(\mathbf{r}, t)E_b^B(\mathbf{r}, t). \tag{21}$$

For fixed time, one may now consider the curved space in which there are vector conjugate pairs of fields $\{\mathbf{A}_a, \mathbf{E}_a\}$ wherein $\mathbf{A}_a$ is constructed as a linear combination of the spatial connection coefficients γ^a_{bc} and extrinsic spatial curvature tensors.

Under the above circumstances, the theory of gravity takes on the appearance of a typical gauge theory but with exactly soluble constraint equations. On the other hand, the exactly soluble Gauss law constraints do *not* yield the usual Newtonian gravitational law of force. We thereby judge that the theory is mathematically interesting but physically irrelevant. One must have a limit in which Newtonian gravity with all of its astronomical success is recovered. Unfortunately, such a limit does not appear in this approach and the gauge quantization of gravity has not yet proved physical.

6. Low-Order Perturbation Theory

The graviton is the quantum which propagates gravitational disturbances in much the same way in which the photon propagates electromagnetic disturbances. Gravitons appear in the conventional [1] quantized gravitational theory. Here we outline the quantum gravity computation for the emission and absorption of gravitons from an object having a length scale L. We further presume a low frequency limit ($\omega L/c) \ll 1$.

As a first picture of a gravitational wave, consider the schematic drawing in Fig. 2. Space–time with or without a gravitational wave may be chosen in the synchronous gauge to have a proper time

$$c^2\, d\tau^2 = c^2\, dt^2 - d\sigma^2,$$
$$d\sigma^2 = |d\mathbf{r}|^2 + 2d\mathbf{r} \cdot \mathsf{u}(\mathbf{r}, t) \cdot d\mathbf{r}. \tag{22}$$

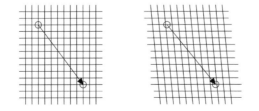

Figure 2. Shown is a schematic drawing of small sections, respectively, of "flat" and "curved" space. In flat space, the distance between the exhibited neighbouring points is $d\sigma_0^2 = |d\mathbf{r}|^2 = dx^2 + dy^2 + dz^2$. In curved space the distance is modified by a gravitational wave to read $d\sigma^2 = |d\mathbf{r}|^2 + 2\,d\mathbf{r} \cdot \mathsf{u} \cdot d\mathbf{r}$. The strain field of gravitational wave distorted space is $\mathsf{u}(\mathbf{r}, t)$. Proper time in the presence of a gravitational wave is $c^2\, d\tau^2 = c^2\, dt^2 - d\sigma^2$.

The gravitational wave itself may be treated as a strain $u(r, t)$ which distorts space according to the transverse traceless gauge conditions. The gravitational wave is a transverse traceless stain on physical space which does not change volume elements. The gravitational wave is "small", that is, the norm $|u| \equiv \sup_{(ij)} |u_{ij}| \ll 1$. The graviton is a spin 2 massless particle quantum associated with a gravitational wave.

Classical sources of gravitational waves in the mass quadrupole approximation emit gravitational energy at a mean rate [13]

$$\dot{\mathcal{E}} = \left(\frac{G}{45c^5} \right) \overline{\frac{d^3 D}{dt^3} : \frac{d^3 D}{dt^3}}, \tag{23}$$

wherein D is the mass quadrupole moment tensor of the antenna. For a single frequency, the energy emitted is given by

$$\dot{\mathcal{E}} = \left(\frac{2G\omega^2}{45c^5} \right) \ddot{D}_\omega : \ddot{D}_\omega^*. \tag{24}$$

Consider the decay of a quantum state with the emission of a graviton of frequency ω_g. The lowest-order transition rate Γ, where $\dot{\mathcal{E}} = \hbar\omega\Gamma$, for a spontaneous decay in lowest-order perturbation theory is given by

$$\Gamma\big[|i\rangle \rightarrow |f\rangle + g\big] = \left(\frac{2G\omega_g}{45\hbar c^5} \right) \langle i| \ddot{D} |f\rangle : \langle f| \ddot{D} |i\rangle. \tag{25}$$

With only the quantum mechanical results discussed above, it is theoretically possible to compute the detection efficiency of gravitational wave antennae.

7. Gravitational Wave Antennae

From a conceptual viewpoint, the most simple form of gravitational wave detector is a large metallic bar held up by its two ends by thin but strong wires. If the x-axis lies on the axis of the cylinder of length L, then the lowest frequency acoustic longitudinal displacement $\xi(x, t)$ in the bar obeys the acoustic wave equation. The lowest acoustic frequency of the bar is $\Omega = (\pi u / L)$; see Fig. 3. The manner in which the gravitational wave detector works may be described as follows: a gravitational wave (graviton beam) of frequency ω will be absorbed by exciting the fundamental mode of the bar with a strong peak at the acoustic resonance frequency $\omega \approx \Omega$. The absorption cross-section for an incident graviton is the central quantity describing the mechanical efficiency of the bar for detecting gravitational waves.

The absorption of gravitons by mechanical antennae may be discussed in terms of the Breit–Wigner [6] theory of resonant scattering. To understand how this works we begin by considering the inverse process by which a phonon (quantum particle of the acoustic mode) decays into a graviton (quantum particle of the gravitational wave). We may first suppose that the resonant acoustic phonon mode has a width $\gamma = (\Omega/Q)$ where Q is the mode quality factor. If one denotes a single phonon by ϕ, then the width (inverse lifetime of the phonon) is determined by the total probability per unit time for the phonon to decay into any and all final products. The total phonon decay rate is $\gamma = \sum_X \Gamma(\phi \rightarrow X)$ wherein X represents the final products of the decay. Only in a small

Metal cylinder

Gravitational wave detector

Vibrational mode frequency Ω

Figure 3. A metallic cylindrical Weber bar. The lowest resonant acoustic mode of frequency Ω is described by longitudinal displacement $\xi(x,t)$ obeying $\{(\partial/\partial t)^2 - u^2(\partial/\partial x)^2\}\xi = 0$ where u is the longitudinal sound velocity.

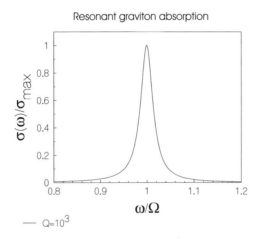

Resonant graviton absorption

Figure 4. The absorption cross-section is plotted for a graviton of frequency ω to produce a bound state antenna phonon of frequency Ω. The phonon has a width $\gamma = (\Omega/Q)$ and $Q = 10^3$ has been chosen for the purpose of illustration.

fraction $\eta = \Gamma(\phi \to g)/\gamma$ of phonon decay events will the final product be equal to a single graviton. The total fixed graviton cross-section near resonance, $\omega \sim \Omega$, may be written as

$$\sigma(\omega) = \frac{\Omega^2 \gamma^2 \sigma_{\text{max}}}{(\omega^2 - \Omega^2)^2 + \omega^2 \gamma^2},$$

$$\sigma_{\text{max}} \approx \left(\frac{4\pi c^2}{\Omega^2}\right) \eta. \tag{26}$$

$$\eta = \frac{2}{45}\left(\frac{G\Omega}{\hbar \gamma c^5}\right) \langle 0|\ddot{\mathbf{D}}|\Omega\rangle : \langle \Omega|\ddot{\mathbf{D}}|0\rangle,$$

where $|0\rangle$ is the ground state of the antenna and $|\Omega\rangle$ is the state with one phonon excitation in the fundamental acoustic mode. The variation of resonant cross-section with graviton frequency is exhibited in Fig. 4.

The calculation of the absorption cross-section is thus reduced to the calculation of the branching ratio η. While η is known to involve quantum mechanical matrix elements [18], the usual theories [13,18] invoke a purely classical macroscopic approximation. The difficulty with a classical computation is that both the graviton and the phonon are strongly coupled to the electronic degrees of freedom in the metal. Furthermore, electronic degrees of freedom are inherently quantum-mechanical in nature. This need for quantum-mechanical computation is evident from the expression for the low temperature sound wave damping rate [4]

$$\gamma \approx \frac{4}{3}\left(v_{el}k^2\right) \quad \text{where } k = \frac{\pi}{L} \tag{27}$$

and v_{el} is the kinematic viscosity of the *electronic* fluid. Kinematic viscosity has "diffusion units" of cm^2/sec which describe how vortices in electronic fluid flows decay in time. Such diffusion must be distinguished from the electron density diffusion coefficient D_{el} which determines electronic conductivity. Electrons diffuse much more rapidly than the vortices of the electron fluid. Experimentally, for (say) aluminium at low temperatures

$$\left[\frac{D_{el}}{(4v_{el}/3)}\right]_{\text{experiment}} \approx 10^4 \quad \text{(aluminium).} \tag{28}$$

In computing the quantum branching ratio we find that [17,19]

$$\eta_{\text{quantum}} \approx \left[\frac{D_{el}}{(4v_{el}/3)}\right]\eta_{\text{classical}} \tag{29}$$

so that there is a considerable theoretical amplification in predicted gravitational wave detection efficiency of mechanical antennae.

For recent analyses of gravitational wave detection via optical means, the reader may consult references [15,16].

8. Conclusion

Quantum mechanics must be employed in treating the interactions between condensed matter and gravity. This is an experimental necessity since materials derive their properties from quantum mechanical motions. As a matter of principle, the foundations of quantum mechanics cannot be formulated properly without including general relativity at least in the form of the equivalence principle. This fact was under discussion between Bohr and Einstein from the very beginning of quantum mechanical theory. Problems at the exceeding small length scale of Planck

$$\Lambda = \sqrt{\frac{\hbar G}{c^3}} \tag{30}$$

cause a fundamental divergence in all physically reasonable formulations of quantum gravity. Nevertheless, low-order perturbation theory results may be considered to be reliable and are very useful for analysing experiments. The large amplification factor implied by quantum mechanics for the detection of gravitational waves in mechanical antennae has been surprising to some workers.

References

[1] Arnowitt R., Deser S., Misner C. 1962: "The Dynamics of General Relativity", in L. Witten (ed.), *Gravitation: An Introduction to Current Research*, Wiley, New York.

[2] Ashtekar A. 1986: "New variables for classical and quantum gravity", *Phys. Rev. Lett.* **57**, pp. 2244–7.

[3] Ashtekar A. 1987: "New Hamiltonian formulation of general relativity", *Phys. Rev.* **D36**, pp. 1587–1602.

[4] Bhatia A.B. 1967: *Ultrasonic Aborption*, Oxford University Press (Chapter 12).

[5] Bohr N. 1970: "Discussions with Einstein on Epistemological Problems in Atomic Physics", in P.A. Schilpp (ed.), *Albert Einstein Philosopher-Scientist*, La Salle (Ill.), Open Court.

[6] Breit G., Wigner E.P. 1936: "Capture of Slow Neutrons", *Phys. Rev.* **49**, pp. 519–31.

[7] Einstein A. 1950: *The Meaning of Relativity*, Princeton University Press.

[8] Gambini R., Pullin J. 1996: *Loops, Knots, Gauge Theories and Quantum Gravity*, Cambridge University Press.

[9] Landau L.D., Lifshitz E.M. 1999: *The Classical Theory of Fields*, Oxford, Butterworth Heinemann, pp. 286–91.

[10] Lee T.D. 1983: *Particle Physics and Introduction to Field Theory*, Hardwood Academic Publishers, Chur (Switzerland).

[11] Lesniewski A. 1997: "Noncommutative Geometry", *Notices of the AMS* **44**, pp. 800–5.

[12] Messiah A. 1999: *Quantum Mechanics*, Mineola, Dover (Chapter 8).

[13] Ohanian H., Ruffini R. 1994: *Gravitation and Spacetime*, New York, Norton (Chapter 5).

[14] Pound R.V., Rebka G.A. 1960: "Apparent Weight of Photons", *Phys. Rev. Lett.* **4**, pp. 337–41.

[15] Sivasubramanian S., Srivastava Y.N., Widom A.: "Gravitational Wave Detection with Michelson Interferometers", arXiv:gr-qc/0307085.

[16] Sivasubramanian S., Srivastava Y.N., Widom A.: "Gravitational Waves and the Sagnac Effect", arXiv:gr-qc/0311070 (presented at the II Space Part Conference, Washington, DC, December 2003; to be published in *Nuclear Physics B Proceedings Supplement*).

[17] Srivastava Y.N., Widom A. Pizzella G.: "Electronic Enhancement in the Detection of Gravitational Waves by Metallic Antennas", arXiv:gr-qc/0302024.

[18] Weinberg S. 1972: *Gravitation and Cosmology*, New York, Wiley (Chapter 10).

[19] Widom A., Drosdoff D., Sivasubramanian S., Srivastava Y.N.: "Electronic Detection of Gravitational Disturbances and Collective Coulomb Interactions", arXiv:gr-qc/0402097.

Physics Before and After Einstein
M. Mamone Capria (Ed.)
IOS Press, 2005
© 2005 The authors

Chapter 12

Superluminal Waves and Objects: Theory and Experiments. A Panoramic Introduction[1]

Erasmo Recami

*Facoltà di Ingegneria, Università statale di Bergamo, Dalmine (BG), Italy;
and INFN – Sezione di Milano, Milan, Italy*

1. Introduction

The question of superluminal ($V^2 > c^2$) objects or waves, has a long history, starting perhaps in 50 BC with Lucretius' *De Rerum Natura*.[2] Still in pre-relativistic times, one meets various related works, from those by J.J. Thomson to papers by A. Sommerfeld.

With special relativity, however, since 1905 the conviction spread that the speed c of light in vacuum was the *upper* limit of any possible speed.

For instance, in 1917 R.C. Tolman believed that he had shown by his "paradox" that the existence of particles endowed with speeds larger than c would have allowed information to be sent into the past. Such a conviction blocked for more than half a century – aside from an isolated paper (1922) by the Italian mathematician G. Somigliana – any research about superluminal speeds. Our problem started to be tackled again in the 1950s and 1960s, in particular after the papers by E.C. George Sudarshan *et al.* [7], and, later on, [38] by E. Recami, R. Mignani *et al.* (who, in their works at the beginning of the 1970s, brought the expressions subluminal and superluminal into popular use), as well as by H.C. Corben and others (to confine ourselves to the *theoretical* researches). The first experiments looking for faster-than-light objects were performed by T. Alväger *et al.* [38].

Superluminal objects were called *tachyons*, T, by G. Feinberg, from the Greek word $\tau\alpha\chi\acute{\upsilon}\varsigma$, quick (and this induced the present author in 1970 to coin the term "bradyon", for ordinary subluminal ($v^2 < c^2$) objects, from the Greek word $\beta\rho\alpha\delta\acute{\upsilon}\varsigma$, slow). Finally, objects travelling exactly at the speed of light are called "luxons".

In recent years, terms as "tachyon" and "superluminal" fell unhappily into the (cunning, rather than crazy) hands of pranotherapists and mere cheats, who started squeezing money out of simple-minded people; for instance by selling plasters (!) that should cure

---

[1]This work was partially supported by INFN and Murst/Miur (Italy). E-mail for contacts: recami@mi.infn.it.
[2]Cf. e.g. book 4, line 201: "Quone vides *citius* debere et longius ire/Multiplexque loci spatium transcurrere eodem/Tempore *quo Solis* pervolgant *lumina* coelum?"

various illnesses by "emitting tachyons"... We are dealing with tachyons here, however, since at least four different experimental sectors of physics seem to indicate the actual existence of superluminal motions (thus confirming some long-standing theoretical predictions [31]).

In the first part of this chapter (after a brief, non-technical theoretical introduction, which can be useful since it informs about an original, still scarcely known approach) we mention the various experimental sectors of physics in which superluminal motions seem to appear. In particular, a bird's-eye view is presented of experiments with evanescent waves (and/or tunnelling photons), and with "localized superluminal solutions" (SLS) to the wave equation, like the so-called X-shaped waves; the shortness of this review is compensated for by a number of references, sufficient in some cases to provide interested readers with reasonable bibliographical information.

2. General Concepts

Let us premise that special relativity (SR), abundantly confirmed by experience, can be built on two simple postulates:

(1) that the laws (of electromagnetism and mechanics) be valid not only for a particular observer, but for the whole class of the "inertial" observers;
(2) that space and time be homogeneous and space be moreover isotropic.

From these postulates one can theoretically *infer* that one, and only one, *invariant* speed exists: and experience tells us such a speed to be the one, c, of light in vacuum (namely, 299,792,458 km/s). Indeed, ordinary light possesses the peculiar feature of always presenting the same speed in vacuum, even when we run towards or away from it. It is just that feature, of being invariant, that makes the speed c quite exceptional: no bradyons, and no tachyons, can enjoy the same property.

Another (known) consequence of our postulates is that the total energy of an ordinary particle increases when its speed v increases, tending to infinity when v tends to c. Therefore, infinite forces would be needed for a bradyon to reach the speed c. This fact generated the popular opinion that speed c can be neither achieved nor overcome.

However, as speed c photons exist which are born, live and die at the speed of light (without any need for acceleration from rest to the light speed), so objects can exist [39] always endowed with speeds V larger than c (see Fig. 1). This circumstance has been picturesquely illustrated by George Sudarshan (1972) with reference to an imaginary demographer studying the population patterns of the Indian subcontinent:

> Suppose a demographer calmly asserts that there are no people North of the Himalayas, since none could climb over the mountain ranges! That would be an absurd conclusion. People of central Asia are born there and live there: they did not have to be born in India and cross the mountain range. So with faster-than-light particles.

Let us add that, still starting from the above two postulates (as well as a third postulate, even more obvious),[3] the theory of relativity can be generalized [31,39] in such a way as to accommodate superluminal objects, a large part of such an extension being contained

---

[3] Namely, the assumption that particles do not exist – regularly travelling forward in time – endowed with negative energies.

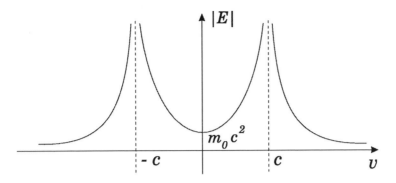

Figure 1. Energy of a free object as a function of its speed [38,31,39].

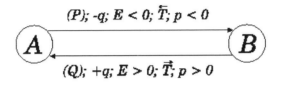

Figure 2. Depicting the "switching rule" (or reinterpretation principle) by Stueckelberg–Feynman–Sudarshan–Recami [31,39,35]: Q will appear as the antiparticle of P. See text.

in a series of works which date back to the 1960s–70s. Also within "extended relativity" [31] the speed c, besides being invariant, is a limiting velocity: but every limiting value has two sides, and one can *a priori* approach it both from the left and from the right.

Actually, the ordinary formulation of SR has been restricted too much. For instance, *even leaving aside superluminal speeds*, it can be easily widened to include antimatter [35]. Then, one finds space–time to be *a priori* populated by normal particles P (which travel forward in time carrying positive energy), *and* by dual particles Q "which travel backwards in time carrying negative energy". The latter shall appear to us as antiparticles, i.e. as particles – regularly travelling forward in time with positive energy, but – with all their "additive" charges (e.g. the electric charge) reversed in sign: see Fig. 2.

To clarify this point, we can recall what follows: We, as macroscopic observers, have to move in time along a single, well-defined direction, to such an extent that we cannot even see a motion backwards in time ... and every object like Q, travelling backwards in time (with negative energy), will be *necessarily* reinterpreted by us as an anti-object, with opposite charges but travelling forward in time (with positive energy): cf. Fig. 2 and [31,39,35].

But let us forget about antimatter and go back to tachyons. A strong objection against their existence is based on the opinion that by using tachyons it would be possible to send signals into the past, owing to the fact that a tachyon T which, say, appears to a first observer O as emitted by A and absorbed by B, can appear to a second observer O' as a tachyon T' which travels backwards in time with negative energy. However, by applying (as it is obligatory to do) the same "reinterpretation rule" or switching procedure seen above, T' will appear to the new observer O' just as an antitachyon $\overline{T}$ emitted by B and absorbed by A, and therefore travelling forward in time, even if in the contrary *space* direction. In such a way, every instance of travel towards the past, and every negative energy, disappears [31,39,35].

Starting from this observation, it is possible to solve [35] the so-called causal para-doxes associated with superluminal motions: paradoxes which are the more instructive and amusing, the more sophisticated they are, but that cannot be re-examined here.[4]

Let us mention here just the following. The reinterpretation principle, according to which signals are carried only by objects which appear to be endowed with positive energy, eliminates any information transfer backwards in time; but this has a price: that of abandoning the ingrained conviction that judgement about what is cause and what is effect is independent of the observer. In fact, in the case examined above, the first observer O considers the event at A to be the cause of the event at B. By contrast, the second observer O' will consider the event at B as causing the event at A. All observers will, however, see the cause happen *before* its effect.

Taking new objects or entities into consideration always forces us to review our prejudices. If we require the phenomena to obey the *law* of (retarded) causality with respect to all observers, then we cannot also demand the *description* "details" of the phenomena to be invariant: namely, we cannot also demand in that case the invariance of the "cause" and "effect" *labels* [38,36].

To illustrate the nature of our difficulties in accepting that e.g. the parts of cause and effect depend on the observer, let us cite an analogous situation that does not imply present-day prejudices:

> For ancient Egyptians, who knew only the Nile and its tributaries, which all flow South to North, the meaning of the word "south" coincided with the one of "upstream", and the mean-ing of the word "north" coincided with the one of "downstream". When Egyptians discovered the Euphrates, which unfortunately happens to flow North to South, they passed through such a crisis that it is mentioned in the stele of Tuthmosis I, which tells us about *that inverted water that goes downstream (i.e. towards the North) in going upstream* [Csonka, 1970].

In the last century theoretical physics led us in a natural way to suppose the exis-tence of various types of objects: magnetic monopoles, quarks, strings, tachyons, besides black holes: and various sectors of physics could not go on without them, even if the existence of none of them is certain (also because attention has not yet been paid to some links existing among them: e.g. a superluminal electric charge is expected to behave like a magnetic monopole; and a black hole *a priori* can be the source of tachyonic matter). According to Democritus of Abdera, everything that was thinkable without meeting con-tradictions had to exist somewhere in the unlimited universe. This point of view – which was given the name "totalitarian principle" by M. Gell-Mann – was later expressed by T.H. White in the humorous form "Anything not forbidden is compulsory" ...

3. A Glance at the Experimental State-of-the-Art

Extended relativity can allow a better understanding of many aspects of *ordinary* physics, even if tachyons would not exist in our cosmos as asymptotically free objects. As already said, we are dealing with superluminal motions, however, since this topic is coming back into fashion, especially because at least three or four different experimental sectors of physics seem to suggest the possible existence of faster-than-light motions. Our first aim is to put forth in the following some information (mainly bibliographical) about the

---

[4]Some of them have been proposed by R.C. Tolman, J. Bell, F.A.E. Pirani, J.D. Edmonds and others [31,36].

experimental results obtained in a couple of these different sectors, with a mere mention of the others.

3.1. Neutrinos

A long series of experiments, begun in 1971, seems to show that the square $m_0{}^2$ of the mass m_0 of muon-neutrinos, and more recently of electron–neutrinos too, is negative. This, if confirmed, would mean that (when using a naïve language, commonly adopted) such neutrinos possess an "imaginary mass" and are therefore tachyonic, or mainly tachyonic [31,3]. In extended relativity, the dispersion relation for a free superluminal object becomes[5]

$$\omega^2 - k^2 = -\Omega^2, \quad \text{or} \quad E^2 - p^2 = -m_0^2,$$

and there is *no* need therefore of imaginary masses . . .

3.2. Galactic Micro-Quasars

As to the *apparent* superluminal expansions observed in the core of quasars [48] and, recently, in so-called galactic microquasars [18], we shall not deal with that problem here, because it is far from the other topics of this chapter: not to mention that for those astronomical observations there exist orthodox interpretations, based on Ref. [40], that – even if "statistically" weak – are accepted by the majority of astrophysicists.[6]

    Here, let us mention only that simple geometrical considerations in Minkowski space show that a *single* superluminal light source would appear [31,33]: (i) initially, in the "optical boom" phase (analogous to the acoustic "boom" produced by a plane travelling at a constant supersonic speed), as an intense source which suddenly comes into view; and (ii) would afterwards seem to split into two objects receding one from the other with speed $V > 2c$ (all of this being similar to what is actually observed, according to [18]).

3.3. Evanescent Waves and "Tunnelling Photons"

Within quantum mechanics (and, more accurately, in the *tunnelling* processes), it had been shown that the tunnelling time – firstly evaluated as simple "phase time" and later on calculated through the analysis of the wavepacket behaviour – does not depend on the barrier width in the case of opaque barriers ("Hartman effect") [22]. This implies superluminal and arbitrarily large (group) velocities V inside long enough barriers: see Fig. 3.

    Experiments that may verify this prediction by, say, electrons are difficult. Luckily enough, however, the Schroedinger equation in the presence of a potential barrier is mathematically identical to the Helmholtz equation for an electromagnetic wave propagating, for instance, down a metallic waveguide along the x-axis (as shown, e.g., by R. Chiao *et al.* [10]); and a barrier height U bigger than the electron energy E corresponds (for a given wave frequency) to a waveguide of transverse size lower than a cut-

---

[5]We put $c = 1$, whenever convenient, throughout this chapter.

[6]For a theoretical discussion, see [33].

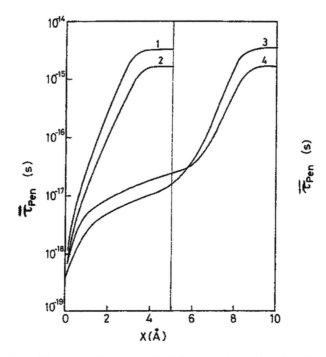

Figure 3. Behaviour of the average "penetration time" (in seconds) spent by a tunnelling wavepacket, as a function of the penetration depth (in åangstroms) down a potential barrier (from Olkhovsky *et al.* [22]). According to the predictions of quantum mechanics, the wavepacket speed inside the barrier increases in an unlimited way for opaque barriers, and the total tunnelling time does *not* depend on the barrier width [22].

off value. A segment of "undersized" guide – to continue with our example – therefore behaves as a barrier for the wave (photonic barrier) [10], as well as any other photonic band-gap filters. The wave assumes therein – like an electron inside a quantum barrier – an imaginary momentum or wave-number and gets, as a consequence, exponentially damped along x. In other words, it becomes an *evanescent* wave (going back to normal propagation, even if with reduced amplitude, when the narrowing ends and the guide returns to its initial transverse size). Thus, a tunnelling experiment can be simulated [10] by having recourse to evanescent waves (for which the concept of group velocity can be properly extended [34]).

The fact that evanescent waves travel with superluminal speeds (cf. e.g. Fig. 4) has actually been verified in a series of famous experiments, performed since 1992 onwards by R. Chiao, P.G. Kwiat and A. Steinberg's group at Berkeley [44], by G. Nimtz *et al.* at Cologne [20], by A. Ranfagni and colleagues at Florence [30], and by others at Vienna, Orsay and Rennes [30], which verified that "tunnelling photons" travel with superluminal group velocities.[7] Let us add also that extended relativity had predicted [50] evanescent waves endowed with faster-than-c speeds; the whole matter therefore appears to be theoretically consistent. The debate in the current literature does not refer to the experimental results (which can be correctly reproduced by numerical elaborations [8,4]

---

[7]Such experiments also raised a great deal of interest [49] within the non-specialized press, and were reported by *Scientific American, Nature, New Scientist, Newsweek*, etc.

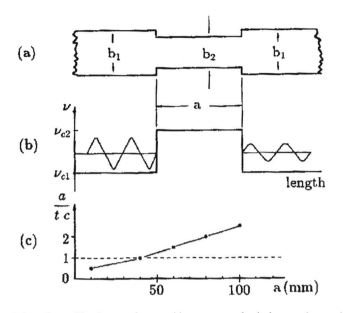

Figure 4. Simulation of tunnelling by experiments with evanescent classical waves (see text), which were predicted to be superluminal on the basis of extended relativity [31,39]. The figure shows one of the measurement results in [44]; that is, the wavepacket average speed while crossing the evanescent region (= segment of undersized waveguide, or "barrier") as a function of its length. As theoretically predicted [22, 50], such an average speed exceeds *c* for long enough "barriers".

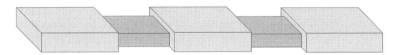

Figure 5. The very interesting experiment along a metallic waveguide with two barriers (undersized guide segments), i.e. with two evanescence regions [21]. See text.

based on Maxwell equations only), but rather to the question whether they allow, or do not allow, the sending of signals or information with superluminal speed [34,4,17].

In the above-mentioned experiments one meets a substantial attenuation of the considered pulses during tunnelling (or during propagation in an absorbing medium). However, by employing a "gain doublet", it has recently been reported the observation of undistorted pulses propagating with superluminal group-velocity with a *small* change in amplitude [46].

Let us emphasize that some of the most interesting experiments of this series seem to be the ones with two "barriers" (e.g. with two gratings in an optical fibre, or with two segments of undersized waveguide separated by a piece of normal-sized waveguide: Fig. 5). For suitable frequency bands – i.e. for "tunnelling" far from resonances – it was found that the total crossing time does not depend on the length of the intermediate (normal) guide: namely, that the wavepacket speed along it is infinite [21,13]. This agrees with what was predicted by quantum mechanics for non-resonant tunnelling through two successive opaque barriers under suitable hypotheses (the tunnelling *phase time*, which depends on the entering energy, has been shown by us to be independent of the distance between the two barriers [24]); something that has been accepted and generalized in

Aharonov *et al.* [24]. Such a prediction has been verified a second time, taking advantage of the circumstance that quite interesting evanescence regions can be constructed in the most varied manners, such as by means of different photonic band-gap materials or gratings (it being possible to use multilayer dielectric mirrors, or semiconductors, to photonic crystals...). And indeed very recent confirmation came – as already mentioned – from an experiment having recourse to two gratings in an optical fibre [13].

We cannot skip a further topic – which, being delicate, should not appear in a brief overview like this one – since the last experimental contribution to it (performed at Princeton by J. Wang *et al.* [46] and published in *Nature* on 20 July 2000) aroused the interest of the world's press.

Even if in extended relativity all the ordinary causal paradoxes seem to be solvable [31,36], nevertheless one has to bear in mind that (whenever an object, $\mathcal{O}$, travels with superluminal speed) one may have to deal with negative contributions to the *tunnelling times* [22,25]: and this should not be regarded as unphysical. In fact, whenever an "object" (particle, electromagnetic pulse, etc.) $\mathcal{O}$ *overcomes* the infinite speed [31,36] with respect to a certain observer, it will afterwards appear to the same observer as the "*anti*-object" $\overline{\mathcal{O}}$ travelling in the opposite *space* direction [31,36].

For instance, when going on from the lab to a frame $\mathcal{F}$ moving in the *same* direction as the particles or waves entering the barrier region, the object $\mathcal{O}$ penetrating through the final part of the barrier (with almost infinite speed [22,4,24,25], as in Fig. 3) will appear in the frame $\mathcal{F}$ as an anti-object $\overline{\mathcal{O}}$ crossing that portion of the barrier *in the opposite space-direction* [31,36]. In the new frame $\mathcal{F}$, therefore, such an anti-object $\overline{\mathcal{O}}$ would yield a *negative* contribution to the tunnelling time, which could even be negative. For clarifications, see [23]. What we want to stress here is that the appearance of such negative times is predicted by relativity itself, on the basis of the ordinary postulates [31,36,4,23]. (In the case of a non-polarized wave, the wave anti-packet coincides with the initial wave packet; if a photon is, however, endowed with helicity $\lambda = +1$, the anti-photon will bear the opposite helicity $\lambda = -1$.)[8]

Let us add here that, via quantum interference effects it is possible to obtain dielectrics with refraction indices very rapidly varying as a function of frequency, also in three-level atomic systems, with almost complete absence of light absorption (i.e. with quantum induced transparency) [1]. The group-velocity of a light pulse propagating in such a medium can decrease to very low values, either positive or negatives, with no pulse distortion. Experiments have been performed both in atomic samples at room temperature and in Bose–Einstein condensates, which showed the possibility of reducing the speed of light to a few metres per second. Similar, but negative, group velocities, implying a propagation with superluminal speeds thousands of times higher than the previously mentioned ones, have also been recently predicted in the presence of such an "electromagnetically induced transparency", for light moving in a rubidium condensate [2], while corresponding experiments are being carried out, for instance, at the LENS laboratory, Florence.

Finally, let us recall that faster-than-c propagation of light pulses can also be (and was, in same cases) observed by taking advantage of anomalous dispersion near an absorbing line, or nonlinear and linear gain lines – as already seen –, or nondispersive di-

---

[8]From a theoretical point of view, besides Refs [31,36,22,4,25,23], see Ref. [9]. On the (quite interesting!) experimental side, see papers [47], the first one having already been mentioned above.

electric media, or inverted two-level media, as well as in some parametric processes in nonlinear optics (cf. e.g. G. Kurizki *et al.*'s works).

3.4. *Superluminal Localized Solutions (SLS) to the Wave Equations. The "X-Shaped Waves"*

The fourth sector is no less important. It came back into fashion when some groups of capable scholars in engineering (for sociological reasons, most physicists had abandoned the field) rediscovered by a series of clever works that any wave equation – to fix the ideas, let us think of the electromagnetic case – also admit solutions as much subluminal as superluminal (besides the ordinary waves endowed with speed c/n).

Let us recall that, starting with the pioneering work by H. Bateman, it had slowly become known that all homogeneous wave equations (in a general sense: scalar, electromagnetic, spinorial, ...) admit wavelet-type solutions with subluminal group velocities [6]. Subsequently, also superluminal solutions started to be written down.[9] An important feature of some of these new solutions (which attracted much attention for possible applications) is that they propagate as localized, non-diffracting pulses: namely, according to Courant and Hilbert's terminology [6], as "undistorted progressive waves". It is easy to realize the practical importance, for instance, of a radio transmission carried out by localized waves, independently of their being sub- or superluminal. But non-diffractive wave packets can be of use even in theoretical physics for a reasonable representation of elementary particles [42]; and so on.

Within extended relativity since 1980 it had been found [5] that – while the simplest subluminal object conceivable is a small sphere, or a point as its limit – the simplest superluminal objects turns out to be instead an "X-shaped" wave (see Refs [5], and Figs 6 and 7 of this chapter), or a double cone as its limit, which moreover travels without deforming – i.e. rigidly – in a homogeneous medium [31]. It is not without meaning that the most interesting localized solutions happened to be superluminal, and with a shape of that kind. Even more, since from Maxwell equations under simple hypotheses one goes on to the usual *scalar* wave equation for each electric or magnetic field component, one can expect the same solutions to exist also in the fields of acoustic waves and of seismic waves (and of gravitational waves too).

Actually, such waves (as suitable superpositions of Bessel beams [12]) were mathematically constructed for the first time, by Lu *et al.* [14], *in acoustics*: and later on by Recami [32] for electromagnetism; they were then called "X-waves" or, rather, X-shaped waves. In the Appendix we briefly show how X-shaped solutions to the wave equation (in particular, the "classical" X-wave) can be constructed.

It is more important for us that the X-shaped waves have indeed been produced in experiments both with acoustic and with electromagnetic waves; that is, X-waves were produced that travel undistorted faster than the speed of sound, in the first case, and than light, in the second case. In acoustics, the first experiment was performed by Lu *et al.* [15] in 1992 at the Mayo Clinic (and their papers received the 1992 first IEEE award). In the electromagnetic case, which is more intriguing, superluminal localized X-shaped solutions were first mathematically constructed (cf. e.g. Fig. 8) in Refs [32], and later experimentally produced by Saari *et al.* [41] in 1997 at Tartu by visible light (Fig. 9),

---

[9]This was done in [45] and, independently, in [11] (in one case just by the mere application of a superluminal Lorentz "transformation" [31,37]).

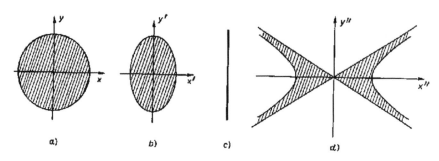

a)   b)   c)   d)

Figure 6. An intrinsically spherical (or pointlike, at the limit) object appears in the vacuum as an ellipsoid contracted along the motion direction when endowed with a speed $v < c$. By contrast, if endowed with a speed $V > c$ (even if the c-speed barrier cannot be crossed, neither from the left nor from the right), it would appear [42] no longer as a particle, but rather as an "X-shaped" wave [42] travelling rigidly (occupying the region delimited by a double cone and a two-sheeted hyperboloid – or as a double cone, at the limit – moving superluminally and without distortion in the vacuum, or in a homogeneous medium).

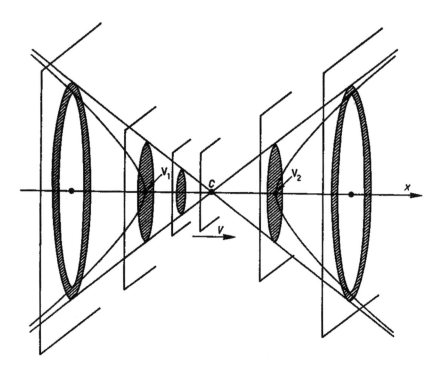

Figure 7. Here we show the intersections of an "X-shaped wave" [42] with planes orthogonal to its motion line, according to extended relativity [38,31,39]. Examination of this figure suggests how to construct a simple dynamic antenna for generating such localized superluminal waves (such an antenna was in fact adopted, independently, by Lu *et al.* [14] for the production of such non-diffractive waves).

and more recently by Mugnai, Ranfagni and Ruggeri at Florence by microwaves [19] (paper appeared in *Phys. Rev. Lett.* of 22 May 2000). Further experimental activity is in progress, while in the theoretical sector the activity has been growing so rapidly that it is not possible to quote here the relevant recent literature; we might recall, e.g., the papers devoted to building up new analogous solutions with finite total energy or more suitable

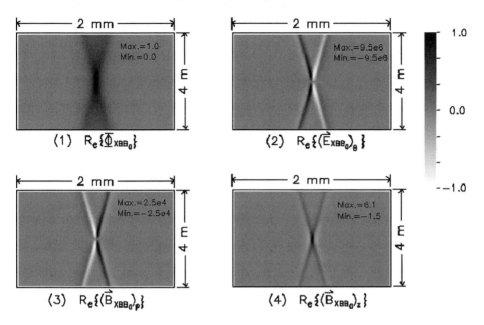

Figure 8. Theoretical prediction of the superluminal localized "X-shaped" waves for the electromagnetic case (from Lu, Greenleaf and Recami [32], and Recami [32]).

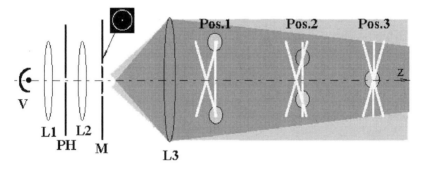

Figure 9. Scheme of the experiment by Saari *et al.*, who announced (*Physical Review Letters* of 24 Nov. 1997) the production in optics of the waves depicted in Fig. 8. In this figure one can see what was shown by the experiment, i.e. that the superluminal "X-shaped" waves run after and catch up with the plane waves (the latter regularly travelling with speed c). An analogous experiment has been performed with microwaves at Florence by Mugnai, Ranfagni and Ruggeri (*Physical Review Letters* of 22 May 2000).

for high frequencies, on one hand, and localized solutions superluminally propagating even along a normal waveguide, on the other hand [26,27]; or attempts at focusing X-shaped waves, at a certain instant, in a small region [28]. But we cannot avoid mentioning that suitable superpositions of Bessel beams can produce a *stationary* intense wave-field: confined within a tiny region, and static; while the field intensity outside the tiny region is negligible [29]; such "frozen waves" can have very many important applications, of course (a patent is pending).

Before continuing, let us eventually touch on the problem of producing an X-shaped superluminal wave like the one in Fig. 7, but truncated – of course – in space and in time (by the use of a finite antenna, radiating for a finite time): in such a situation, the wave will keep its localization and superluminality only along a certain "depth of field", decaying abruptly afterwards [12,32].

We can become convinced about the possibility of realizing it, by imagining the simple ideal case of a negligibly sized superluminal source S endowed with speed $V > c$ in a vacuum and emitting electromagnetic waves W (each travelling at the invariant speed c). The electromagnetic waves will be internally tangent to an enveloping cone C having S as its vertex, and as its axis the propagation line x of the source [31].

This is analogous to what happens for a plane that moves in the air with constant supersonic speed. The waves W interfere mostly negatively inside the cone C, and constructively on its surface. We can place a plane detector orthogonally to x, and record the magnitude and direction of the W waves that hit on it, as (cylindrically symmetric) functions of position and of time. It will be enough, then, to replace the plane detector with a plane antenna which *emits* – instead of recording – exactly the same (axially symmetric) space–time pattern of waves W, for constructing a cone-shaped electromagnetic wave C that will propagate with the superluminal speed V (of course, without a source any longer at its vertex): even if each wave W travels at the invariant speed c.[10]

Here let us only remark that such localized superluminal waves appear to keep their good properties only as long as they are fed by the waves arriving (at speed c) from the antenna. Taking account of the time needed for fostering such superluminal pulses (i.e. for the arrival of the feeding speed-c waves coming from the aperture), one concludes that these localized superluminal waves are probably unable to transmit *information* faster than c. However, they don't seem to have anything to do with the illusory "scissors effect", even if the energy feeding them appears to travel with the speed of light. In fact, the spot – endowed, as we know, with superluminal group-velocity – is able to get, for instance, two (tiny) detectors at a distance L to click after a time *smaller* than L/c. A lot of discussion is still going on about the possible differences among group-velocity, signal-velocity and information speed.

As we mentioned above, the existence of all these X-shaped superluminal (or "supersonic") waves seems to constitute at the moment, together, e.g. with the superluminality of evanescent waves, nothing but confirmations of extended relativity: a theory, let us repeat, based on the ordinary postulates of SR and which consequently does not appear to violate any of its fundamental principles. It is curious, moreover, that one of the first applications of such X-waves (which takes advantage of their propagation without deformation) is in progress in the field of medicine; more precisely, ultrasound scanners [16].

Before ending, let us remark that a series of new SLSs to the Maxwell equations, suitable for arbitrary frequencies and arbitrary bandwidths, have recently been constructed by us, some of them being endowed with *finite* total energy. Among the others, we have set forth an infinite family of generalizations of the classical X-shaped wave; and shown how to deal with the case of a *dispersive* medium. Results of this kind may find application in other fields in which an essential role is played by a wave-equation (such as acoustics, seismology, geophysics, gravitation, elementary particle physics, etc.).

---

[10]For further details, see the first of Refs [32].

Acknowledgements

The author acknowledges continuous, stimulating discussions with M. Mamone Capria and M. Zamboni Rached. For useful discussions, he is also grateful to J.D. Bekenstein, Ray Chiao, Gianni Degli Antoni, J.R. Fanchi, Flavio Fontana, Larry Horwitz, Peter Milonni, Alwyn van der Merwe and Marco Villa, as well as to F. Bassani, A. Bertin, L. Cifarelli, R. Collina, G.C. Costa, J.H. Eberly, M. Fleishhauer, A. Gigli, J.-y. Lu, G. Marchesini, M. Mattiuzzi, D. Mugnai, G. Nimtz, V.S. Olkhovsky, M. Pernici, A. Ranfagni, R.A. Ricci, P. Saari, G. Salesi, A. Shaarawi, L.S. Shulman, D. Stauffer, A. Steinberg, M.T. Vasconselos and A. Vitale.

Appendix

In this Appendix we want to show how localized superluminal solutions (SLS) to the wave equations, in particular the X-shaped ones, can be mathematically constructed. Here for simplicity we shall consider only the case of a dispersionless medium such as a vacuum, and of free space (without boundaries).

It has been known for over a century that a particular axially symmetric solution to the wave equation in vacuum ($n = n_0$) is, in cylindrical coordinates, the function $\psi(\rho, z, t) = J_0(k_\rho \rho) \, e^{+ik_z z} \, e^{-i\omega t}$ with $k_\rho^2 = n_0^2(\omega^2/c^2) - k_z^2$; $k_\rho^2 \geq 0$, where J_0 is the zeroth-order ordinary *Bessel function*, k_z and k_ρ are the axial and the transverse wavenumbers respectively, ω is the angular frequency and c is the light velocity. Using the transformation

$$\begin{cases} k_\rho = \dfrac{\omega}{c} n_0 \sin \theta, \\[2ex] k_z = \dfrac{\omega}{c} n_0 \cos \theta \end{cases} \tag{1}$$

the solution $\psi(\rho, z, t)$ can be rewritten in the well-known *Bessel beam* form:

$$\psi(\rho, \zeta) = J_0\left(n_0 \frac{\omega}{c} \rho \sin \theta\right) e^{+in_0(\omega/c)\zeta \cos \theta}, \tag{2}$$

where $\zeta \equiv z - Vt$ while $V = c/(n_0 \cos \theta)$ is the phase velocity, quantity θ ($0 < \theta < \pi/2$) being the cone angle of the Bessel beam.

More generally, SLSs (with axial symmetry) to the wave equation will be the following ones [14,43]:

$$\psi(\rho, \zeta) = \int_0^\infty S(\omega) J_0\left(\frac{\omega}{V} \rho \sqrt{n_0^2 \frac{V^2}{c^2} - 1}\right) e^{+i(\omega/V)\zeta} \, d\omega, \tag{3}$$

where $S(\omega)$ is the adopted frequency spectrum.

Indeed, such solutions result in pulses propagating in free space without distortion and with the superluminal velocity $V = c/(n_0 \cos \theta)$. The most popular spectrum $S(\omega)$ is that given by $S(\omega) = e^{-a\omega}$, which provides the ordinary ("classical") X-shaped wave

$$X \equiv \psi(\rho, \zeta) = \frac{V}{\sqrt{(aV - i\zeta)^2 + \rho^2(n_0^2(V^2/c^2) - 1)}}. \tag{4}$$

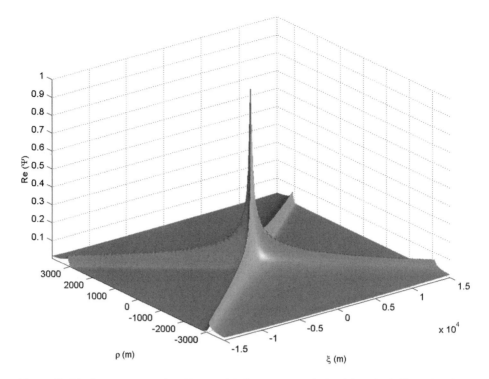

Figure 10. The figure represents (in arbitrary units) the square magnitude of the "classical", X-shaped superluminal localized solution (SLS) to the wave equation, with $V = 5c$ (and $a = 0.1$): see [14,32]. It has been shown elsewhere that (infinite) families of SLSs exist, however, which generalize this classical X-shaped solution.

Because of its non-diffractive properties and its low frequency spectrum,[11] the X-wave is being particularly applied in fields such as acoustics [15]. The "classical" X-wave is represented in Fig. 10.

References

[1] G. Alzetta, A. Gozzini, L. Moi and G. Orriols, *Nuovo Cimento B*, vol. 36, 5 (1976).

[2] M. Artoni, G.C. La Rocca, F.S. Cataliotti and F. Bassani, *Phys. Rev. A* (in press).

[3] Cf. M. Baldo Ceolin, "Review of neutrino physics", invited talk at the *XXIII Int. Symp. on Multiparticle Dynamics* (Aspen, CO; Sept. 1993); E.W. Otten, *Nucl. Phys. News*, vol. 5, 11 (1995). From the theoretical point of view, see, e.g., E. Giannetto, G.D. Maccarrone, R. Mignani and E. Recami, *Phys. Lett. B*, vol. 178, 115–120 (1986), and refs. therein; S. Giani, "Experimental evidence of superluminal velocities in astrophysics and proposed experiments", CP458, in *Space Technology and Applications International Forum 1999*, ed. by M.S. El-Genk (A.I.P.; Melville, 1999), pp. 881–888.

[4] A.P.L. Barbero, H.E.H. Figueroa and E. Recami, "On the propagation speed of evanescent modes", *Phys. Rev. E*, vol. 62, 8628 (2000), and refs. therein.

[5] A.O. Barut, G.D. Maccarrone and E. Recami, *Nuovo Cimento A*, vol. 71, 509 (1982); P. Caldirola, G.D. Maccarrone and E. Recami, *Lett. Nuovo Cim.*, vol. 29, 241 (1980); E. Recami and G.D. Maccarrone, *Lett. Nuovo Cim.*, vol. 28, 151 (1980). See also E. Recami, refs. [3,4,41], and E. Recami, M.Z. Rached

---

[11] Let us emphasize that this spectrum starts from zero, is suitable for low-frequency applications, and has the bandwidth $\Delta\omega = 1/a$.

and C.A. Dartora: "The X-shaped, localized field generated by a Superluminal electric charge", *Phys. Rev. E*, vol. 69 (2004) no. 027602.

[6] H. Bateman, *Electrical and Optical Wave Motion* (Cambridge Univ. Press; Cambridge, 1915); R. Courant and D. Hilbert, *Methods of Mathematical Physics* (J. Wiley; New York, 1966), vol. 2, p. 760; J.N. Brittingham, *J. Appl. Phys.*, vol. 54, 1179 (1983); R.W. Ziolkowski, *J. Math. Phys.*, vol. 26, 861 (1985); J. Durnin, J.J. Miceli and J.H. Eberly, *Phys. Rev. Lett.*, vol. 58, 1499 (1987); A.O. Barut et al., *Phys. Lett. A*, vol. 143, 349 (1990); *Found. Phys. Lett.*, vol. 3, 303 (1990); *Found. Phys.*, vol. 22, 1267 (1992); P. Hillion, *Acta Applicandae Matematicae*, vol. 30, 35 (1993).

[7] See, e.g., O.M. Bilaniuk, V.K. Deshpande and E.C.G. Sudarshan, *Am. J. Phys.*, vol. 30, 718 (1962).

[8] H.M. Brodowsky, W. Heitmann and G. Nimtz, *Phys. Lett. A*, vol. 222, 125 (1996).

[9] R.Y. Chiao, A.E. Kozhekin and G. Kurizki, *Phys. Rev. Lett.*, vol. 77, 1254 (1996); E.L. Bolda et al., *Phys. Rev. A*, vol. 48, 3890 (1993); C.G.B. Garret and D.E. McCumber, *Phys. Rev. A*, vol. 1, 305 (1970).

[10] See, e.g., R.Y. Chiao, P.G. Kwiat and A.M. Steinberg, *Physica B*, vol. 175, 257 (1991); A. Ranfagni, D. Mugnai, P. Fabeni and G.P. Pazzi, *Appl. Phys. Lett.*, vol. 58, 774 (1991); Th. Martin and R. Landauer, *Phys. Rev. A*, vol. 45, 2611 (1992); Y. Japha and G. Kurizki, *Phys. Rev. A*, vol. 53, 586 (1996). Cf. also G. Kurizki, A.E. Kozhekin and A.G. Kofman, *Europhys. Lett.*, vol. 42, 499 (1998); G. Kurizki, A.E. Kozhekin, A.G. Kofman and M. Blaauboer, paper delivered at the VII Seminar on Quantum Optics, Raubichi, Belarus (May, 1998).

[11] R. Donnelly and R.W. Ziolkowski, *Proc. Roy. Soc. London A*, vol. 440, 541 (1993); I.M. Besieris, A.M. Shaarawi and R.W. Ziolkowski, *J. Math. Phys.*, vol. 30, 1254 (1989); S. Esposito, *Phys. Lett. A*, vol. 225, 203 (1997); J. Vaz and W.A. Rodrigues, *Adv. Appl. Cliff. Alg.*, vol. S-7, 457 (1997).

[12] J. Durnin, J.J. Miceli and J.H. Eberly, *Phys. Rev. Lett.*, vol. 58, 1499 (1987); *Opt. Lett.*, vol. 13, 79 (1988).

[13] S. Longhi, P. Laporta, M. Belmonte and E. Recami, *Phys. Rev. E*, vol. 65 (2002) no. 046610.

[14] J.-y. Lu and J.F. Greenleaf, *IEEE Trans. Ultrason. Ferroelectr. Freq. Control*, vol. 39, 19 (1992).

[15] J.-y. Lu and J.F. Greenleaf, *IEEE Trans. Ultrason. Ferroelectr. Freq. Control*, vol. 39, 441 (1992): In this case the wave speed is larger than the *sound* speed in the considered medium.

[16] J.-y. Lu, H.-h. Zou and J.F. Greenleaf, *Ultrasound in Medicine and Biology*, vol. 20, 403 (1994); *Ultrasonic Imaging*, vol. 15, 134 (1993).

[17] Cf., e.g., P.W. Milonni, *J. Phys. B*, vol. 35, R31–R56 (2002); G. Nimtz and A. Haibel, *Ann. der Phys.*, vol. 11, 163–171 (2002); R.W. Ziolkowski, *Phys. Rev. E*, vol. 63 (2001) no. 046604; A.M. Shaarawi and I.M. Besieris, *J. Phys. A*, vol. 33, 7227–7254, 7255–7263 (2000); E. Recami, F. Fontana and R. Garavaglia, *Int. J. Mod. Phys. A*, vol. 15, 2793 (2000), and refs. therein.

[18] I.F. Mirabel and L.F. Rodriguez, "A superluminal source in the Galaxy", *Nature*, vol. 371, 46 (1994) [with an editorial comment, "A galactic speed record", by G. Gisler, at page 18 of the same issue]; S.J. Tingay et al., "Relativistic motion in a nearby bright X-ray source", *Nature*, vol. 374, 141 (1995).

[19] D. Mugnai, A. Ranfagni and R. Ruggeri, *Phys. Rev. Lett.*, vol. 84, 4830 (2000).

[20] G. Nimtz and A. Enders, *J. de Physique-I*, vol. 2, 1693 (1992); vol. 3, 1089 (1993); vol. 4, 1379 (1994); *Phys. Rev. E*, vol. 48, 632 (1993); H.M. Brodowsky, W. Heitmann and G. Nimtz, *J. de Physique-I*, vol. 4, 565 (1994); *Phys. Lett. A*, vol. 222, 125 (1996); vol. 196, 154 (1994). For a review, see, e.g., G. Nimtz and W. Heitmann, *Prog. Quant. Electr.*, vol. 21, 81 (1997). See also Ref. [4].

[21] G. Nimtz, A. Enders and H. Spieker, in *Wave and Particle in Light and Matter*, ed. by A. van der Merwe and A. Garuccio (Plenum; New York, 1993); *J. de Physique-I*, vol. 4, 565 (1994). See also A. Enders and G. Nimtz, *Phys. Rev. B*, vol. 47, 9605 (1993).

[22] V.S. Olkhovsky, E. Recami and J. Jakiel, "Unified time analysis of photon and particle tunnelling", *Phys. Reports*, vol. 398, pp. 133–178 (2004); V.S. Olkhovsky and E. Recami, *Phys. Reports*, vol. 214, 339 (1992), and refs. therein: in particular T.E. Hartman, *J. Appl. Phys.*, vol. 33, 3427 (1962); L.A. MacColl, *Phys. Rev.*, vol. 40, 621 (1932). See also V.S. Olkhovsky, E. Recami, F. Raciti and A.K. Zaichenko, *J. de Phys.-I*, vol. 5, 1351–1365 (1995); G. Privitera, E. Recami, G. Salesi and V.S. Olkhovsky, "Tunnelling Times: An Elementary Introduction", *Rivista Nuovo Cim.*, vol. 26 (2003), monographic issue no. 4.

[23] V.S. Olkhovsky, E. Recami, F. Raciti and A.K. Zaichenko, ref. [22], page 1361 and refs. therein. See also refs. [31,36], and E. Recami, F. Fontana and R. Garavaglia, ref. [34], page 2807 and refs. therein.

[24] V.S. Olkhovsky, E. Recami and G. Salesi, *Europhysics Letters*, vol. 57 (2002) 879–884; "Tunneling through two successive barriers and the Hartman (Superluminal) effect", e-print quant-ph/0002022; Y. Aharonov, N. Erez and B. Reznik, *Phys. Rev. A*, vol. 65 (2002) no. 052124. See also E. Recami, "Superluminal tunneling through successive barriers: Does QM predict infinite group-velocities?", *Journal of Modern Optics*, vol. 51, 913–923 (2004).

[25] V.S. Olkhovsky, E. Recami and G. Salesi, refs. [26]; S. Esposito, *Phys. Rev. E*, vol. 67 (2003) no. 016609.

[26] M.Z. Rached, E. Recami and H.E.H. Figueroa, "New localized Superluminal solutions to the wave equations with finite total energies and arbitrary frequencies", *European Physical Journal D*, vol. 21, pp. 217–228 (2002); M.Z. Rached, K.Z. Nóbrega, H.E.H. Figueroa and E. Recami: "Localized Superluminal solutions to the wave equation in (vacuum or) dispersive media, for arbitrary frequencies and with adjustable bandwidth", *Optics Communications*, vol. 226, 15–23 (2003); M.Z. Rached, E. Recami and F. Fontana, "Superluminal localized solutions to Maxwell equations propagating along a waveguide: The finite-energy case", *Physical Review E*, vol. 67 (2003) no. 036620.

[27] M.Z. Rached, E. Recami and F. Fontana, "Localized Superluminal solutions to Maxwell equations propagating along a normal-sized waveguide", *Phys. Rev. E*, vol. 64 (2001) no. 066603; M.Z. Rached, K.Z. Nobrega, E. Recami and H.E.H. Figueroa, "Superluminal X-shaped beams propagating without distortion along a co-axial guide", *Physical Review E*, vol. 66 (2002) no. 046617; I.M. Besieris, M. Abdel-Rahman, A. Shaarawi and A. Chatzipetros, *Progress in Electromagnetic Research (PIER)*, vol. 19, 1–48 (1998).

[28] M.Z. Rached, A.M. Shaarawi and E. Recami, "Focused X-shaped pulses", *Journal of the Optical Society of America A*, vol. 21, pp. 1564–1574 (2004), and refs. therein.

[29] M.Z. Rached, "Stationary optical wave-fields with arbitrary longitudinal shape", *Optics Express*, vol. 12, pp. 4001–4006 (2004).

[30] A. Ranfagni, P. Fabeni, G.P. Pazzi and D. Mugnai, *Phys. Rev. E*, vol. 48, 1453 (1993); Ch. Spielmann, R. Szipocs, A. Stingl and F. Krausz, *Phys. Rev. Lett.*, vol. 73, 2308 (1994), Ph. Balcou and L. Dutriaux, *Phys. Rev. Lett.*, vol. 78, 851 (1997); V. Laude and P. Tournois, *J. Opt. Soc. Am. B*, vol. 16, 194 (1999).

[31] E. Recami, *Rivista N. Cim.*, vol. 9(6), 1–178 (1986), and refs. therein.

[32] E. Recami, *Physica A*, vol. 252, 586 (1998); J.-y. Lu, J.F. Greenleaf and E. Recami, "Limited diffraction solutions to Maxwell (and Schroedinger) equations" [Lanl Archives # physics/9610012], Report INFN/FM–96/01 (I.N.F.N.; Frascati, Oct. 1996). See also R.W. Ziolkowski, I.M. Besieris and A.M. Shaarawi, *J. Opt. Soc. Am., A*, vol. 10, 75 (1993); *J. Phys. A*, vol. 33, 7227–7254 (2000); A.T. Friberg, A. Vasara and J. Turunen, *Phys. Rev. A*, vol. 43, 7079 (1991).

[33] E. Recami, A. Castellino, G.D. Maccarrone and M. Rodonò, "Considerations about the apparent Superluminal expansions observed in astrophysics", *Nuovo Cimento B*, vol. 93, 119 (1986). Cf. also R. Mignani and E. Recami, *Gen. Relat. Grav.*, vol. 5, 615 (1974).

[34] E. Recami, F. Fontana and R. Garavaglia, *Int. J. Mod. Phys. A*, vol. 15, 2793 (2000), and refs. therein.

[35] E. Recami, in *I Concetti della Fisica*, ed. by F. Pollini and G. Tarozzi (Acc. Naz. Sc. Lett. Arti; Modena, 1993), pp. 125–138; E. Recami and W.A. Rodrigues, "Antiparticles from Special Relativity", *Found. Physics*, vol. 12, 709–718 (1982); vol. 13, E533 (1983).

[36] E. Recami, *Found. Physics*, vol. 17, 239–296 (1987). See also *Lett. Nuovo Cimento*, vol. 44, 587–593 (1985); P. Caldirola and E. Recami, in *Italian Studies in the Philosophy of Science*, ed. by M. Dalla Chiara (Reidel; Boston, 1980), pp. 249–298; A.M. Shaarawi and I.M. Besieris, *J. Phys. A*, vol. 33, 7255–7263 (2000).

[37] See also E. Recami and W.A. Rodrigues Jr., "A model theory for tachyons in two dimensions", in *Gravitational Radiation and Relativity*, ed. by J. Weber and T.M. Karade (World Scient.; Singapore, 1985), pp. 151–203, and refs. therein.

[38] See E. Recami and R. Mignani, *Rivista N. Cim.*, vol. 4, 209–290, E398 (1974), and refs. therein. Cf. also E. Recami (editor), *Tachyons, Monopoles, and Related Topics* (North-Holland; Amsterdam, 1978); and T. Alväger and M.N. Kreisler, *Phys. Rev.*, vol. 171, 1357 (1968), and refs. therein.

[39] See, e.g., E. Recami, in *Annuario '73, Enciclopedia EST*, ed. by E. Macorini (Mondadori; Milano, 1973), pp. 85–94; and *Nuovo Saggiatore*, vol. 2(3), 20–29 (1986).

[40] M.J. Rees, *Nature*, vol. 211, 46 (1966); A. Cavaliere, P. Morrison and L. Sartori, *Science*, vol. 173, 525 (1971).

[41] P. Saari and K. Reivelt, "Evidence of X-shaped propagation-invariant localized light waves", *Phys. Rev. Lett.*, vol. 79, 4135–4138 (1997).

[42] A.M. Shaarawi, I.M. Besieris and R.W. Ziolkowski, *J. Math. Phys.*, vol. 31, 2511 (1990), Sect. VI; *Nucl Phys. (Proc. Suppl.) B*, vol. 6, 255 (1989); *Phys. Lett. A*, vol. 188, 218 (1994). See also V.K. Ignatovich, *Found. Phys.*, vol. 8, 565 (1978); and A.O. Barut, *Phys. Lett. A*, vol. 171, 1 (1992); vol. 189, 277 (1994); *Ann. Foundation L. de Broglie*, Jan. 1994; and "Quantum theory of single events, Localized de Broglie-wavelets, Schroedinger waves and classical trajectories", preprint IC/90/99 (ICTP; Trieste, 1990).

[43] H. Sõnajalg and P. Saari, "Suppression of temporal spread of ultrashort pulses in dispersive media by Bessel beam generators", *Opt. Letters*, vol. 21, pp. 1162–1164 (August 1996).

[44] A.M. Steinberg, P.G. Kwiat and R.Y. Chiao, *Phys. Rev. Lett.*, vol. 71, 708 (1993), and refs. therein; *Scient. Am.*, vol. 269(2), 38 (1993). For a review, see, e.g., R.Y. Chiao and A.M. Steinberg, in *Progress in Optics*, ed. by E. Wolf (Elsevier; Amsterdam, 1997), p. 345. Cf. also Y. Japha and G. Kurizki, *Phys. Rev. A*, vol. 53, 586 (1996).

[45] J.A. Stratton, *Electromagnetic Theory* (McGraw-Hill; New York, 1941), p. 356; A.O. Barut et al., *Phys. Lett. A*, vol. 180, 5 (1993); vol. 189, 277 (1994). For a review-article about Localized Superluminal Waves, see E. Recami, M.Z. Rached, K.Z. Nóbrega, C.A. Dartora and H.E.H. Figueroa, "On the localized superluminal solutions to the Maxwell equations", *IEEE Journal of Selected Topics in Quantum Electronics*, vol. 9(1), 59–73 (2003).

[46] L.J. Wang, A. Kuzmich and A. Dogariu, *Nature*, vol. 406, 277 (2000).

[47] L.J. Wang, A. Kuzmich and A. Dogariu, *Nature*, vol. 406, 277 (2000); M.W. Mitchell and R.Y. Chiao, *Phys. Lett. A*, vol. 230, 133–138 (1997). See also S. Chu and W. Wong, *Phys. Rev. Lett.*, vol. 48, 738 (1982); B. Segard and B. Macke, *Phys. Lett. A*, vol. 109, 213–216 (1985); B. Macke et al., *J. Physique*, vol. 48, 797–808 (1987); G. Nimtz, *Europ. Phys. J., B* (to appear as a Rapid Note).

[48] See, e.g., J.A. Zensus and T.J. Pearson (editors), *Superluminal Radio Sources* (Cambridge Univ. Press; Cambridge, UK, 1987).

[49] *Scientific American* (Aug. 1993); *Nature* (21 Oct. 1993); *New Scientist* (Apr. 1995); *Newsweek* (19 June 1995).

[50] Ref. [31], p. 158 and pp. 116–117. Cf. also D. Mugnai, A. Ranfagni, R. Ruggeri, A. Agresti and E. Recami, *Phys. Lett. A*, vol. 209, 227 (1995).

Physics Before and After Einstein
M. Mamone Capria (Ed.)
IOS Press, 2005
© 2005 The authors

Chapter 13

Standard Cosmology and Other Possible Universes

Aubert Daigneault

*Professeur honoraire, Département de mathématiques et de statistique,
Université de Montréal*

Introduction

The phenomenon of the extragalactic redshift was barely known and the existence of galaxies was a disputed subject at the time Einstein published his epoch-making 1917 seminal paper [22].

Since then, cosmology has evolved in ways unforeseen by this celebrated author at that time under the impetus of these discoveries. To this day, the observable fact of the redshift remains the foremost determining factor of all competing cosmological theories. Its interpretation gives rise to *healthy disagreement* between scientists of diverse breeds.

Amongst several contenders, the so-called Big Bang cosmology, here labelled BBC, has acquired a largely dominant position which justifies calling it "Standard Cosmology" even if it is at odds with the original theory proposed by Einstein. Although its proponents and the media present it as definitive science, a number of other views are put forward by respected scientists and it is the purpose of the present chapter to describe, in addition to the pros and the cons of BBC, the arguments in favour or against a few of the competing hypotheses.

In so doing, we will endeavour to be as objective as brevity allows, avoiding in particular the kind of sarcasm unfortunately so widespread in this area of science which is an eminently arguable subject. Jean-Claude Pecker, a vigilant man, regretfully writes in his excellent book [79, p. 525], one of several recommended readings and a most open minded source of documentation: "There is undoubtedly a sort of intellectual dictatorship of the cosmological establishment, which has its Gospels, its Paradise and its Hell". That being said, the defenders of conventional wisdom, often quite haughty, certainly do not have a monopoly of disrespectful or acrimonious words. In consulting the literature, the reader would be well advised to disregard irrelevant passages of this nature which do not contribute to a better understanding of this fascinating subject.

The competition between cosmological theories after Einstein revolves mostly around their respective ability to account for the observational data relating to three phenomena: 1 – the redshift of electromagnetic radiation from extragalactic objects; 2 – the *cosmic background radiation* (CBR); 3 – the relative *abundances* of light elements in the universe.

However the main challenge is discovering the nature of the extragalactic redshift. What complicates matters is that it would seem to have more than one cause. Sometimes, in the expanding universe theories such as BBC, its explanation is postulated and the theory is built around this postulate; other times, as in *chronometric cosmology* (CC), it is derived from the curvature of space, a second possibility originally contemplated also by Hubble. Still other times, as in some *tired-light* mechanisms, it appears as a consequence of photon collisions. The *Quasi-Steady-State Cosmology* (QSSC) ascribes more than one cause to the redshift. In the *Plasma Universe* of Alfvén an antimatter theory is put forward with some reservations to explain the phenomenon. Halton Arp attributes it to intrinsic properties of matter and the numerous so-called *anomalous redshifts* that he has discovered appear to seriously defy all cosmologies. These findings are also called *discordant redshifts* and they seem to be quite real, despite the denials of most advocates of BBC, and the question remains to what extent they may be statistically significant.

In the same order of ideas, one may wonder about the impact on statistical studies of observations if multiple images of the same object are seen in the night sky, be it in the context of CC or in a variant of BBC in which the cosmological principle is denied giving rise to multiply-connected finite spaces.

Not all aspects of the theories presented are in contradiction with one another. For instance, *plasma cosmology* (PC) appears to the present author essentially compatible with CC except for the flatness and infiniteness of space which seem to be inessential features of PC. The two cosmologies CC and PC may be complementary to a large extent: the first providing kinematics and the second dynamics. The dominant role that electromagnetism plays in the plasma universe and the fact that the Einstein universe is the proper setting for Maxwell's equations, as Segal has shown, suggest a link between these two theories.

Mostly for lack of space, we have left out theories, such as *ultimate theories of everything*, which for the time being appear metaphysical in the sense of having no phenomenological basis, such as the possibility of the existence of parallel universes or of space–time having more than four dimensions. This is not to say that such questions are devoid of meaning or of interest.

The literature being very abundant, we have listed mostly the references that are explicitly referred to. We have marked with asterisks some titles that we consider to be particularly relevant. Also listed are some websites that we think are useful.

1. Greatness and Miseries of the Big Bang Theory: Its Rise and Predictable Fall?

1.1. The Dogmas of the Big Bang Cosmology and the Arguments in Its Favour

"Big Bang" is a disparaging expression coined by an opponent of BBC, Sir Fred Hoyle. The cautious and respectful Jean-Claude Pecker would rather call it "the primeval fireball hypothesis" or "the general explosion" [79, footnote p. 411].

Here are the major propositions of the BBC creed:

- The universe is expanding and this expansion follows Hubble's law which relates linearly the redshift from an extragalactic object to its distance from us.

- Time did not exist before the Big Bang; there is simply no "before". Likewise, space was born with the Big Bang so the general explosion did not take place in pre-existing space [115, p. 272].

Whether space is finite or infinite is still debated amongst BBC supporters. If space is finite, it was born as just a point and one may perhaps say this is where the Big Bang occurred. If space is infinite, it always was since it was generated and so was never small; some say then that the Big Bang occurred everywhere at the same time in that infinite space (Hubert Reeve, p. 10 in [29]). In any case, space is held to be a three-dimensional manifold with no edge. A widespread mistaken belief as to what BBC means is that the universe was once a small three-dimensional ordinary *ball* (as opposed to a three-dimensional *sphere*) and is now a much larger one still expanding i.e. still pushing its edge further away into nowhere.

If space is infinite, some say open or flat, it contains an infinite quantity of matter; there are then infinitely many galaxies by virtue of its homogeneity.

What does "expanding universe" mean? According to a widespread conception, or misconception, of that phrase, it is space itself that is expanding and not the galaxies in it which are flying away from one another in pre-existing space.

For instance Sten Odenwald, an astronomer with Raytheon ITSS, and Rick Fienberg, Editor in Chief of *Sky & Telescope* magazine, declare:

In Big Bang cosmology, galaxies are located at fixed positions in space. They may perform small dances about these positions in accordance with special relativity and local gravitational fields, but the real 'motion' is in the literal expansion of space between them! This is not a form of movement that any human has ever experienced [77].

In the same vein Wolfgang Rindler writes:

Note that the cosmological redshift is really an *expansion* effect rather than a *velocity* effect [86, p. 213].

From the mathematical definition of the Friedmann–Lemaître–Robertson–Walker (FLRW) space–times (see Chapter 6) which are used as models of BBC and most other cosmologies, these authors seem to be right.

Or is this indeed the way it is? To the question "How is it possible for space, which is utterly empty, to expand? How can nothing expand?", Nobel Prize physicist and foremost BBC enthusiast Steven Weinberg of the University of Texas replies:

Good question. The answer is: space does not expand. Cosmologists sometimes talk about expanding space – but they should know better [17, p. 32].

One way of interpreting Weinberg's words is that, according to him, talk about expansion of *space* is no more than metaphorical, the physical fact being the increase in (cosmological) time of the distances between any two galaxies.

Jeffrey R. Weeks expresses his view on this matter this way:

Houses, people, atoms, and metersticks are not expanding. Planets, stars, and even galaxies are not expanding. Space *is* expanding, and so is the distance between galaxies, but that's about it [115, p. 269].

There are considerable variations in the speed of expansion. It may be much faster than the speed of light as it is supposed to have been in the first zillionth of a second after the Big Bang according to *inflation theory*, a now very popular patch up of the original BBC, which Pecker refers to as the *New Big Bang* [79, p. 486]. After a long sedate period at a much slower pace, it is accelerating again. This last claim, made in 1998 and earlier [33,83], is based on an interpretation of observations of distant supernovae i.e. exploding stars.

In standard BBC, the global topology and geometry of space are determined by the density of the matter it contains in accordance with General Relativity. As a uniform distribution of matter is generally assumed, the local geometry is the same around any point of space and, in particular, space is of constant curvature, i.e. its curvature is the same at all points.

This curvature depends on the density of matter which is to be observationally determined. If this density is above a certain threshold, the curvature is positive, the geometry is elliptic and space is finite; if the density is equal to this dividing value, the curvature is zero and space is Euclidean and hence infinite; if the density is below this value, the curvature is negative, space is again infinite and the geometry is hyperbolic. The ratio of the density to this value is denoted Ω (omega) and called the *density parameter* [80, p. 100].

Conventional BBC affirms the *cosmological principle* also known as the *Copernican principle* asserting the isotropy and hence the homogeneity of the large-scale distribution and composition of matter [32, pp. 570, 617; 69, pp. 714–715; 112]. The exact mathematical definition of this property of space–time is given in [31, pp. 134–136] and, as noted there, it implies the existence of a six-parameter group of isometries of space–time onto itself.

A dissident sect in the BBC chapel challenges this cosmological principle. More about that in a later section.

The *cosmic background radiation* (CBR), also called the *microwave background radiation*, is an echo or afterglow of the Big Bang dating back to the time of *decoupling* of matter and radiation, some hundreds of thousands of years after the Big Bang.

Inflation theory was invented by MIT physicist Alan Guth in the early 1980s to eliminate till then unnoticed or ignored, contradictions in the theory. The *inflationary scenario* resolves three or four problems involving times when the universe was much less than a second old: the '*flatness*' problem, the '*horizon*' problem, the '*smoothness*' problem and the '*entropy*' problem [15, pp. 178–179; 32, pp. 610–612; 30]. A consequence of inflation is that the density parameter Ω is 1 and, as already noted, space is Euclidean. Without inflation the value 1 of the density parameter is very unstable: it rapidly tends to 0 should it be below 1 by any infinitesimal amount; it tends as rapidly to infinity, should it exceed 1, again by any amount. Inflation stabilizes this value 1: should the parameter be any different from 1 one way or the other, it would be brought back to that value in no time [46, pp. 34, 158, 159, 165; 30].

No data whatsoever exists to corroborate this inflation hypothesis. The notorious astrophysicist P.J.E. Peebles declares: "But inflation is not tested, and it is not easy to see how it could be falsified, so it is not part of the standard model [of BBC]" [80, p. 7].

The long decried cosmological constant Λ (lambda) first introduced by Einstein in 1917 makes a comeback in the 1990's [40,39]; it is needed to resolve the universe age problem arising from the fact that some stars are found to be older than the universe if the constant is nil. Assuming it to be 0 and upholding the cosmological principle, the global topology of space i.e. its infinite or finite character is determined by the density of matter as already noted: if there is enough matter, space is closed and the expansion will reverse itself and lead to a big crunch. Otherwise the expansion will continue forever. However, if lambda is nonzero, as some revisionist BBC supporters are now claiming, no such conclusion can be drawn [31, pp. 137, 139].

If moreover the cosmological principle is downgraded from a global to a local principle, the infinite or finite character of space becomes independent of the density of matter though the fate of the universe is presumably determined as before. But we are anticipating on a later section of this chapter.

All matter was created at the time of the Big Bang. The parameter η (eta), the universal ratio of nucleons to photons, has been restricted to a narrow range on the basis of the *observed abundances* of the light chemical elements in the universe. This accomplishment is held to be one of the three main "proofs" establishing BBC, the other two being the claimed empirically verified Hubble law governing the redshift and the existence of the CBR conveniently interpreted.

1.2. The Problems with BBC

Perhaps the most decisive assault on BBC is the large set of rigorous statistical studies of Irving Ezra Segal and J.F. Nicoll of the *Institute for Defense Analysis* in Alexandria, VA, USA ([90,97,93,98] and references therein) and also of V.S. Troitsky[1] [108,107] based on all available astronomical data which show that the distance-redshift relation is not linear but quadratic. This goes to the root of the problem as the linear law, a direct consequence of the expanding universe hypothesis, represents its principal empirically falsifiable implication. Various attacks on these studies have all been sharply responded to.

The hypothesis of *evolution* of galaxies and quasars, another consequence of BBC, is as well shattered by the same kind of statistical analyses which at the same time establish the chronometric cosmology (CC) of Segal (to be discussed in the next section) and which is exempt of any adjustable parameter. Followers of the *evolution* hypothesis use whole adjustable functions to demonstrate its existence whereas the CC hypothesis of no evolution fits the data just as well, with no parameters entering the CC formulae.

To the question; "Does the observable universe show traces of evolution?", Pecker replies: "... the statistical data are no more than suggestive. They are inconclusive in favour of some evolution, within the range of measured z" [79, p. 474].

The proliferation of adjustable parameters in BBC flies in the face of the canons of genuine scientific knowledge. Aside from the deceleration parameter, now turned into an acceleration parameter, the density parameter, the Hubble constant (which, perhaps as other 'constants' may vary with time), and now the cosmological constant, a whole array of parameters is introduced about the evolution of galaxy and quasar populations since the Big Bang to account for the redshift and luminosity data for distant objects. Other parameters are needed for the calculations of the abundances of light chemical elements.

About this last panoply of parameters, Geoffrey Burbidge and his companions write in [14, p. 39]:

> Supporters of BBC gain for themselves a large bag of free parameters that can subsequently be tuned as the occasion may require. [...] We do not think science should be done that way.

---

[1]V.S. Troitsky died on 5 June 1996. He was from the *Radiophysical Research Institute* N. Novgorod, Russia.

Moreover, the long elusive Hubble constant, the recuperated cosmological constant and the deceleration parameter now become not only unknown but unrestricted unknown functions of time. For instance, Lawrence M. Krauss, an astrophysicist from Case Western University, speaking of the cosmological constant, writes [39, p. 59]: "It might not in fact be constant." Krauss speculates further about the vagaries of the expansion itself ([39, p. 58]): "Perhaps the universe is just now entering a new era of inflation, one that may eventually come to an end."

A theory with so much built-in freedom escapes refutation and has little predictive power if any.

In 1989, the editor of the prestigious weekly science periodical, *Nature*, John Maddox, wrote in an editorial entitled, "Down with the Big Bang," [59]: "Apart from being philosophically unacceptable, the Big-Bang is an over-simple view of how the Universe began, and it is unlikely to survive the decade ahead." (!)

BBC *needlessly* forces our minds to make an 'agonizing reappraisal' of our fundamental ways of thinking. Apart from having to adapt to hard-to-swallow notions such as there being no time nor space before the Big Bang our brains have to renounce asking questions such as "What does space expand into?" which become simply inadmissible [80, p. 6]. Such mental exercises should be entertained only after less dramatic hypotheses have been exhausted.

One problem with inflation is that the observational value of the density parameter is considerably below 1. Hence the *dark matter problem*: where is the *missing mass*? If, on the other hand, one renounces inflation, one has to live with the several problems that it resolves and which plague BBC.

Big Bang cosmology requires that all the matter in the observable universe be created in one single moment occurring 10^{-36} second after the Big Bang itself [15, p. 107]. Yet Einstein's well-known equation expressing the equivalence of matter and energy suggests otherwise (and see §3.2 of this chapter). Other cosmologies assert the possibility of converting energy into matter or are compatible with this.

The CBR is not uniquely indicative of a Big Bang, as claimed by BBC. All the other cosmologies can explain it just as well.

For instance, even before any Big Bang ideas, a theoretical prediction of the existence of such a radiation had already been made in 1953–1954, eleven years before its detection, by E. Findlay-Freundlich and Max Born on the basis of a stationary universe. In his famous 1926 book [21], Sir Arthur Eddington calculated the minimum temperature any body in space would cool to, given that it is immersed in the radiation of distant starlight. With no adjustable parameters, he obtained 3 K, later refined to 2.8 K by E. Regener.

The famous physicist Walther Nernst (1864–1941) who received the Nobel Prize in Chemistry in 1920 for his third law of thermodynamics (1906) attempted to compute the temperature of extragalactic space in 1938. He found the lower value of 0.75 K. The works of Eddington, Regener and Nernst make essential use of Stefan–Boltzmann's law, which is characteristic of a black body radiation as that of the CBR at 2.7 K [12].

It is argued that there is just not enough time in the less than 20 billion years since decoupling, the time of separation of matter and radiation, to form the large-scale structure of the universe, given the measurable maximum speeds of galaxies and the size of the superclusters of galaxies and hence the distance matter would have had to travel [46, pp. 23, 25, 28]. BBC man Edward L. Wright [123] objects to this by arguing that distances were smaller in the past by virtue of expansion and therefore less time was needed.

Standard BBC and some other cosmologies assume homogeneity in the distribution and composition of matter in the universe. About this Pecker has this to say [79, p. 462]: "One of the a priori principles of mathematical cosmology is the homogeneity of the Universe. The evidence is opposite to that principle." French astronomer Gérard de Vaucouleurs (1918–1996) describes the distribution of matter in the universe as hierarchical and fractal rather than homogeneous [79, pp. 412, 463, 464]. Pecker writes moreover:

> "Actually, the larger the volume in which density is measured, the smaller is that density. [...] Hence the usual value of this density has to be looked at with the utmost care and suspicion" [79, pp. 463, 465].

An abundant literature does exist on the fractal structure of the distribution of matter. Some writers, while recognising this, nevertheless defend its large-scale smoothness [120].

Just as Einstein did in his original 1917 paper, BBC completely ignores electromagnetism in its *fluid of galaxies* idealisation of the content of the universe. For instance, one reads in the textbook [69, p. 712]: "The magnetic fields [...] are unimportant for large-scale cosmology, except perhaps very near the 'Big Bang beginning' of the universe." In [31, p. 70], some attention is paid to electric charge which is later left out of the picture [31, p. 136], where the energy-momentum tensor to be used is that of a fluid defined solely by two functions of time: the energy density and the pressure of this gas of galaxies.

Conveying what would seem to be the majority view amongst BBC advocates, Jeremiah Ostriker of Princeton University comments [46, p. 53]: "There is no observational evidence that I know of that indicates electric and magnetic forces are important on cosmological scales."

Segal declares his considerable skepticism about the extrapolation of General Relativity from a theory of gravitation valid on the galactic scale to one on which the dynamics of the entire universe must be based in [95, p. 189], where he adds

> Probably still less justified physically is the application of general relativistic hydrodynamics to extragalactic questions such as the mass density and the stability of the entire Cosmos. The approximation of the distribution of galaxies by a fluid is quite uncontrolled and open-ended; at best, conclusions drawn in this way are merely suggestive.

BBC devotees themselves are not so sure about their calculations of the *primordial abundances* of chemical light elements. For instance, Gary Steigman of Ohio State University says [105, p. 312]:

> Abundances are not *observed*. Abundances are *derived* from the observational data, often following a long and tortuous path involving theory. [...] Errors (or uncertainties), often systematic, may be introduced at many steps in the overall process of deriving abundances from observational data. Furthermore we are here concerned with primordial abundances. Even if present day universal abundances were known to arbitrary accuracy (which they are not!), we still would have to employ theory and observation to extrapolate back to obtain primordial (or at least pregalactic) abundances. Additional errors (uncertainties) are surely introduced here too.

Now listen to N. Yu Gnedin and Jeremiah Ostriker, who are also Big Bang supporters [27]:

Light element nucleosynthesis has been a central pillar supporting the standard FLRW hot Big Bang cosmological model. [...] But there are several confusing and apparently inconsistent elements in the canonical picture which have led to 'patches' which are quite ad hoc and are accepted only because of our familiarity with them and our basic belief that the underlying standard model is accurate.

Readers are referred to the website [123] of astronomer Edward L. Wright from UCLA, a steadfast defender of BBC, to read contrary views on several of the above criticisms.

2. Einstein Static Spherical Universe Revisited: The Chronometric Cosmology of Mathematician Irving Ezra Segal

2.1. The Causality Relation

Irving Ezra Segal, who died on August 30, 1998 at age 79, was an MIT professor of mathematics who thought that the universe is exactly as Albert Einstein had first suggested in 1917: an eternal static three-dimensional sphere. His *chronometric cosmology* takes Special Relativity (SR) for granted but is largely unrelated to General Relativity (GR) about which Segal expresses reservations as a foundation for cosmology.[2]

Space–time defined as the totality of all events – past, present and future – is, before anything else, a partially ordered set: the relation $p < q$ between two events means that p precedes q. This relation, known as the relation of *causality*, of *temporal precedence*, or of *anteriority*, is the most immediate observational data. It conceptually and psychologically precedes the measurement of distances and duration, and is independent of any observer.

This innocuous observation is indeed the starting point of the cosmology of mathematician Segal which he calls *chronometric cosmology* (CC). The FLRW space–times are all endowed with such a causality relation, which derives from their time-oriented Lorentzian metrics. Indeed, such a metric defines at each point, i.e. each event p, a future cone in the tangent space at that point. An event p *temporally precedes* an event q if there exists an oriented curve going from p to q at every point of which the forward pointing tangent belongs to the future cone at that point. Such a relation is well known in Minkowski space–time and determines its ordinary Lorentzian metric to within a constant strictly positive factor. This latter fact is a nontrivial but fundamental theorem established in 1953 by the Russian mathematicians A.D. Alexandrov and V.V. Ovchinnikova and rediscovered a decade later by E.C. Zeeman. It entails that causality preserving maps between FLRW spaces are the same as conformal maps.[3]

Minkowski space–time (M) and the Einstein universe (EU) are two such FLRW space–times. The latter has the distinguishing property that all others can be imbedded into it by essentially unique causality-preserving maps, though the metric of time or space, and the factorisation of space–time into time and space, may not be preserved by such imbeddings. In particular, topologically, space in the imbedded space–time may or may not be compact. This property bestows on EU the name of *universal cosmos*. To a

---

[2]Reference [28] lists all Irving Ezra Segal's publications.
[3]A clear and detailed exposition of it appears in Gregory L. Naber's excellent book [72, pp. 64–74].

mathematician's mind, this fact alone gives EU a very special status and calls it to his or her attention. But admittedly, this may leave most astrophysicists indifferent.

These two models M and EU are related to each other somehow just as the complex plane is related to the Riemann sphere. On the one hand M is causally embedded into EU by a relativistic generalization of stereographic projection, and on the other hand it is tangent to EU at the point of observation, as is the case for any FLRW space–time, or, more generally, for any Lorentzian manifold. We will come back to this point.

Amongst several other distinguishing properties of EU, probably more physically relevant, one has to do with Maxwell's equations. These equations governing light and, more generally, all electromagnetic radiation, which is basically all that is observable in large scale astronomy, are defined primarily in Minkowski space–time M but they extend uniquely, as well as their solutions of finite Minkowskian energy, to the larger universe EU into which M is causally immersed. This mathematical fact designates EU as the proper arena for the interpretation of electromagnetic phenomena.

To our knowledge these properties of EU were unfamiliar to Einstein himself who arrived at EU by an altogether different route related to his *General Relativity* (GR) which, according to some [26, pp. 7, 373, 393, 395] is a misnomer for what should be known only as (Einstein's) *Theory of Gravitation*.

In Segal's opinion, the Einstein universe and Minkowski space–time are the only two space–times satisfying general conditions embodying three fundamental physical principles: the isotropy and global homogeneity of space, i.e. the absence of a preferred direction at any point of space and the absence of a preferred point in space; second, the "principle of inertia", i.e. the statement that there is no preferred timelike direction, which means the equivalence between observers in relative motion at the same point; and third, the possibility to *globally* factor space–time as *time × space* and *temporal homogeneity* with respect to this factorization. This is to be understood as requiring not only the absence of a preferred moment on the time axis, but also that time translations with respect to this factorization make up a group of conformal or causal automorphisms of space–time, the *temporal group* belonging to this factorization. The infinitesimal generator of this group is conventionally identified with the energy, thus giving rise to energy conservation laws both in Minkowski and Einstein space–times.[4]

Alexander Levichev from Boston University and the Sobolev Institute of Mathematics in Novosibirsk, a foremost defender of chronometry, calls CC "the crowning accomplishment of special relativity" in the opening paragraph of a paper [50] which remains the best concise introduction to the mathematics of CC.

2.2. The Main Tenets of Chronometric Cosmology

In this theory, the universe is not expanding and is eternal in both directions.

Space is a three-dimensional sphere of fixed radius.[5]

---

[4]Segal repeatedly claims (beginning in [102, pp. 53, 58]) that de Sitter space–time fails to satisfy the third condition. Though de Sitter space–time is homeomorphic to the product of the real line with a three-sphere it is not *globally* factorizable as *time × space* in the above sense. Thereby it admits, in Segal's view, no natural definition of energy. This is in spite of the fact that its isometry group is of dimension 10 (which is the dimension of the Poincaré group and hence the maximal number compatible with general relativity) enabling that many conservation laws in the ordinary sense.

[5]The reader is referred to Chapter 6 for a discussion of the three-dimensional sphere also known as the hypersphere.

The theory predicts the redshift phenomenon in EU but with a quadratic law for small distances instead of the linear Hubble law.

CC is a purely kinematical theory; it offers no specific dynamics of the universe though it is essentially compatible with ideas of some other cosmologies in that respect.

For instance, the treatment of nucleosynthesis is similar in CC and in BBC to first order:

> The difference is only that a stochastic sequence of mini Bangs, associated with, e.g.: the formation of galaxy clusters, replaces the unique Big Bang. Cluster formation would be expected to be accompanied by extremely high temperatures, which, as in BBC, would be productive of light elements [103, p. 323].

The idea of a series of '*minibangs*' replacing the single Big Bang is also put forward by Narlikar [74, p. 29] and is also found in the plasma universe [46, p. 217] as well as in Arp's cosmology.

However, CC has no need for dark matter, inflation or other scenarios found in BBC. The matter density in the universe which in BBC plays a major role in determining the shape of space is irrelevant in CC.

Homogeneity and isotropy in CC are postulated only for an empty space. No homogeneous or isotropic distribution of matter is ever asserted. For the phenomenological justification of CC, *luminosity uniformity* (LU) but no *spatial uniformity* (SU) in the distribution of matter in space is postulated. LU states that the intrinsic luminosity of objects is statistically independent of the distance, i.e. of the redshift. This means the absence of evolution of the statistical characteristic of the population of galaxies. This is contrary to the case in BBC where one talks of largely unknown statistical evolution since the Big Bang. Of course individual objects do evolve.

In everyday experience, Euclidean geometry is used. Euclidean space is homogeneous and isotropic. No one claims that these geometrical properties belong to the distribution of matter in space.

Ever since the advent of GR, a widely held belief is that the global geometry and topology of space, including its finiteness or infiniteness, is determined by its energetic content, i.e. the matter and the energy it contains. In CC, the energetic content determines at best only a preferred factorisation of space–time as a product of '*space*' with '*time*' which in turn determines a Lorentzian metric on space–time. Unlike what is the case in Minkowski space–time where there is a unique Lorentzian metric, in EU, there is one for each such factorisation. However, time is eternal and space is a three-sphere of a fixed radius which is the same radius in all such factorisations. But, as is the case in ordinary special relativity, CC pays no attention to matter to begin with, and the Einstein universe makes its appearance quite independently of considerations about gravity contrary to what was the case in Einstein seminal cosmological 1917 paper [22].

As to the cosmic background radiation, Segal writes:

> The observed blackbody form of the cosmic microwave background is simply the most likely disposition of remnants of light on a purely random basis, assuming the classic principle of the conservation of energy, and is not at all uniquely indicative of a Big Bang.

As Jean-Claude Pecker says in [79, p. 511], similar views are often expressed in an infinite-eternal universe which makes the CBR result from an equilibrium between photons and the rest. For instance, Burbidge, Hoyle and Narlikar as well as Arp have comparable explanations of the CBR.

According to CC, the redshift–distance relation is quadratic rather than linear for small redshifts and the theory specifies a formula valid for all distances, as we shall see next.

2.3. The Chronometric Redshift Theory

According to CC, Einstein's model EU is the correct one to understand the universe as a whole, except that there are two kinds of time: a cosmic (or Einstein) time t, and a local (or Minkowski) time x_0.

This two-time situation arises from the essentially uniquely defined conformal immersion of Minkowski space $M = R \times R^3$ (the Cartesian product of a time axis with three-dimensional Euclidean space) into $EU = R \times S^3$ (the Cartesian product of a time axis with a three-sphere). This immersion is a relativistic variant of stereographic projection. Minkowski space M can be thought of as being tangent to EU just as the complex plane is tangent to the Riemann sphere. This immersion of M into EU preserves causality but does not preserve the time coordinate nor the space coordinate in the factorizations of these space–times as a Cartesian product of 'time' with 'space'.

In Segal's words,

> The key point is that time and its conjugate variable, energy, are fundamentally different in the *EU* from the conventional time and energy in the local flat Minkowski space *M* that approximates the *EU* at the point of observation.

Simply put, Einstein's cosmic time t is the "real" one, whereas Minkowski's time is only an approximation of t. Which time and space coordinates, those of EU or those of M, are actually measured in observations is empirically immaterial, since the two differ by unobservably small amounts except for extragalactic observations.

Using appropriate units such that the speed of light $c = 1$ and denoting by r the radius of the three-sphere, Einstein's and Minkowski's time coordinates are related by the equation

$$x_0 = 2r \tan\left(\frac{t}{2r}\right) \tag{1}$$

which may be called the chronometric two-times formula, from which the relation

$$z = \tan^2\left(\frac{t}{2r}\right) \tag{2}$$

of an observed redshift z to time of propagation t, or equivalently, geodesic distance on the 3-sphere, may be derived essentially by simple differentiation.

Assuming formula (1), we derive formula (2). For a wave of frequency v and of observed redshifted smaller frequency v', the redshift z is defined by the quotient $(v - v')/v'$. Letting dt stand for a small interval of cosmic time, dx_0 for the corresponding small interval of Minkowskian time and f for the number of oscillations during that duration, we have, using formula (1), that

$$\frac{v}{v'} = \frac{f/dt}{f/dx_0} = \frac{dx_0}{dt},$$

$$z = \frac{v - v''}{v'} = \frac{v}{v'} - 1 = \frac{dx_0}{dt} - 1 = \sec^2\left(\frac{t}{2r}\right) - 1 = \tan^2\left(\frac{t}{2r}\right).$$

Equation (2), called the *chronometric redshift formula*, may also be derived [102,103] in (and above) the observable frequency range by a rigorous analysis based on Maxwell's equations. It follows from this analysis that a free photon will experience a redshift when propagated over a very long period according to Einstein time.

Equation (1) implies that as t varies from $-\pi r$ to $+\pi r$, ordinary time goes from minus infinity to plus infinity. Hence, eternity (past and future) in the ordinary sense corresponds to a finite interval of cosmic time, which cosmic time t, nevertheless, varies over the whole real line. As for Eq. (2), it reveals that for small values of t (or, equivalently, of the distance), the redshift varies as the square of t, in contradiction with Hubble's law, which is linear. From (2) we also see that as r tends to infinity, z tends to 0. Hence, as envisaged as a possibility by Hubble himself, the curvature of *space* is the reason for the cosmic redshift in CC.

If the choice of units is completed in such a way that the fixed radius of space r is 1, in addition to the choice $c = 1$ made earlier, which makes t equal to the distance d, then the chronometric redshift–distance relation (2) is seen to be entirely parameter-free and is thus quite vulnerable. This is quite unlike the similar relation in BBC which involves the deceleration parameter, the cosmological constant, the curvature parameter and the present radius of space as parameters.[6]

As Lerner points out at the very end of his book [46] the importance, for cosmology, and for physics more generally, of having the correct relation between the extragalactic redshift and the distance cannot be overemphasized.

2.4. Elementary Particles in Chronometric Theory

Segal's chronometric theory not only covers cosmology, it also includes a successful theory of fundamental particles based on the Einstein universe instead of on the Minkowski space–time. As a matter of fact, the chronometric theory was originally developed for elementary particle applications.

In the chronometric elementary particle theory, each particle, including the photon, has an unobservably small, but theoretically very important, "*bare*" mass, in addition to a considerably more substantial empirically observed "*clothed*" or *gravitational*, or else *Machian*, mass deriving from its interactions throughout the universe. The two together make up the *inertial mass* [91, p. 853; 101, p. 175]. This is not unlike the Narlikar–Arp *variable mass hypothesis* to be mentioned in a subsequent section.

In a forthcoming paper [49], Levichev writes:

> In Segal's chronometry, the entire list of known particles is derived mathematically. One chronometric particle (the "exon") has not yet been experimentally identified [94].

2.5. Observational Tests of Chronometric Cosmology

Using a sophisticated bootstrap statistical technique that they call ROBUST, which takes into account the so-called "observational cutoff bias" making faraway celestial objects less likely to be observed than closer ones since they are apparently less luminous and hence harder to 'see', Segal and collaborators have demonstrated in several articles that the quadratic redshift–distance law predicted by CC fits all available experimental data

---

[6]See, for instance, Eq. (29.16), p. 781 coupled with (29.2), p. 772 in [69].

very well whereas the Hubble law fails miserably on every count. The more general CC redshift–distance formula valid for all distances is also experimentally confirmed without appealing to an hypothetical largely unknown evolution as in BBC.

In his address to the Colloquium on Physical Cosmology, sponsored by the National Academy of Sciences of the USA in March 1992, Segal stated:

> The good news is that there is a simple redshift–distance relation that appears consistent with observations in complete objectively defined samples in the infrared and X-ray wave bands, as well as the optical. The bad news is that at low redshifts it doesn't at all resemble the Hubble Law, which appears simply irreconcilable with these observations.

Quite independently of Segal, the Russian author, V.S. Troitsky, has established a quadratic redshift–distance law in a 1996 paper [107, pp. 94, 105] on the basis of a statistical analysis of a very considerable sample of more than 73197 galaxies.

2.6. About Supernovae, 'Time Dilation' and 'Dark Energy'

In 1996, acknowledging that widely accepted experimental proof of the universal expansion hypothesis was lacking, B. Leibundgut *et al.* claim [45] that the observation of a particular supernova provides strong evidence for this assumption and is incompatible with a static universe or with tired-light theories. The evidence, held to be a "clear vindication of an expanding universe", consists in observing that the light curve of this distant supernova is stretched by a factor of $1 + z$ as, they say in their abstract, is "prescribed by cosmological expansion". This means that the rise and fall of the light intensity of that distant supernova is over a longer time span than similar events in nearby supernovae. A year later this interpretation of the observation has been challenged both by Segal on the one hand and by Narlikar and Arp on the other from the standpoints of their respective theories.

Segal [90][7] shows that the time dilation factor $1 + z$ is also *prescribed* in the context of chronometric cosmology. This dilation effect in CC is essentially obtained by simple differentiation of our formula (1) which expresses Minkowskian time in terms of cosmic time, or equivalently the cosmic distance ρ on the 3-sphere of radius r as we assume the velocity of light c to be 1, while taking our formula (2) into account. Indeed, one readily obtains

$$\frac{\partial x_0}{\partial t} = \sec^2\left(\frac{t}{2r}\right) = 1 + \tan^2\left(\frac{t}{2r}\right) = 1 + z. \tag{2.1}$$

In the same issue of the *Astrophysical Journal*, Narlikar and Arp [75] claim to establish the same dilation factor $1 + z$ in the context of their "variable mass hypothesis".

In 2001, Leibundgut acknowledges these contradictory views. He writes on p. 79 of [44]: "The SN [= supernova] result, however, has also been interpreted in other, nonexpanding cosmologies [75,90]." And on p. 89:

> In a quasi-steady state cosmology, a combination of acceleration due to the formation of matter and the particular, gray dust proposed by Aguirre (1999b) can reproduce the SN observa-

---

[7]A misprint occurs in (the first part of) formula (2) of [90]: "$\cos t$" there should be replaced by "1" as in formula (3) of the same paper. So corrected, that formula (2) reads $x_0 = 2\sin t/(1 + \cos t)$ and becomes equivalent to our formula (1) with $r = 1 = c$ and $t = \rho$.

tions (Banerjee et al. 2000). Because this world model has an oscillating scale parameter, the SN result would not be extraordinary in such a framework.

The observed '*time dilation*' phenomenon concerning distant supernovae is the root of the *accelerating universe hypothesis* [44, p. 67].

The belief in an accelerating universe, often attributed to some 'dark energy' related to a nonzero cosmological constant, stems from the discovery that distant supernovae are less luminous than they are expected to be on the basis of their measured redshifts as interpreted in the framework of BBC.

Robert P. Kirshner from the Harvard-Smithsonian Center for Astrophysics, in Cambridge, USA explains it thus:

> If the universe had been decelerating – in the way it would if it contained the closure density of matter, that is, if $\Omega_m = 1$ – then the light emitted at redshift $z = 0.5$ by a SN Ia would not have travelled as far, compared with a situation where the universe had been coasting at a constant rate – characteristic of an empty universe, where $\Omega_m = 0$. For a universe with $\Omega_m = 1$, the flux from the distant supernova therefore would be about 25% brighter. But the distant supernovae are not brighter than expected in a coasting universe, they are dimmer. For this to happen, the universe must be accelerating while the light from the supernova is in transit to our observatories [37, p. 4226].

Adam G. Riess from the Space Telescope Science Institute in Baltimore, puts it somewhat differently:

> A more illuminating way to quantify the evidence for an accelerating universe is to consider how the SN Ia distances depart from decelerating or 'coasting' models. The average high-redshift SN Ia is 0.19 mag dimmer or about 10% farther than expected for a universe with no cosmological constant and negligible matter [85, p. 1287].

More recently the findings of the Wilkinson Microwave Anisotropy Probe (WMAP) [136] (a NASA satellite whose mission is measuring the temperature of the CBR over the full sky) are thought to confirm the existence of this 'dark energy' which together with 'hot' or 'cold dark matter' would make up most of the material in the universe. The influential *Science* magazine pompously saluted this discovery as the Number 1 breakthrough of the year in December 2003 [104].

However, several articles such as '*Things fall apart*' [121], cite the work of a number of unconvinced cosmologists who busy themselves pouring cold water on hot or cold dark matter and predicting a dark future for dark energy.

It is not suggested that the '*accelerating universe*' implies the speed of expansion has always been increasing. For instance Riess et al write in [84, Sect. 1.1]:

> If the cosmological acceleration inferred from SNe Ia is real, it commenced rather recently, at $0.5 < z < 1$. Beyond these redshifts, the universe was more compact and the attraction of matter dominated the repulsion of dark energy. At $z > 1$, the expansion of the universe should have been decelerating.

An accelerating universe means that the Hubble 'constant' is in fact a decreasing function of the distance at least for the values of z in the interval $0.5 < z < 1$. This would seem to be in contradiction with chronometric cosmology as one easily computes, by differentiating formula (2) with respect to t, that cz, which may be considered as the *speed of expansion* in CC, is an increasing function of t (which is the same as the distance with $c = 1$).

In any case, the following passage from a 1997 paper by Segal [90, pp. 70–71] leaves little doubt as to what he thought, and presumably would still think, of the observation of supernovae as a basis for establishing Hubble's law:

Today there is a new wave of claims for the validation of the Hubble law, on the basis of observations of another quite non-generic type of object, namely supernovae. Bold, if not somewhat disingenuous, claims for "measurement" of the distances to supernovae are made, notwithstanding that the crucial difficulty in extragalactic astronomy is that the distance to a source can never be measured in a truly model-independent way. The directly observable content of the Hubble law in no way involves putative distances, but is rather to the effect that the apparent magnitude m and the redshift z are related by the equation $m = 5 \log z + M$, where M is a random variable that is independent of z. The 'distances' of supernovae are, like the 'standard candle' character of the Bright Cluster Galaxies, theorised rather than observed. Because of their transience, irregularity, scarcity and difficulty of classification into appropriate types, the use of supernovae as primary sample objects for cosmological testing would probably serve to moot the redshift–distance relation indefinitely.

The next year, Segal wrote in a joint paper with Nicoll submitted shortly before his sudden death:

More recently, the linear law has been derived from observations on large-redshift supernovae of type Ia. It is now acknowledged that the peak luminosities of these objects have a substantial dispersion and that an apparent linear law requires extensive model-dependent corrections [98, p. 510].

This last declaration is well supported in 2001 by Leibundgut, a staunch advocate of the accelerating universe hypothesis:

The observational characteristics of nearby Sne Ia show some differences from event to event. Despite their considerable range in observed peak luminosity, they can be normalized by their light-curve shape. Through this normalization, SNe Ia can be used as exquisite distance indicators. [...] SNe Ia have long been proposed as good distance indicators for cosmology, first through their standard candle character, i.e., identical peak luminosity, and later normalized by corrections from light-curve shapes. [...] Different methods for the light-curve shape corrections, however, do not compare well with each other; significant differences in the implementations of the corrections are found [44, pp. 67–69].

2.7. Phantom Galaxies?

Could it be that the night sky were a family album of the living and the dead celestial objects, each of them being depicted a large number of times? There would then be far fewer objects than there appears to be.

As we will see later, this view is now defended by a small group of Big Bang supporters. It would seem that this possibility arises in the context of chronometric cosmology as well.

Already in 1920, Hermann Weyl wrote [116, p. 278]:

If the world is closed, spatially, it becomes possible for an observer to see several pictures of one and the same star. These depict the star at epochs separated by enormous intervals of time (during which light travels once entirely round the world).

In 1974, Segal wrote:

In view of the apparent transparency of intergalactic space, the residual radiation should typically make many circuits of space before being ultimately absorbed by matter.

The following excerpt from a 1995 paper by Segal and Zhou seems to imply that
this theoretical possibility is in fact a prediction of CC which nevertheless has not been
explicitly stated so far. Indeed in the concluding paragraph of [103] one reads:

Finally, the transparency of cosmic space implies that photons in the Einstein universe EU
will typically make many circuits of space (i.e. of the 3-sphere) before being absorbed or
undergoing interaction. A free photon will be infinitely redshifted at the antipode of S^3 to its
point P of emission, but on returning to P it will be in its original state, as a consequence of
the periodicity of free photon wave function in EU.

2.8. Objections to CC

On the whole, CC has been ignored or despised by mainstream cosmologists and by
some others as well.

But, disregarding the often inappropriate rhetoric on more than one side, I do not
know of any objection to CC, from anybody, that has not been answered in print, to my
mind satisfactorily, by Segal except for a somewhat astonishing but minor blemish of
little consequence found in one of his numerous papers [19].

We have provided in [18] an answer to a 1997 critical study of CC on the basis of
a sharp and extensive rebuttal found in an unpublished manuscript submitted by Segal
and his collaborator Jeff Nicoll, but which was rejected by the *Astrophysical Journal* in
1998, shortly before Segal's death.

It is sometimes argued that CC is a purely kinematical theory that says nothing of
the dynamics of the universe. True, but this only shows that the theory is incomplete and
not that it is wrong. It may incorporate, for instance, the essential elements of the *Plasma
Universe* to be discussed in a later section.

The most serious challenge still facing CC, shared by other cosmologies, is posed
by Arp's discordant redshifts to be discussed in the next section.

3. Is the Redshift a Distance Indicator? The Anomalous Redshifts of Halton Arp

3.1. Arp's Enigmatic Observational Discoveries

A distinguished observational astronomer, Halton Arp was for 29 years a staff member
of the observatories known originally as the Mt. Wilson and Palomar Observatories. He
is currently at the Max-Planck-Institute in Munich. His 1998 book *Seeing Red* [11] and
also his 1987 book *Quasars, Redshifts and Controversies* [10] expound the details of his
findings and tell the vicissitudes of his scientific life and publications. His most recent
book is a *Catalogue of Discordant Redshift Associations* [8].

The most important thing in what Arp contends is that the redshift cannot be a distance indicator as he lists numerous "discordant redshifts", also called *anomalous redshifts*, consisting of some celestial objects having very different redshifts which must
nevertheless be at the same distance from us since Arp argues that they are obviously in
interaction.

Arp's observations have generally been disregarded or denied by mainstream cosmologists.

In Arp's view, there is no expansion of the universe, extragalactic redshifts are interpreted predominantly as an intrinsic property of matter measuring its age, i.e. the time elapsed since its creation.

Arp shares with Jayant V. Narlikar [73] a belief in the *variable mass hypothesis*, which asserts that all particle masses uniformly scale with epoch, and which he claims "fits the data better than the Big Bang mantra". He sees in it the reason why the nonexpanding universe does not collapse [11, pp. 231–233].

According to Arp, there is continuous creation of matter; meaning the transformation of previously existing mass–energy. This is suggested by the many ejections of quasars from small dense nuclei of active galaxies that we can see. The quantized properties of the redshifts are linked with the properties of young matter as it is expelled from galaxies by old dense matter [11, p. 228; 79, p. 508].

The younger the object, the larger the redshift [11, p. 77]. Arp suggests that perhaps younger matter emits weaker photons [32, p. 467]. Matter is born with high redshift and *zero* mass. The former decreases and the latter increases (the *variable mass* hypothesis) with the passage of time as the new-born particles interact with an ever larger portion of the universe [11, pp. 108, 238].

These ejected quasars typically move outward with speeds of from a few tenths of c to the speed of light. They move out in pairs in opposite directions along the rotation axis of the parent galaxy. "Opposite ejection of extragalactic material is a ubiquitous process that operates on all scales". But where Arp sees ejections, others see collisions and mergers [11, pp. 245, 191, 71, 72].

In Arp's view, quasars are low luminosity recently created objects and can be quite nearby. This is the opposite of what they are in BBC: far away very luminous objects having existed in the young universe [11, p. 190].

According to Arp, the CBR is formed in the static intergalactic space and has been in equilibrium for billions and billions of years [79, p. 509]. Contrary to what he calls the *establishment astronomy*, he sees in the CBR, and particularly in its long hoped for ripples finally discovered in 1992, the proof that the universe is not expanding [11, pp. 236–238].

As quantitative proof of the dependence of redshift on age rather than on distance, he cites the observation that companion galaxies of the largest galaxy in a group of galaxies have systematically larger redshifts than this dominant central galaxy which they are orbiting and which engendered them [11, pp. 62–64]. In the customary interpretation of the redshift, the distribution of the redshifts of the companions should be approximately evenly distributed about that of the dominant one as some should have an approaching velocity and others a receding one.

In particular, the Andromeda galaxy is the dominant galaxy, the parent galaxy, of the Local Group of which our Milky Way is a member [11, pp. 62, 69]. The companion galaxies are the end point of the evolution of quasars born of the parent galaxy [11, p. 84]. As evidence for the fact that the companion galaxies are the result of ejections from the central galaxy, Arp calls attention to the fact that the satellite galaxies are preferentially along the axis of rotation of the dominant galaxy.

The Magellanic Clouds are also members of the Local Group and would appear to be the Milky Way's younger offspring [11, p. 95]. This statement as well as the preceding one involving the Andromeda galaxy go against conventional wisdom and illustrate the fact that where Arp sees ejections, others may see collisions.

Arp has no taste for, or high opinion of, general relativity and, in particular, for curved space or curved space–time [11, pp. 254–255].

3.2. *"Let There Be Matter"*

Arp's contention, also shared by his friends Hoyle, Burbidge and Narlikar (see below), that matter is continually being 'created' is receiving support from recent experimental results.

"Let there be matter". This phrase, reminiscent of Genesis, is from Jeffrey Winters, (*Discover* magazine, December 1997) who writes:

> But like any equation, $E = mc^2$ works in both directions, at least theoretically. That is, it should be possible to convert energy into matter. Now a team of physicists has accomplished just that: they have transmuted light into matter. 'We're able to turn optical photons into matter', says Princeton physicist Kirk McDonald, co-leader of the team. 'That is quite a technological leap'.
>
> Until now, no one had directly created matter from light. 'Back in 1934 physicists realized that it would be possible to do this in principle' says McDonald, 'but it just wasn't technically feasible'.

The experiment is reported in *Physical Review Letters* (vol. 79, p. 1626). It consists in injecting a beam of very high energy electrons from a linear accelerator into an extremely tightly packed photon beam, slamming some photons backwards into others. The collisions created electrons and their antimatter siblings – positrons.

One may also consult Kirk McDonald's website [133].

Alexander Levichev says that this result is "not a surprise for a mathematician. From the representation theory viewpoint (namely, in its chronometric version) a photon is an electron–positron bond".

3.3. *Confirming Arp's Abnormal Redshifts and Attempts at Explaining them*

3.3.1. *Sympathetic Opinions of HBN and Pecker*

We often use the acronym HBN to designate the three scientists Fred Hoyle, Geoffrey Burbidge and Jayant V. Narlikar to whom we devote a later section.

One of the main arguments held against Arp's conclusion was the discovery of gravitational lensing, which allows us to see several images of a single quasar in the vicinity of a perturbing but much closer galaxy [79, pp. 467, 470].

But Arp denies this [11, pp. 169–171] and HBN also argue lengthily against this possibility beginning on p. 147 of [15]. Pecker shares this opinion; he writes:

> [...] there are gravitational lensing effects, and images of real quasars given by this mechanism. There are, however, other quasars which cannot be accounted for by any effect of that sort. [...] We feel that, at the present time, the evidence for abnormal redshifts, whatever their cause, is truly convincing [79, pp. 470–472].

G. Burbidge and others have conducted broad statistical studies in 1971 and again in 1989 [15, pp. 122–124] which claim to have demonstrated the existence of Arp's associations. Other studies, notably by A. Webster in 1982, concluded otherwise. On the basis of the computer search of 1989 based on the catalogue of all quasars known in 1987 (numbering about 3000), HBN assert "without any possibility of doubt" that the overwhelming majority of the 400 pairs consisting of a quasar and a galaxy separated by less than 10 minutes of arc are physically associated and hence that the quasar in each case is at the same distance as the associated galaxy, requiring the large observed redshift of the quasar to have a dominant intrinsic property.

HBN believe that the effect of anomalous redshifts must be rare for otherwise if it were widespread in galaxies as they apparently are for quasars, there would be a large scatter in the Hubble diagram and no Hubble relation would be found for normal galaxies [15, p. 327].

3.3.2. Emil Wolf's New Optical Redshift Mechanism

An important discovery made in 1986 (see, for instance, [117] and [118]) that its finder calls *correlation-induced spectral changes* might explain Arp's discordant redshifts. It was made by Emil Wolf, professor of optical physics at the University of Rochester, and appears to have been generally ignored or incorrectly explained. Wolf is coauthor with Max Born of the monumental work on optics [13]. According to Wolf's theory, in some well defined circumstances one may "generate shifts of spectral lines which are indistinguishable from those that would be produced by the Doppler effect" [118, p. 48]. These theoretical predictions were subsequently verified by experiments conducted by two of Wolf's colleagues, G.M. Morris and D. Faklis [70,71].

Wolf writes:

> [...] contrary to the usual claims there is a mechanism, rooted in statistical optics, which may give rise to Doppler-like shift of spectral lines, even though the source and the observer may be at rest relative to each other. Whilst this mechanism does not necessarily challenge the Big Bang theory it may resolve a long standing controversy relating to pairs of astronomical objects (e.g. certain galaxies and quasars) which have very different redshifts and yet appear to be connected.

He adds:

> It has also been shown that scattering on a fluctuating medium whose correlation function is strongly anisotropic may generate shifts of spectral lines which are indistinguishable from those that would be produced by the Doppler effect [118, pp. 41, 48].

> The shifts may be arbitrarily large [36].

> Red shifts as well as blue shifts can be produced by this mechanism, depending on the scattering geometry [35].

In his *Plasma cosmology* article of 1992 [82], Anthony L. Peratt gives an instructive description of the Wolf mechanism:

> A mechanical analogue of Wolf's discovery is a pair of tuning forks with nearly identical resonant frequencies (pitches). If these forks are connected together by, say, a sounding board, the coupling is strong and the resonant frequencies tend to get "dragged down" to lower ones. In other words, the wavelength is lengthened or redshifted. This phenomenon has been verified experimentally with light waves and for sound waves for coupled speakers. [...] This mechanism can be extended from the case of two radiating point sources to that of a whole collection of such objects, for example a plasma cloud. Wolf and his colleagues have shown that such a cloud can produce shifts that closely mimic the Doppler effect.

3.3.3. Are They Compatible with Chronometric Cosmology?

In [99, p. 2951], Segal and his co-writers comment on Arp's anomalous redshifts and the quantization of redshifts from the viewpoint of their chronometric cosmology as follows:

> In connection with phenomenological indications for possible discordant redshifts and the quantization of redshifts, it should be noted that the chronometric redshift $z = \tan^2(t/2)$ is a function of the *duration* t of the time interval between the emission of the photons and its absorption in the process of observation. In particular, a photon in a localized *photon trap*, if such exist, would experience the redshift given by this equation while traversing a negligible distance. The effect of temporary trapping of a source at distance r would modify the redshift–distance relation in accordance with the equation $z = \tan^2[(r + \delta t)/2]$, where δt is the time spent in the trap. In chronometric theory, the difference between the global energy of a photon and its locally observable linear component (i.e. its Minkowskian energy) is essentially gravitational, from a general theoretical standpoint, representing an attractive force that scales like the Newtonian potential, etc. The trapping of photons would thus be analogous to the gravitational binding of massive particles, and it would not be surprising if such traps exist in regions of extreme physical conditions (J.A. Wheeler, *Geometrodynamics*, 1962). By virtue of limitation to such regions, the effect on overall quasar statistics would be marginal and hardly detectable until substantially larger complete samples had been observed; but on occasion, discordant redshifts of the type proposed by Arp would be expected to result.

In a 1992 preprint of a paper that was to appear the next year under the title of "The redshift–distance relation", with the following passage deleted, Segal expresses his view about Arp's anomalous redshifts and the quantization of redshifts thus:

> The existence of discordant redshifts of the type proposed by Arp is subject to objective statistical test by their expected impact on the distribution of dispersion in apparent magnitude in large complete samples. They should appear as an underestimate of this dispersion when this is predicted in random samples by CC. The quantization of redshifts proposed e.g. by Burbidge could result from interfaces between particle and antiparticle regions in CC (in which particle–antiparticle symmetry is a priori natural), and could be similarly tested directly by analyses of observed V/V_m (a spatial uniformity test described in [102, pp. 163, 164]) distributions in complete samples.

4. Challenging the Cosmological Principle: Finite Topological Spaces with Holes

4.1. Mirror Images in the Night Sky?

A small group of astrophysicists and mathematicians has emerged who, while supporting most of BBC, deny the cosmological principle and propose unanticipated shapes for the space we live in. More specifically, the cosmological principle is demoted from the global geometry to the local geometry. Mathematically, local homogeneity means that for any two points of space there exists neighbourhoods that are isometric; while local isotropy means that for any two directions from a point of space, there exists an isometry of some neighbourhood of that point onto itself which leaves that point fixed (i.e. a local isometry) and maps one direction on the other [115, pp. 96, 264, 265; 31, p. 136; 119, p. 381].

This dissent opens the door to the possibility of space being finite irrespective of the density of matter in the universe though this density, as in standard BBC, still determines the constant curvature of space and thereby, presumably, the fate of the universe. Perhaps more spectacularly, this downgrading of the Copernican principle also suggests the possibility that space may be multiply-connected, i.e. not simply connected, the way a doughnut surface is and hence that light originating from far away may reach us via several paths thereby generating several images of the same object, including of our own galaxy, at different epochs.

Just as it is with the three-sphere as a model of space which goes back at least to the mathematician Riemann in 1854, this idea of a multiply-connected universe is not new, since as early as 1890, the mathematician Felix Klein imagined that space might be a 3-torus which is a flat three-dimensional analogue of a doughnut surface (which is not flat). The astrophysicist Karl Schwarzschild [89] briefly mentioned this idea in 1900 [115, p. 280]. The idea that the homogeneity of space might be illusory with the above mentioned consequential hypotheses as to the shape of space and the multiplicity of images of a single object was later pursued, for instance in a 1986 paper [24] by G.F.R. Ellis and G. Schreiber.

A subgroup of this school of thought, now making the headlines [52,25,114] appears convinced that the density of matter-energy marginally favors a positively curved space model, and has found reasons in the analysis of anisotropies in the CBR to announce that the space we live in has the unexpected shape of a small dodecahedron whose pairs of opposite faces are somehow identified and which is known as Poincaré space.

Janna Levin ends the first chapter of her book [51], *How the Universe Got Its Spots: Diary of a Finite Time in a Finite Space*, with a plea for understanding:

> I'll try to tell you my reasons for believing the universe is finite, unpopular as they are in some scientific crowds, and why a few of us find ourselves at odds with the rest of our colleagues.

Others [1] think that the geometry is Euclidean i.e. the curvature is zero and from known geometrical results they can list all such possibilities. There are ten such, of which four are infinite. The simplest of the remaining six is the flat 3-torus obtainable from a cube by identifying or, one says, by abstractly gluing, the opposite faces of all three pairs of parallel faces. A two-dimensional analogue is the flat 2-torus obtainable from a square by identification of its parallel edges. This is topologically the same object as the doughnut-surface torus but geometrically the two must be distinguished as the first is flat and the second is curved. The (topological) 2-torus is a surface with one hole. Of course three-dimensional manifolds are harder to visualize.[8]

The flat 3-torus may be compared to a room all six faces of which would be mirrors. This exemplifies the possibility of a large number of images of a celestial object being visible should space result from the identification of some points at the edge of a fundamental domain such as a cube in the case of the flat 3-torus.

---

[8]The recommended book by Jeffrey R. Weeks *The Shape of Space* [115] and the article by William P. Thurston *How to see 3-manifolds* [106] will be found quite helpful in that respect. One may also read Chapters V and VIII in the excellent book by Richard Osserman *Poetry of the Universe* [78].

Mathematically speaking, such a "small universe", i.e. a typical such compact model of space, is obtainable from the spatial part of any FLRW space–time by suitable identifications under a discrete group of isometries. The original space can be recovered from the small universe as its universal covering space which is simply connected. An observer in the small universe "sees" the universal covering space – his observations will be exactly the same as those of an observer in the covering space with an exactly repeating set of galaxies [24, pp. 98, 99]. This would be the root of the illusion of space being homogeneous.

As experimental corroboration of these ideas, two methods are being used. They have tantalizing names: the *cosmic crystallographic method* and the *circles-in-the-sky method*.

Cosmic crystallography looks at the 3-dimensional observed distribution of high redshift sources (e.g. galaxy clusters, quasars) using catalogues of cosmic objects in order to discover repeating patterns in their distribution, much like the repeating patterns of atoms observed in crystals. This seems to be independent of BBC and could perhaps also be used in the context of chronometric cosmology. This method appears to have yielded few results [43; 115, Sect. 21].

The second method appears to be more promising and it is on its basis that the fantastic dodecahedron announcement was made in October 2003. It relies on an analysis of the irregularities in the Cosmic Background Radiation thought of as an echo of the Big Bang. The temperature fluctuations of the CMB may be decomposed into a sum of spherical harmonics, much like the sound produced by a musical instrument may be decomposed into ordinary harmonics. It is on the basis of an analysis of the CMB temperature fluctuations as measured by the satellite WMAP that the Poincaré space has been selected as a model for our physical space [52].

Poincaré space may be represented by a dodecahedron (a regular polyhedron with 12 pentagonal faces) whose opposite faces are glued after a 36 degree clockwise twist; such a space is positively curved, and is a multiply connected variant of the hypersphere, with a volume 120 times smaller [115, Sect. 16].[9]

Of course, all this rests essentially on some BBC hypotheses and the reliability of these conclusions cannot exceed that of these assumptions! At the time of writing considerable uncertainty still surrounds this analysis of the observed CMB temperature fluctuations [114, pp. 617–618]. WMAP data find that certain features of CMB harmonics align with the ecliptic plane, the plane of the earth's orbit, at 99.9% confidence level, calling into question the presumed cosmic origins of these features and suggesting instead some hitherto unknown solar system contributions to the CMB. It is appropriate here to quote the following excerpt from a letter of I.E. Segal published in *The New York Times* of May 13, 1992 shortly after the discovery of the CMB fluctuations which were desperately hoped for to help take BBC from the hook:

> The marginally observable fluctuations in the cosmic background radiation are likely to be confirmed if only because such fluctuations would be a concomitant of almost any known type of possible physical origin for this radiation. They are not at all uniquely indicative of a Big Bang.

---

[9]One can download free of charge, interactive 3D software to explore this space and other 3-manifolds from Weeks' Topology and Geometry software website [137].

5. The Steady-State Cosmology Resurrected: The Quasi-Steady-State Cosmology of Three Venerable Maverick Stalwarts

5.1. Teachings of the QSSC

Here comes, from three continents, a resolute triumvirate comprising the most well-known dissenters from the conventional wisdom, i.e. BBC: Fred Hoyle, Geoffrey Burbidge and Jayant V. Narlikar.

Sir Fred Hoyle, who died on August 20, 2001, was professor of astronomy at the University of Cambridge, Geoffrey Burbidge is professor of physics at the University of California at San Diego, and Jayant V. Narlikar is at the Inter-University Center for Astronomy and Astrophysics in Pune, India.

The original steady-state cosmological model was developed in 1948 by Bondi and Gold, and Hoyle. This model was based on two interrelated postulates. First, the *perfect cosmological principle* according to which the universe has always and will always look the same to any observer. Second, there is no such thing as a Big Bang and matter is created, emerging spontaneously out of nowhere, at a uniform rate determined by the expansion, rather than being created at time $t = 0$ as in BBC [38, p. 142].

Apparently definitely defeated by BBC in the 1960s, the original steady state cosmology has made a comeback in the 1990s under the name of *quasi-steady state cosmology* (QSSC) under active development since 1993 by its initiators.

The geometry of QSSC is a Friedmann–Lemaître–Robertson–Walker space–time in which the time-dependent scalar factor $R(t)$ is the product of a periodic sinusoidal pulsation function and a slow exponential. This gives rise to a bouncing universe whose *radius* $R(t)$ (no finiteness of space intended by the use of the word *radius* which is a common name for $R(t)$) oscillates between a maximum size and a *nonzero* minimum size, with these extremes and the amplitude slowly increasing with time but with a fixed frequency.[10]

Near an oscillatory maximum, the universe is sufficiently diffuse that light propagation is essentially free. This, together with the long time scale of a maximum phase and the large-scale homogeneity and isotropy in the distribution of galaxies, causes the radiation to also acquire a high degree of homogeneity that persists through subsequent cycles. Thus an explanation of the remarkable uniformity of the CBR is obtained [14, p. 41].

A scalar field analogous to the one that appears in inflationary models of BBC permits new matter to appear in an already existing universe [14, p. 39].

While BBC hypothesises that a typical galaxy centre is occupied by a massive black hole surrounded by an accretion disk which emits only gravitational energy, QSSC teaches that galactic nuclei are near-black holes in which mass and energy are created. The authors assert, as Arp does, that there is overwhelming evidence of quasar ejection from active galactic nuclei. Some quasars, in their view, are quite nearby.

They regard the observed phenomenon of the ejection of condensed objects from excited galaxy nuclei as *prima facie* evidence for galaxy formation [14, p. 43].

The authors reject several assertions of BBC for which they think there is no *primary* observational evidence.

---

[10]The graph of the scale factor is illustrated in [16, p. 732].

Although they concede the existence of some baryonic dark (i.e. nonluminous) matter in galaxies on the basis of the flat rotation curves of spiral galaxies they contend that some clusters of galaxies are not gravitationally bound and that their masses are exaggerated as conventionally computed. They argue that the gravitational merger hypothesis is overemphasised and that the innumerable observational evidences in favour of ejections and explosions of galaxies are almost disregarded [14, pp. 38, 42, 44].

On the basis of our understanding of stellar evolution leading to the conclusion that in general matter will only be tied up in luminous stars for a comparatively small part of its life, they assert that there is every reason to believe that a significant fraction of all the matter in the universe is dark, much of it as diffuse gas ejected at the end of stars' lives and the rest made of dead stellar remnants, namely, slowly cooling white dwarfs, neutron stars and near-black holes [15, p. 281].

They deny the existence of nonbaryonic dark matter invoked for large-scale structure scenarios, or to attain the closure density. They talk of "continuing failure of attempts to identify this nonbaryonic dark matter [14, p. 38].

Contrary to Big Bang cosmology which requires that all the matter in the observable universe be created in one single moment occurring 10^{-36} second after the Big Bang itself, they relate matter creation to conditions within near-black holes. It is a process that takes place in many locations [15, p. 107]. The products of creation expand in a "universal sea" in BBC but in "separated fireballs" in QSSC [15, p. 109].

The light elements, like all other isotopes in the periodic table, are of an astrophysical origin, i.e. are produced by astrophysical processes. This again is contrary to the prevailing idea that the light elements were formed almost entirely at the time of the Big Bang [15, p. 107].

In 1957, it had already been shown that of the 320 isotopes of the chemical elements all but eight were synthesised by nuclear processes in stellar interiors. This short inventory of problematic elements consisted of light nuclei and HBN claim that, of this list, deuterium is the last survivor which still poses a difficulty, i.e. the only one for which primordial nucleosynthesis might be needed. Yet they also conclude that this last case is uncertain and that it would seem that any nucleus heavier than the proton has been synthesised by processes associated with stars [14, p. 41].

The CBR can be explained by the hypothesis that all helium-4 is synthesised from hydrogen in stars as opposed to Big Bang nucleosynthesis. This alone defeats two major claims justifying BBC [14, p. 38].

Although helium is known to be produced from hydrogen inside stars, BBC argues that stellar synthesis can account for only a negligible contribution to the observed abundance of helium because of lack of time since the Big Bang for such slow astrophysical processes to have produced most of the observed helium. Most of the production is explained in BBC by primordial synthesis in the early universe [15, p. 96].

Observations over many years have accumulated good statistical evidence that many high-redshift quasars are physically associated with galaxies having much smaller redshifts, sometimes with a luminous bridge connecting these objects. This suggests, in agreement with Arp's interpretation, that a quasar ejected from a low-redshift parent galaxy possesses an intrinsic redshift component not associated with any recessional motion [14, p. 43; 15, p. 122].

HBN's theory of the observed redshift z_0 recognises that it has three components: one due to the cosmological expansion denoted z_c, a Doppler component due to the motion of the object in question denoted z_d, and finally one of intrinsic origin denoted z_i. They relate them with the following formula: $(1 + z_0) = (1 + z_c)(1 + z_d)(1 + z_i)$. From an analysis of this formula they conclude [15, p. 332] that the quasars from a population which show sharp peaks in its redshift distribution (the phenomenon of redshift quantization) must be comparatively local objects: "The peaks are most prominent in quasars which can be clearly associated with nearby galaxies with very small values of z_c."

One prediction the QSSC which could be tested with future very large telescopes is that galaxies belonging to a preceding cycle are indeed blue shifted [79, p. 501].

HBN introduce a creation function. The properties of homogeneity and isotropy for the whole universe for their creation function are rejected as this would in their view require that the curvature k be 0. The cases $k = +1$ and $k = -1$ arise in regions of the universe as consequences of inhomogeneities in the creation rate. A high local creation rate leads to $k = +1$ and to a bouncing region of the universe i.e. one in which phases of expansion and contraction alternate, matter creation episodes taking place at the oscillatory minima each producing some increase in the scale of the maxima. A low local creation rate gives rise to an ever-expanding region of the universe, a *growing hole*. They regard the entire observable universe as a $k = +1$ region into the surrounding universe to which their Friedmann–Robertson–Walker metric does not strictly apply. They speak of the observable universe as being a *near-black hole* as seen from outside. Within it there are regions of condensation and also expanding holes, but never strictly formed in a topological sense [15, pp. 194–196]. We are further referred to [76].

HBN have much in common with their friend Arp, coauthoring some papers. One point that keeps them apart is Arp's contention that there is no expansion of the universe while HBN claim that the matter creation episodes are causing it to expand. Arp says that if HBN accepted that matter be created with *zero* mass instead of having the particles produced with terrestrial masses, they would do away with the need for an unstable expansion i.e. their bouncing universe [11, p. 238].

6. The Eternal Self-Sustaining Plasma Universe: Hannes O.G. Alfvén Rehabilitates Electromagnetism

6.1. A Major Role for Electromagnetism

Electromagnetism is the main character in the plasma universe scenario. General relativity is downplayed but special relativity is maintained [3, p. 597]. Plasma cosmology holds that 99.999% of the volume of the universe is made up not of "invisible matter", but rather of matter in the plasma state. Electrodynamic forces in electric plasmas far exceed the gravitational force and therefore are the main actors that shape the cosmos all the way to superclusters of galaxies. Gravity is the only force that conventional cosmology based on general relativity takes into account with its metric tensor coupled with fluid dynamics. The *fluid of galaxies* idealisation defined by the two hydrodynamical parameters *density* and *pressure* ignores electromagnetism, as we have noted earlier.

The proponents of this theory assert that the universe is incomprehensible without taking into account the huge electrical currents and magnetic fields that permeate it and whose existence is denied or ignored by mainstream cosmology. Hannes Alfvén[11], an electrical power engineer and a Swedish Nobel laureate, has founded modern plasma physics, i.e. the physics of electrically conducting gases, which he has been studying since the 1930s, and has coined the phrase '*plasma universe*' to describe the view of the universe he espouses. Eric Lerner from Lawrenceville Plasma Physics in New Jersey, has championed and popularised this idea in [46].

The electromagnetic force has a far greater range than the gravitational one since the former varies as the inverse of the distance whereas the latter varies as the inverse of the *square* of that distance.

"In this theory, a galaxy, spinning in the magnetic field of intergalactic space, generates electricity, as any conductor does when it moves in a magnetic field" [46, p. 46].

Past and future eternity are postulated for the plasma universe. The shape of space is not emphasized: space is assumed to be infinite but we think the theory could accommodate a finite space as well [46, p. 388 and its footnote; and also p. 279].

6.2. What is a Plasma?

Anthony L. Peratt from the Los Alamos National Laboratory, defines 'plasma' as follows in his book *Physics of the Plasma Universe* [81, p. 1]:

> Plasma consists of electrically charged particles that respond collectively to electromagnetic forces. The charged particles are usually clouds or beams of electrons or ions, or a mixtures of electrons or ions, but also can be charged grains or dust particles. Plasma is also created when a gas is brought to a temperature that is comparable to or higher than that in the interior of stars.
>
> Plasma is a fourth state of matter, different from a solid, liquid or gas, but most closely resembling the last [...] Because of its free electrons, a plasma is a good conductor of electricity, much better than copper, silver or gold. Lightning offers one of the most dramatic manifestations of this property [82, p. 136].

If, as it is claimed, 'plasma makes up more than 99 percent of the visible universe', it is clear that understanding the dynamics of plasma is a key to a better comprehension of the universe.

Stars, for example, are gravitationally bound plasmas, while all of interstellar and intergalactic space is plasma.

Whenever plasmas exist, they produce prodigious amounts of electromagnetic radiation.

6.3. Observational Support for the "Plasma Universe"

Plasma electric currents were first imagined by the Norwegian scientist Kristian Birkeland (1867–1949) [34] in his attempt to understand the aurora borealis. Although supported both by his own observations and by his experiments in the laboratory, where he was able to simulate the aurora, Birkeland's theories failed to gain widespread acceptance until essentially confirmed by satellite evidence in the 1970s.

Modern plasma cosmologists have been heavily influenced by Birkeland's earlier research. In 1950, Hannes Alfvén, who later won a Nobel Prize in physics for his so-

---

[11]Hannes Alfvén died in April 1995; he worked at the Royal Institute of Technology, Stockholm and the University of California, San Diego.

lar studies, proposed that streams of electrons move at nearly the speed of light along magnetic-field lines not only in the Earth's magnetosphere and above the Sun, but also throughout the cosmos. If so, sheets and ropes of electric current should criss-cross the universe in ever-increasing sizes. These currents, Alfvén thought, should give the universe a cellular and filamentary structure.

Astronomers accepted Alfvén's notion of widespread synchrotron radiation but refused to believe that electric currents give rise to the large-scale structure of the universe.

However, the filamentary structure of the distribution of galaxies is confirmed by the three-dimensional maps of Tully and Fischer which show that nearly all the two thousand galaxies in their *Atlas of nearby galaxies* [109] are concentrated into an interconnecting network of a few large filaments called superclusters [46, p. 21].[12]

Peratt's book [81] is a presentation of the mathematical laws of physics that govern the behaviour of plasmas. It provides the fundamental argument for why electrical effects cannot be ignored in any modern study of the cosmos.

Peratt uses a large computer to apply Maxwell equations governing the forces produced by, and the interactions between, electric and magnetic fields to each of a huge ensemble of charged particles. He calls this *Particle in Cell* (PIC) simulations. His results are almost indistinguishable from astroimages of actual galaxies.

Some simulations concern the behaviour of two interacting Birkeland plasma electric currents. The computer images so obtained bear striking resemblance to real ones depicting the full range of galaxy types. It is very important to note that electromagnetic processes rather than gravitational ones create these images. As already pointed out, the electromagnetic force between the two currents, which falls off in direct proportion to the distance between them, is therefore much stronger than the gravitational force which falls off as the square of the distance. Just compare Newton's law of universal attraction to Ampère's law giving the strength of the magnetic field created by an electrical current in a straight wire.[13]

Computer simulations to understand the cosmos through plasma physic are still being conducted on some of the most powerful machines in existence at the Los Alamos National Laboratory [130].

6.4. The CBR, Nucleosynthesis and the Redshift in the Plasma Universe

Eric J. Lerner has developed a Plasma theory of nucleosynthesis and of the CBR. In a soon to be published (but already web accessible [140]) review article [48] he expounds and updates them, and also compares them quite favourably to the ones of BBC.

Alfvén [4, p. 9] and Lerner [46, pp. 52, 278–280, 425–430] express doubts about the nature of the redshift. They speculate that matter/antimatter collisions created an explosion or a big bang in one part of the universe thus creating the Hubble expansion in that small corner of the infinite universe that we can observe: "But this was in no way a Big Bang that created matter, space and time."

Thus there is perhaps no general expansion, only local occasional ones in an eternal universe. Apart from the conjectured infinity of space, this looks much like what Segal

---

[12] A color illustration of this is visible on a page of [132]: "http://home.pacbell.net/skeptica/structure.html". On Brent Tully's website [139] at the Institute for Astronomy of the University of Hawaii, one can find a '*Flight through the Local Supercluster*' amongst other interesting software. Tully supports BBC.

[13] Cf. Richard P. Feynman's *Lectures on Physics* (1964), Vol. II, pp. 13–5.

says in the context of CC. An alleged weakness of the Einstein Universe is that it would be unstable from the point of view of general relativistic hydrodynamics. As already noted, Segal considers this argument naïve. In that connection it is worth noting that Gerald S. Hawkins of the Harvard-Smithsonian Observatories writes in [32, p. 197]: "Hannes Alfvén and his colleagues have shown that a nonexpanding universe can be stabilized by electric and magnetic forces."

Edward L. Wright offers criticism of the Plasma Universe on his website [123] as he does of most cosmologies other than BBC.

7. The New-Tired Light Theory of the Redshift: The Nonexpanding Universe of Paul Marmet

7.1. Old Tired-Light Theories

Tired-light theories attempt to account for the cosmological redshift with variants of the hypothesis that the photons lose energy on their long journey to us generally through encounters with known or hypothetical particles or fields, or else, atoms or molecules.

A tired-light mechanism based on a photon-photon interaction was proposed in the 1950s by E. Findlay-Freundlich and Max Born. Later, Jean-Claude Pecker and Jean-Pierre Vigier with several co-authors could explain many facts on the basis of an interaction between photons of nonzero mass with a hypothetical particle [79, pp. 510–511].

7.2. Paul Marmet's New Tired-Light Theory

Paul Marmet is a Canadian physicist. He distinguishes himself from all other scientists we have mentioned so far by his rejection, not only of general relativity [63] but also of special relativity, as expounded in his books [61,65] and in several papers most of which are available on his website [126], sometimes in updated versions.

In his 1988 paper [60], he introduced his *New Non-Doppler Redshift* based on inelastic photon–molecule collisions. This leads to a new tired-light mechanism for the cosmic redshift which he developed further the next year with Grote Reber in [66], where the authors explain that:

> In this model the redshift is produced by inelastic collisions of photons on atoms and molecules. Some scientists reject this mechanism because they are not aware that most photon–molecule collisions do not lead to any significant angular dispersion of photons in all directions.

The 1999 discovery by Valentijn and van der Werf [110] of large amounts of relatively *warm* hard to detect *molecular* hydrogen in a galaxy, which Marmet and Reber had foreseen in [66] and [62], leads Marmet to forecast the detection of large amounts of (*colder* and hence still harder to spot) *molecular* hydrogen throughout space. It is with these molecules, he thinks [64], that photons collide inelastically and with little scattering on their long journey from galaxies to the earth.

This 1999 finding of *molecular* hydrogen seems to resolve the problem of the constancy of tangential velocities of matter in galaxies according to the discoverers and also according to what Marmet and Reber had conjectured in the above papers. Valentijn declares:

Our results give a much stronger footing for the 'ordinary matter' simple solution of the dark matter problem, in the form of massive clouds in the disks of galaxies.

However, Marmet argues that this discovery of *molecular* hydrogen has other significant corollaries. To support his new tired-light mechanism he writes in [64]:

The recent discovery of an enormous quantity of *molecular* hydrogen not only solves the problem of missing mass; it also solves the problem of the redshift, in a non-expanding unlimited universe [...]. We know that light interacts with a transparent medium, because its velocity is reduced, without scattering, as calculated and observed using the simple index of refraction of gases. Cosmic light, moving across billion of light years, suffers an almost unimaginable number of collisions with those transparent molecules of hydrogen in the universe.

Marmet continues:

[...] as a result of the large amount of *atomic* hydrogen already observed in space, and the extreme stability of *molecular* hydrogen, the chemical equilibrium giving the relative abundance between *atomic* hydrogen and *molecular* hydrogen in space, strongly favors the formation of the diatomic form (H_2) over the monoatomic form. We must thus conclude that the recent discovery of H_2 is no surprise, and should have been expected from the known facts concerning the natural equilibrium between H_2 and H. It is expected that much more colder H_2 will also be discovered.

Marmet arrives at the Hubble law in a nonexpanding universe:

Because *atomic* and *molecular* hydrogen have an approximately homogenous distribution in the universe, this induces a non-Doppler redshift, which is proportional to the distance of the light source.

7.3. Discordant Judgments

However, discordant voices are heard, even from non-conventional cosmologists, about Marmet's redshift mechanism. In *The Big Bang Never Happened*, Lerner contends (pp. 428–429) that there cannot be as much intergalactic matter as Marmet claims there must be to justify his redshift theory, for otherwise "Such a high matter density would have enormous gravitational effects that simply aren't observed." For his part, Arp rejects all tired-light explanations of the redshift since he rejects the notion that the redshift be a distance indicator. In addition, he calls attention to the fact that within our own galaxy:

as we look to lower galactic latitudes, we see objects through an increasing density of gas and dust until they are almost totally obscured and no increase of redshift has ever been demonstrated for objects seen through this increased amount of material [11, p. 97].

8. Mordehai Milgrom's Modified Newtonian Dynamics: The Embryo of a New Cosmology Without Any Dark Matter

The *dark matter problem*, also known as the *missing mass problem*, apparently partly resolved by the 1999 discovery of *molecular* hydrogen in galaxies and which arose in the early 1980's by the observation of the *flat rotation curves* in spiral galaxies is also offered another solution, this one put forward in 1983 by Mordehai Milgrom from the Weizmann Institute in Rehovot, Israel. It is known by its acronym MOND which stands for *Modified Newtonian Dynamics* or sometimes for *Modified Nonrelativistic Dynamics*. In [68], Milgrom attempts to lay the foundations of *a MOND-inspired cosmology, which, at any event, would start with no dark matter.*

The *dark matter problem* may be summed up thus. If the mass of a galaxy were concentrated near its center from which its luminosity falls off rapidly with distance, the stars in the galaxy at increasing distances from the center would have decreasing tangential orbital velocities. Contrary to that expectation it was found that this velocity is essentially independent of distance. One possible conclusion, which appears to be corroborated by the molecular hydrogen discovery, is that the light distribution in a galaxy is not at all a guide to the mass distribution [88,87].

Instead of looking for this missing mass MOND attempts to resolve the problem by challenging Newton's second law of motion usually written as $F = ma$; F being any force, m the mass acted upon by F and a the resulting acceleration. Although this law is well established in laboratories and in the solar system, it is argued that it may no longer be valid when small accelerations, such as the ones found at larger distances in galaxies, are involved.

MOND introduces a constant a_0 which has the dimensions of an acceleration and replaces $F = ma$ by $F = ma^2/a_0$ but only when $a \ll a_0$. As a consequence one obtains a modified universal attraction law which yields larger accelerations when the Newtonian acceleration is much smaller than a_0:

> The basic point of MOND, from which follow most of the main predictions, can be simply put as follows: a test particle at a distance r from a large mass M is subject to the acceleration a given by $a^2/a_0 = MGr^{-2}$ when $a \ll a_0$, instead of the standard expression $a = MGr^{-2}$, which holds when $a \gg a_0$.

The theory is basically nonrelativistic. Milgrom writes on his website [128]:

> Several relativistic theories incorporating the MOND principle have been discussed in the literature, but none is wholly satisfactory.

The theory has its supporters: for instance Tom van Flandern [111] writes:

> Milgrom's model (the alternative to "dark matter") provides a one-parameter explanation that works at all scales and requires no "dark matter" to exist at any scale.

HBN, on the other hand, are not enthusiastic [15, p. 286]:

> Milgrom's proposal [...] is an *ad hoc* modification of Newton's law designed to suit the particular phenomenon under consideration.

9. Epilogue

9.1. Sound Science, Beliefs and Fantasies

Cosmology is in imminent danger of deserting the ground of sound science for the terrain of beliefs and fantasies. It is shameful that in most accounts of contemporary cosmology all options other than BBC are mentioned, if ever, only to be dismissed without discussion. It is appropriate here to remind some basic canons of rational knowledge as opposed to articles of faith of an established orthodoxy. Too many cosmologists have forgotten Occam's razor, the epistemological principle which normally underlies all scientific theory building: one should not make more assumptions than the minimum needed. Also at stake is disrespect for disquieting observed facts and their observers: the case of Arp's abnormal redshifts and its discoverer's forced early retirement come to mind. On the other hand pure speculations are routinely presented as facts: *exotic nonbaryonic dark matter* and *dark energy* make up most of the universe!

There is an urgent need for more open-mindedness; for readiness to abandon long held views in the face of contrary observational evidence; for willingness to revise fundamental assumptions; for no sweeping of facts under the carpet; and for a free flow of information. Attempts at preventing publication of views contrary to the conventional wisdom are absolutely scandalous: astronomers of the *USA National Academy of Science* doing their best to prevent Segal from publishing in the *Proceedings of the NAS*; Alfvén's papers and those of other plasma physicists routinely rejected by astrophysical journals for being in contradiction with common thinking. Documented objections to widespread beliefs must be met and not brushed aside or ignored: even opposition to special relativity deserves a response.

9.2. Apology of Chronometric Cosmology

My view is that Irving Ezra Segal is the true continuator of Einstein's genius in cosmology. All adepts of the universal expansion hypothesis would be well advised to look carefully at Segal's chronometric cosmology. Its redshift–distance relation has met the test of all available reliable data whereas the Hubble law does not hold water. CC faces none of the tribulations afflicting BBC.

9.3. Will Big Bang Cosmology Ever Fall?

Unless BBC somehow resolves the enormous difficulties it faces, it should rationally be forsaken but, in view of its present social status and a subservient popular press, it would take more than a small flock of missionaries to have any of the other contending theories (or a coalition thereof) also facing their own problems, replace it. Mainstream science may remain on the wrong track for a long time for reasons that have little to do with rationality. Meanwhile this will not divert all valiant minds, amongst them many believers in BBC, from the pursuit of truth.

In the meantime a downturn in their arrogant tone, that borders on fanaticism, would be a welcome change in the discourse of many advocates of BBC [6].

Acknowledgments

I am most thankful, in the first place, to Marco Mamone Capria who is doing a superb job as editor of this volume. His very careful readings and many thoughtful suggestions have been very useful to me. I also owe many thanks to Danièle Dubé, Alexander Mittelmann,[14] Alexander Levichev and Arturo Sangalli for their corrections and ideas from reading earlier versions of this chapter.

---

[14]"Cold Creation", www.coldcreation.com.

Appendix. The *Minkowskian-Cosmic Times Formula of CC*

For the benefit of the more mathematically minded reader we present here an outline of a proof of Eq. (1) i.e. the *Minkowskian-cosmic times formula of CC* mostly based on [92]. For simplicity's sake we assume, at first, the three-sphere radius r to be unity. As we also assume the speed of light c to be 1, the distance travelled on the three-sphere is equal to the cosmic time elapsed. The formula to be established then becomes simply $x_0 = 2\tan(s/2)$ in which s is cosmic time and $x_0(s)$ is the corresponding Minkowskian time. Two main characters in this drama are Minkowski space–time M and the unit three-sphere S^3. Their roles will be played respectively by $H(2)$, the set of 2×2 complex Hermitian matrices and $SU(2)$, the special unitary group made up of the 2×2 unitary matrices of determinant 1 as we now explain.

Minkowski space M is mapped biuniquely onto $H(2)$ by sending $x = (x_0, x_1, x_2, x_3)$ onto the matrix

$$A = \begin{pmatrix} x_0 + x_3 & x_1 + ix_2 \\ x_1 - ix_2 & x_0 - x_3 \end{pmatrix}. \tag{3}$$

This is an isomorphism between two real vector spaces. The closed future light-cone at the origin O of M including light rays is the set of points

$$\{(x_0, x_1, x_2, x_3) \mid x_0^2 - x_1^2 - x_2^2 - x_3^2 \geq 0; x_0 \geq 0\}. \tag{4}$$

This light-cone defines the causal structure on M, the light-cones at other points of M being obtained by translations. The above isomorphism maps this light-cone on the set of positive definite Hermitian matrices. Taking this as the light-cone at the origin turns $H(2)$ into a causal manifold isomorphic to M. The real vector space $H(2)$, more exactly, the real vector space $iH(2)$ of antihermitian matrices, to which the causal structure on $H(2)$ can be transferred by the obvious real isomorphism, is the Lie algebra of the unitary group $U(2)$, the multiplicative group of 2×2 unitary matrices. This Lie algebra can be thought of as the tangent space at the unit element of the Lie group $U(2)$. This group acquires a causal structure by translating the light-cone at the identity element of $U(2)$. It matters not whether one uses right or left translations in $U(2)$.

The unit three-sphere S^3 can be identified with the special unitary group $SU(2)$ by virtue of the fact that the generic element of that group is the matrix

$$U = \begin{pmatrix} a + ib & c + id \\ -c + id & a - ib \end{pmatrix} \tag{5}$$

where, the determinant being 1, we must have

$$a^2 + b^2 + c^2 + d^2 = 1 \tag{6}$$

which is the equation of the unit sphere in a Euclidean four-dimensional space.

The point $p = (a, b, c, d)$ of S^3, the unit sphere at the origin in a four dimensional space, corresponds to the above matrix U in this identification. The group $SU(2)$, inherits through this identification, the Riemannian metric denoted ds^2 on the unit sphere induced by the surrounding four space. The group of symmetries of S^3 gives rise to a group of inner automorphisms of the group $SU(2)$ which preserve this metric. This metric is the only one on this group which is invariant under right and left translations.

Another main character in the plot is the Einstein universe $E = R^1 \times S^3$ which, by definition, is endowed with the Lorentzian metric $dt^2 - ds^2$ where dt^2 is the ordinary metric on the real time line R^1 and ds^2 is as above. Its role is played by the product of the additive group R^1 and the multiplicative group $SU(2)$. This group $R^1 \times SU(2)$ is the universal covering group of $U(2)$. The projection map sends an element (s, V) onto the unitary matrix $e^{is} V$. In view of the fact that $e^{is} V = e^{(is+\pi)}(-V)$ this projection factors through the two sheet cover $S^1 \times SU(2)$ where S^1 is the unit circle, i.e. the multiplicative group of complex numbers of module 1.

Now thinking of $M = H(2)$ as the tangent real linear space at the identity of the group $U(2)$, the 0 matrix as a point of $H(2)$ coinciding with the identity matrix I as a point of the Riemannian manifold $U(2)$ we define a causal embedding of $H(2)$ into $U(2)$ using the *Cayley transform* which is a generalisation of the inverse of the stereographic projection. This maps a Hermitian matrix A in $H(2)$ onto the unitary matrix U defined as

$$U = \left(I + \frac{iA}{2}\right)\left(I - \frac{iA}{2}\right)^{-1}. \tag{7}$$

The reverse mapping which generalises the stereographic projection is defined for all U in $U(2)$ except when $U + I$ has determinant 0, by the equation

$$A = -2i(U - I)(U + I)^{-1}. \tag{8}$$

We still need lifting the group $U(2)$ to a section of its universal cover $R^1 \times SU(2)$ as to complete the causal imbedding of Minkowski space M in the Einstein universe E by the sequence of causal mappings

$$M \to H(2) \to U(2) \to R^1 \times SU(2) \to R^1 \times S^3. \tag{9}$$

This continuous map sends an element $U = e^{it} V$ of $U(2)$ where $-\pi < t < \pi$ and V is in $SU(2)$ onto (t, V) in such a way that the identity matrix I is sent to $(0, I)$. Although each element U of $U(2)$ can be written in this form in two ways by virtue of the equation $e^{is} V = e^{i(s+\pi)}(-V)$ the requirement of continuity eliminates the ambiguity in the definition of the map.

In what follows we identify M with $H(2)$ and E with $R^1 \times SU(2)$ keeping in mind the above four steps imbedding of M into E. Hence each point in M has four names: one x in M proper, an Hermitian matrix A given by (5), a unitary matrix U given by (9) and (t, V) as a member of E where

$$U = e^{it} V \tag{10}$$

as above. The stage is now set for the proof of the two-times formula.

One must think of a photon being emitted somewhere on the three sphere and being observed later elsewhere after some cosmic time s. It is important to distinguish between three events all of which belonging to the image of M in E as we suppose that the photon is observed after less than one half-tour of the three sphere merry-go-round which it is circling along a grand circle. The three events are: the emission (t, W) of the photon at cosmic time t at the point W of space $SU(2)$; the observation $(t + s, V)$ of the photon. s cosmic time later at the point V of $SU(2)$; and the event (t, V) when a patient observer starts waiting at V for the arrival of the photon at the moment it is emitted.

Let A be the Hermitian matrix corresponding to the event (t, V) with $t = x_0$ and U be its Cayley transform defined by (7). Let $A(s)$ be the Hermitian matrix corresponding to the event $(t + s, V)$. One obtains

$$A(s) = -2i\left(e^{is}U - I\right)\left(e^{is}U + I\right)^{-1}, \tag{11}$$

where U is defined by (12). This follows from the fact that the Einsteinian temporal translation T_s, the isometry of E which maps any (t, V) onto $(t + s, V)$, once interpreted in the notation of $U(2)$, maps any U onto $U(s) = e^{is}U$. From this one obtains after some calculations

$$A(s) = 2(2aI + bA)(2bI - aA)^{-1}, \tag{12}$$

where $a = \sin(s/2)$ and $b = \cos(s/2)$.

We may safely assume that the observation takes place at the origin of M, and that the cosmic time of emission is $t = 0$ so that $A = 0$, $U = I = V$. It then follows immediately from (12) that

$$A(s) = 2\tan\left(\frac{s}{2}\right)I. \tag{13}$$

By definition the matrix $A(s)$ can be written

$$A(s) = \begin{pmatrix} x_0(s) + x_3(s) & x_1(s) + ix_2(s) \\ x_1(s) - ix_2(s) & x_0(s) - x_3(s) \end{pmatrix}. \tag{14}$$

In view of the fact that $x_j(s) = 0$ for $j = 1, 2, 3$, one obtains from (13) and (14) the desired conclusion $x_0(s) = 2\tan(s/2)$.

A better understanding of the two-times and the redshift formulae is achieved at the cost of intensifying the calculations. This consists in looking at the effect of the temporal translation T_s on some small neighbourhood *NGR* (in the image of M in E) of the origin of M as a point of E instead of this effect just on the origin of M. This is done through the differential approximation dT_s of T_s. For a point $x = (x_0, x_1, x_2, x_3)$ other than the origin in *NGR* we no longer have $A = 0$ nor $x_j(s) = 0$ for $j = 1, 2, 3$ but we still have (11), (12) and (14).

The linear transformation dT_s maps the tangent space at the origin of M onto the tangent space at the image of the origin by T_s. The matrix of this linear transformation expressed in Minkowskian coordinates is the *Jacobian*

$$\left(\frac{\partial x_i(s)}{\partial x_j}\right) \tag{15}$$

evaluated at the origin of M. A computer calculation shows that this is the diagonal matrix all elements of the main diagonal being equal to $\sec^2(s/2)$.[15] This means that the approximate effect of T_s on any point x of *NGR* is to map it on a point whose Minkowskian coordinates are those of x magnified by the same factor $\sec^2(s/2)$. Perhaps surprisingly, this is reminiscent of BBC except that here Minkowskian time also is expanding. This

---

[15]Jean-Marc Terrier, an expert in the computer program *Mathematica*, has kindly verified this result as well as the formulae (8) and (12). The formulae for $x_j(s)$, $j = 0, 1, 2, 3$, given on p. 11115 of [92] appear to be wrong but, in any case, they are not needed here.

does not contradict the fact that T_s being an isometry of E it maps NGR isometrically onto its image. Wavelengths which are small relative to the radius of the universe may be assumed to fall within NGR. As a result, Segal concludes [92, p. 11115]: "In particular, wavelengths, after time s are observed as magnified by the factor $\sec^2(s/2)$."

Thus if λ is such a wavelength, it becomes $\sec^2(s/2)\lambda$ under T_s. As a result one obtains an equivalent proof of the redshift formula using the definition $z = \Delta\lambda/\lambda$ as follows

$$z = \frac{\sec^2(s/2)\lambda - \lambda}{\lambda} = \text{tg}^2\left(\frac{s}{2}\right). \tag{16}$$

Similarly a small interval ds of cosmic time is magnified by a factor of $\sec^2(s/2)$ into a larger interval $dx_0(s)$ of Minkowskian time under the temporal translation T_s so that $dx_0(s)/ds = \sec^2(s/2)$. Integrating this relation immediately yields the two-times relation $x_0(s) = 2\tan(s/2)$ taking into account the initial condition $x_0(0) = 0$.

Going back now to the general case of a three-sphere of any radius r instead of the unit sphere, we note that all but the last map in (9) remain unchanged whereas the last one, $R^1 \times SU(2) \rightarrow R^1 \times S^3$, must be multiplied by r. As a result both x_0 and t ($= s$) must be divided by r in the formulae we have just established in the case $r = 1$. This immediately yields the original chronometric two-times and redshift formulae (1) and (2).

References

[1] Adams C. and Shapiro J. 2001: "The shape of the universe: Ten possibilities", *American Scientist*, **89**, pp. 443–453.

[2] Albrecht A. 1999: Reply to "A Different Approach to Cosmology", *Physics Today* (April), pp. 44–46.

[3] * Alfven H. 1984: "Cosmology – Myth or science?", *Journal of Astrophysics and Astronomy* (ISSN 0250-6335), **5**, pp. 79–98.

[4] * Alfven H. 1990: "Cosmology in the plasma universe: an introductory exposition", *IEEE Transactions on Plasma Science*, **18** (1), pp. 5–10.

[5] Alfven H. 1980: "Cosmology and recent developments in plasma physics", *The Australian Physicist*, **17**, pp. 161–165.

[6] Arp H. et al. 2004: "An Open Letter to the Scientific Community", *New Scientist*, May 22, available at http://www.cosmologystatement.org/.

[7] Arp H.C. 1966: *Atlas of Peculiar Galaxies* (a collectors' item).

[8] * Arp H.C. 2003: *Catalogue of Discordant Redshift Associations*, Montréal, QC, Apeiron, 234 p.: ill. (some col.); 26 cm, ISBN 0-9683689-9-9.

[9] Arp H.C. and Madore B.F. 1987: *A Catalogue of Southern Peculiar Galaxies and Associations*, Cambridge, New York: Cambridge University Press, 2 v.: ill.; 31 cm.

[10] Arp H.C. 1987: *Quasars, Redshifts, and Controversies*, Berkeley, CA: Interstellar Media, 198 p.: ill.; 24 cm.

[11] * Arp H.C. 1998: *Seeing Red: Redshifts, Cosmology and Academic Science*, Apeiron, Montreal, paperback, 306 pages plus 8 pages of colour plates; ISBN 0-9683689-0-5.

[12] Assis A.K.T. and Neves M.C.D. 1995: "History of the 2.7 K temperature prior to Penzias and Wilson", *Apeiron*, **2** (3), pp. 79–84. Downloadable from: http://www.dfi.uem.br/~macedane/history_of_2.7k.html#note.

[13] Born M. and Wolf E. 1997: *Principles of Optics: Electromagnetic Theory of Propagation, Interference and Diffraction of Light*, Cambridge University Press, xxviii, 808 p., ISBN 0521639212 (pbk.).

[14] Burbidge G., Hoyle F. and Narlikar J.V. 1999: "A different approach to cosmology", *Physics Today*, **52** (4), pp. 38–44.

[15] ∗ Burbidge G., Hoyle F. and Narlikar J.V. 2000: A different approach to cosmology: from a static universe through the Big Bang towards reality, Cambridge [England]; New York: Cambridge University Press, xi, 357 p.: ill., cartes; 26 cm, ISBN 0521662230 (hbk.)

[16] Burbidge G., Hoyle F. and Narlikar J.V. 1994: "Further astrophysical quantities expected in a quasi-steady state universe", *Astronomy and Astrophysics*, **289**, pp. 729–739.

[17] Chown M. 1993: "All you ever wanted to know about the Big Bang…", *New Scientist* (17 April), pp. 32–33.

[18] Daigneault A. 2003: "L'universo si sta espandendo ?", pp. 239–259 in "Scienza e democrazia" / a cura di Marco Mamone Capria, Napoli: Liguori, xxxiv, 551 p.; 24 cm., ISBN 88-207-3495-8, translated from the English original Is the universe expanding? available at http://www.dipmat.unipg.it/~mamone/sci-dem/sci&dem.htm.

[19] Daigneault A. 2002: "Response to Wright and Boerner", *Notices Amer. Math. Soc.*, **49** (1), p. 6.

[20] ∗ Daigneault A. and Sangalli A. 2001: "Einstein's static universe: an idea whose time has come back?", *Notices Amer. Math. Soc.*, **48** (1), pp. 9–16. MR1798927 (2001j:83106). Downloadable from http://www.ams.org/notices/.

[21] Eddington Sir A. 1926: *Internal Constitution of the Stars*, Cambridge University Press, reprinted 1988.

[22] ∗ Einstein A. 1917: "Kosmologische Betrachtungen zur allgemeinen Relativitätstheorie", *Sitzungsberichte der Preussischen Akademie der Wissenschaften*, pp. 142–152. Also with the title "Cosmological considerations on the general relativity theory", pp. 177–188 in The Principle of Relativity, a collection of original memoirs on the special and general theory of relativity, by H.A. Lorentz, A. Einstein, H. Minkowski, and H. Weyl, with notes by A. Sommerfeld; translated by W. Perrett and G.B. Jeffery. Published by Methuen & Co. Ltd., London, 1923; reprinted in 1990 (MR 94a: 01043); also published unaltered and unabridged by Dover Publications, Inc., in 1952.

[23] Ellis G.F.R. 1971: *Gen. Rel. Grav.*, **2**, pp. 7–21.

[24] Ellis G.F.R. and Schreiber G. 1986: "Observational and dynamical properties of small universes", *Physics Letters A*, **115** (3), pp. 97–107.

[25] Ellis G.F.R. 2003: "The shape of the universe", *Nature*, **425**, pp. 566–567.

[26] Fock V. 1966: *The Theory of Space, Time and Gravitation*, 2nd revised edition, Pergamon Press, 448 pages, ISBN 0 08 010061 9.

[27] Gnedin N.Yu. and Ostriker J.P. 1992: "Light element nucleosynthesis: a false clue?", *Astrophysical J.*, **400**.

[28] Gross L. 2002: "In memory of Irving Segal", in special issue dedicated to the memory of I.E. Segal, *J. Funct. Anal.*, **190** (1), pp. 1–14.

[29] Gruhier F. 1993: "Et si le big-bang n'avait jamais eu lieu…", *Le Nouvel Observateur* (3 juin), pp. 4–11.

[30] Guth A.H. 1997: *The Inflationary Universe*, Addison Wesley, 358 pp.

[31] Hawking S.W. and Ellis G.F.R. 1973: *The Large-Scale Structure of Space–Time*, Cambridge Univ. Press.

[32] Hetherington N.S., Ed. 1993: *Encyclopedia of Cosmology*, Garland Publishing, Inc., ISBN 0-8240-7213-8.

[33] ∗ Hogan C.J., Kirshner R.P. and Suntzeff N.B. 1999: "Surveying space–time with supernovae", *Scientific American* (January), pp. 46–51.

[34] Jago L. 2000: The Northern Lights: The True Story of the Man Who Unlocked the Secrets of the Aurora Borealis, 320 pages Knopf.

[35] James D.F.D. and Wolf E. 1994: "A class of scattering media which generate Doppler-like frequency shifts of spectral lines", *Phys. Lett. A*, **188**, pp. 239–244.

[36] James D.F.D. and Wolf E. 1990: "Doppler-like frequency shifts generated by dynamic scattering", *Phys. Lett. A*, **146**, pp. 167–171.

[37] ∗ Kirshner R.P. 1999: "Supernovae, an accelerating universe and the cosmological constant", *Proc. Natl. Acad. Sci. USA*, **96**, pp. 4224–4227.

[38] ∗ Kragh H. 1996: *Cosmology and Controversy: the Historical Development of Two Theories of the Universe*, Princeton University Press.

[39] ∗ Krauss L.M. 1999: "Cosmological antigravity", *Scientific American* (January), pp. 53–59.

[40] Krauss L.M. 1998: "The end of the age problem, and the case for a cosmological constant revisited", *The Astrophysical Journal*, **501**, pp. 461–466.

[41] Lee J.M. 1997: "Riemannian manifolds: an introduction to curvature", volume 176 in: *Graduate Texts in Mathematics*, Springer-Verlag, ISBN 0-387-98322-8, 09/05/1997, http://www.math.washington.edu/~lee/Books/Riemannian/desc.html.

[42] Lehoucq R., Uzan J.-P. and Weeks J. 2003: "Detecting topology of nearly flat spherical universes", *Class. Quant. Grav.*, **20**, pp. 1529–1542 (e-print astro-ph/0209389).

[43] Lehoucq R., Uzan J.-P. and Luminet J.-P. 2000: "Limits of crystallographic methods for detecting space topology", *Astronomy and Astrophysics*, **363** (1) (astro-ph/0005515).

[44] Leibundgut B. 2001: "Cosmological implications from observations of type Ia Supernovae", *Annu. Rev. Astron. Astrophys.*, **39**, pp. 67–98.

[45] Leibundgut B., Schommer R., Phillips M., Riess A., Schmidt B. et al. 1996: "Time dilation in the light curve of the distant type 1a supernova SN 1995K", *Ap. J.*, **466**, pp. L21–24.

[46] Lerner E.J. 1991: The Big Bang never happened, Random House.

[47] Lerner E.J. 1993: "The case against the Big Bang", in: *Progress in New Cosmologies*, H.C. Arp and C.R. Keys, Eds., Plenum Press, New York, pp. 89–104.

[48] Lerner E.J. 2003: "Two world systems revisited: a comparison of plasma cosmology and the Big Bang" (to be published in *IEEE Trans. on Plasma Sci.*). Downloadable from http://www.nobigbang.com/p27.htm.

[49] Levichev A.V. Mathematical description of three agni yoga worlds and of their unity (to be published).

[50] ∗ Levichev A.V. 1993: "On mathematical foundations and physical applications of chronometry", in: *Semigroups in Algebra, Geometry and Analysis* (Oberwolfach), pp. 77–103, de Gruyter Exp. Math., 20, de Gruyter, Berlin, 1995. Downloadable in PostScript at [127].

[51] ∗ Levin J. 2002: *How the Universe Got Its Spots: Diary of a Finite Time in a Finite Space*, Princeton University Press, ISBN: 0-691-09657-0, 224 pp.

[52] ∗ Luminet J.-P. et al. 2003: "Dodecahedral space topology as an explanation for weak wide-angle temperature correlations in the cosmic microwave background", *Nature*, **425**, pp. 593–595.

[53] Luminet J.-P. A finite dodecahedral universe (From Luminet's website).

[54] Luminet J.-P. 2001: A small spherical universe after all (From Luminet' website; December).

[55] Luminet J.-P. La topologie de l'univers; L'univers est-il chiffonné ? (from his web site).

[56] Luminet J.-P. 2002: "L'univers est-il infini ?", *La Recherche*, **358**, pp. 77–80.

[57] ∗ Luminet J.-P., Starkman G.D. and Weeks J.R. 1999: "Is Space finite?", *Scientific American*, **280**, pp. 90–97.

[58] ∗ Luminet J.-P. 2001: *l'Univers chiffonné*, Fayard, Paris, 369 p.

[59] Maddox J. 1989: "Down with the Big Bang", *Nature*, **340**, p. 425.

[60] ∗ Marmet P. 1988: "A new Non-Doppler redshift", Updated from: *Physics Essays*, **1** (1), pp. 24–32. http://www.newtonphysics.on.ca/HUBBLE/Hubble.html.

[61] Marmet P. 1993: *Absurdities in Modern Physics*, Les Éditions du Nordir, ISBN 0-921272-15-4.

[62] Marmet P. 1990: "Big Bang cosmology meets an astronomical death", *21st Century, Science and Technology*, **3** (2), pp. 52–59. http://www.newtonphysics.on.ca/BIGBANG/Bigbang.html.

[63] Marmet P. 1989: Classical description of the advance of the perihelion of Mercury.

[64] ∗ Marmet P. 2000: "Discovery of H_2", in Space Explains Dark Matter and Redshif; *21st Century Science & Technology* (Spring), pp. 5–7. http://www.newtonphysics.on.ca/hydrogen/index.html.

[65] Marmet P. 1997: *Einstein's Theory of Relativity versus Classical Mechanics*, Newton Physics Books, ISBN 0-921272-18-9.

[66] Marmet P. and Reber G. 1989: "Cosmic matter and the nonexpanding universe", *IEEE Transactions on Plasma Science*, **17** (2), pp. 264–269. http://www.newtonphysics.on.ca/UNIVERSE/Universe.html.

[67] ∗ Marmet P. 1995: "The origin of the 3 K radiation", *Apeiron*, **2** (1).

[68] Milgrom M. 1999: "The modified dynamics. A status review", in: *Dark Matter in Astrophysics and Particle Physics*, 1998. Proceedings of the Second International Conference, pp. 443–457, Institute of Physics Publishing, Bristol, UK, ISBN: 0 7503 0634 3.

[69] Misner C.W., Thorne K.S. and Archibald Wheeler J. 1973: *Gravitation*, San Francisco: W.H. Freeman, xxvi, 1279 p.: ill., ISBN 0716703440.

[70] Morris G.M. and Faklis D. 1987: "Effects of source correlation on the spectrum of light", *Optics Communications*, **62** (1), pp. 5–11.

[71] Morris G.M. and Faklis D. 1988: "Spectral shifts produced by source correlations", *Optics Letters*, **13** (1), pp. 4–6.

[72] ∗ Naber G.L. 1992: *The Geometry of Minkowski Spacetime*, 257 p., Springer Verlag, ISBN 3-540-97848-8.

[73] Narlikar J. and Arp H. 1993: "Flat space–time cosmology: a unified framework for extragalactic redshifts", *Astrophysical Journal*, **405**, pp. 51–56.

[74]   * Narlikar J. 1993: "Challenge for the Big Bang", *New Scientist* (June 19), pp. 27–31.

[75]   Narlikar J.V. and Arp H.C. 1997: "Time dilation in the supernovae light curve and the variable mass hypothesis", *Ap. J.*, **482**, pp. L119–120.

[76]   Narlikar J.V. 1973: "Singularity and matter creation in cosmological models", *Nature (Physical Science)*, **242** (122), pp. 135–136.

[77]   Odenwald S. and Fienberg R.T. 1993: "Galaxy redshifts reconsidered", *Sky & Telescope* (February), pp. 31–35. http://cecelia.physics.indiana.edu/life/redshift.html.

[78]   * Osserman R. 1995: *Poetry of the Universe, A Mathematical Exploration of the Cosmos*, Anchor Books, 210 pages, ISBN 0-385-47429-6.

[79]   * Pecker J.-C. 2001: *Understanding the Heavens: Thirty Centuries of Astronomical Ideas from Ancient Thinking to Modern Cosmology*, edited by S. Kaufman, Berlin; New York: Springer, xiii, 597 p. : ill., maps; 24 cm.

[80]   * Peebles P.J.E. 1993: *Principles of Physical Cosmology*, Princeton University Press.

[81]   * Peratt A.L. 1991: *Physics of the Plasma Universe*, Springer-Verlag, New York, 372 pages 208 illus., ISBN: 0-387-97575-6, published.

[82]   Peratt A.L. 1992: "Plasma Cosmology", *Sky & Tel.* p. 136 (Feb.); Downloadable from [130].

[83]   Riess A.G. et al. 1998: "Observational evidence from supernovae for an accelerating universe and a cosmological constant", *Astronomical Journal*, **116** (3), pp. 1009–1038.

[84]   Riess A.G. et al. 2001: "The farthest known supernova: support for an accelerating universe and a glimpse of the epoch of deceleration", *The Astrophysical Journal*, **560**, pp. 49–71.

[85]   Riess A.G. 2000: "The case for an accelerating Universe from Supernovae", *Publications of the Astronomical Society of the Pacific*, **112**, pp. 1284–1299.

[86]   * Rindler W. 1977: *Essential Relativity, Special, General, and Cosmological*, Second edition, Springer Verlag.

[87]   * Rubin V.C. 1983: "Dark matter in spiral galaxies", *Scientific American* (ISSN 0036-8733), **248**, pp. 96–106, 108.

[88]   Rubin V.C. 1983: "The rotation of spiral galaxies", *Science* (ISSN 0036-8075), **220**, pp. 1339–1344.

[89]   Schwarzschild K. 1998: "On the permissible curvature of space", *Class. Quantum Grav.*, **15**, pp. 2539–2544; a translation from *Vierteljahrschrift d. Astronom. Gesellschaft.*, **35** (1900), pp. 337–347.

[90]   Segal I.E. 1997: "Cosmic time dilation", *Ap. J.*, **482**, pp. L115–117.

[91]   Segal I.E. 1982: "Covariant chronogeometry and extreme distances, III. Macro-micro relations (Dirac Symposium, New Orleans, 1981)", *Internat. J. Theoret. Phys.*, **21**, pp. 851–869.

[92]   Segal I.E. 1993: "Geometric derivation of the chronometric redshift", *Proc. Natl. Acad. Sci. USA*, **90**, pp. 11114–11116.

[93]   Segal I.E. 1999: "Is redshift-dependent evolution of galaxies a theoretical artifact?", *PNAS*, **96** (24), pp. 13615–13619.

[94]   Segal I.E. 1991: "Is the cygnet the quintessential baryon?", *PNAS*, **88**, pp. 994–998.

[95]   Segal I.E. 1976: *Mathematical Cosmology and Extragalactic Astronomy*, Academic Press, New York.

[96]   * Segal I.E. 1997: "Modern statistical methods for cosmology testing", in: *Statistical Challenges in Modern Astronomy II*, Springer, 469 pages 72 illus., hardcover, ISBN: 0-387-98203-5, published 1997, Edited by G. Jogesh Babu and E.D. Feigelson.

[97]   Segal I.E. and Nicoll J.F. 1998: "Cosmological implications of a large complete quasar sample", *PNAS*, **95**, pp. 4804–4807.

[98]   Segal I.E. and Nicoll J.F. 2000: "Phenomenological analysis of redshift–distance power laws", *Astrophysical and Space Science*, **274**, pp. 503–512.

[99]   Segal I.E., Nicoll J.F., Wu P. and Zhou Z. 1991: "The Nature of the redshift and directly observed quasar statistics", *Naturwissenschaften*, **78**, pp. 289–296.

[100] * Segal I.E. 1985: "*The Cosmic Background Radiation and the Chronometric Cosmology*", Società Italiana di Fisica, Conference Proceedings, The cosmic background radiation and fundamental physics, **1**, pp. 209–223.

[101] Segal I.E. 1990: "The mathematical implications of fundamental physical principles", in: *The Legacy of John von Neumann* (Hempstead, NY, 1988), pp. 151–178, *Proc. Sympos. Pure Math.*, **50**, Amer. Math. Soc., Providence, RI.

[102] * Segal I.E. 1976: "Theoretical foundations of the chronometric cosmology", *Proc. Nat. Acad. Sci.*, **73**, pp. 669–673.

[103] * Segal I.E. and Zhou Z. 1995: "Maxwell's equations in the Einstein universe and chronometric cosmology", *Astrophys. J. Suppl. Ser.*, **100**, pp. 307–324.

[104] Seife C. 2003: "Breakthrough of the year: Illuminating the dark universe", *Science*, **302** (5653), 19 pp. 2038–2039.

[105] Steigman G. 1988: "Big Bang nucleosynthesis: The standard model", in: *Cosmic Abundances of Matter* (Symposium at University of Minnesota), edited by C. Jake Waddington, *American Institute of Physics Conference Proceedings*, vol. 183, New York, 1989.

[106] * Thurston W.P. 1998: "How to see 3-manifolds", *Class. Quantum Grav.*, **15**, pp. 2545–2571.

[107] Troitskij (= Troitskii = Troitsky) V.S. 1996: "Observational test of the cosmological theory testifies to the static universe and a new redshift-distance relation", *Astrophysics and Space Science*, **240**, pp. 89 –121.

[108] Troitsky V.S. 1996: "On the nature of the redshift in the standard model of cosmology", *Apeiron*, **3** (1).

[109] Tully R.B. and Fisher J.R. 1987: *Nearby Galaxies Atlas*, Cambridge University Press.

[110] * Valentijn E.A. and van der Werf P.P. 1999: "First extragalactic direct detection of large-scale molecular hydrogen in the disk of NGC 891", *Astrophysical Journal, Letters*, **522** (1), pt. 2 (1 Sept.), pp. L29–33.

[111] Van Flandern T. 2002: "The top thirty problems with the Big Bang", *Apeiron* (April); downloadable from http://redshift.vif.com/JournalFiles/V09NO2PDF/V09N2tvf.PD.

[112] Walker A.G. 1944: "Completely symmetric spaces", *Journal of the London Mathematical Society*, **19**, pp. 219–226.

[113] Weeks J.R. 1998: "Circles in the sky: finding topology with the microwave background radiation", *Class. Quantum Grav.*, **15**, pp. 2657–2670.

[114] * Weeks J.R. 2004: "The Poincaré Dodecahedral Space and the Mystery of the Missing Fluctuations", *Notices Amer. Math. Soc.*, **51** (6), pp. 610–619.

[115] * Weeks J.R. 2002: *The Shape of Space*, Second edition, Marcel Dekker. Hardcover, 382 pages, ISBN: 0-8247-0709-5.

[116] Weyl H. 1918: *Space Time Matter*, first edited in 1918, translated into English from the 1920 German edition in 1950 and published by Dover Publications Inc. in 1952.

[117] * Wolf E. 1987: "Non-cosmological redshifts of spectral lines", *Nature*, **326** (6111), pp. 363–365.

[118] * Wolf E. 1998: "The redshift controversy and correlation induced spectral changes", in: *Waves, Information and Foundations of Physics*, R. Pratesi and L. Ronchi eds. *Conference Proceedings*, **60**, pp. 41–49 (Italian Physical Society, Bologna, Italy).

[119] Wolf J. 1984: *Spaces of Constant Curvature*, Fifth Edition, Publish or Perish Inc. 412 pages, ISBN 0-914098-07-1.

[120] Wu K.K.S., Lahav O. and Rees M.J. 1999: "The large scale smoothness of the Univere", *Nature*, **397** (6716), pp. 225–230.

[121] * 2004: "Things fall apart", *The Economist* (Feb. 5th).

A Selection of Websites

[122] Readings from Open Questions in Cosmology <http://www.openquestions.com/oq-cosmo.htm>.

[123] Errors in some popular attacks on the Big Bang according to Edward L. Wright <http://www.astro.ucla.edu/~wright/intro.html>.

[124] Tom Van Flandern: <The Top 30 Problems with the Big Bang> *Apeiron*, Vol. 9, No. 2, April 2002 72 <http://redshift.vif.com/JournalFiles/V09NO2PDF/V09N2tvf.PDF>.

[125] Tom van Flandern: Did the universe have a beginning? <http://www.metaresearch.org/cosmology/DidTheUniverseHaveABeginning.asp>.

[126] Paul Marmet's website <http://www.newtonphysics.on.ca/>.

[127] The chronometric cosmology: Alexander Levichev's website <http://math.bu.edu/people/levit/>.

[128] Modified newtonian dynamics (MOND) of Mordehai Milgrom <http://nedwww.ipac.caltech.edu/level5/Sept01/Milgrom2/Milgrom_contents.html>.

[129] Electrical Plasma <http://www.electric-cosmos.org/electricplasma.htm>.

[130] The Plasma universe; Antony Peratt's website <http://public.lanl.gov/alp/plasma/universe.html>.

[131] The electric-cosmos <http://electric-cosmos.org/>.

[132] A resource for Big Bang skeptics and critics. <http://home.pacbell.net/skeptica/>.

[133] Light creates matter: Kirk McDonald 's website <www.hep.princeton.edu/~mcdonald/>.

[134] Jean-Pierre Luminet's website <http://luth2.obspm.fr/~luminet/PLS.html>.

[135] Robert Gardner's Web Version of his 1999 AAS Poster Presentation <http://www.etsu.edu/math/gardner/aas/aaspstr.htm>.

[136] Wilkinson Microwave Anisotropy Probe website <http://map.gsfc.nasa.gov/>.

[137] Jeffrey R.Weeks' Topology and Geometry software website <http://www.geometrygames.org/>.

[138] The Big Bang <http://www.marxist.com/science/bigbang.html\#Dark\%20Matter?>.

[139] R. Brent Tully homepage <http://www.ifa.hawaii.edu/faculty/tully/>.

[140] The Big Bang Never Happened website <www.bigbangneverhappened.org>.

[141] Halton Arp's website: <http://haltonarp.com/>.